THEORY OF SCIENCE

SYNTHESE HISTORICAL LIBRARY

TEXTS AND STUDIES IN THE HISTORY OF

LOGIC AND PHILOSOPHY

Portrait of Bernard Bolzano in 1839 (oils, 48 × 38 cm, signature: 22.541)
in the National Museum, Prague.

BERNARD BOLZANO

THEORY OF SCIENCE

Edited, with an Introduction, by

JAN BERG

Translated from the German by

BURNHAM TERRELL

D. REIDEL PUBLISHING COMPANY

DORDRECHT-HOLLAND / BOSTON-U.S.A.

First printing: December 1973

Library of Congress Catalog Card Number 72–92524

ISBN 90 277 0248 9

Published by D. Reidel Publishing Company,
P.O. Box 17, Dordrecht, Holland

Sold and distributed in the U.S.A., Canada, and Mexico
by D. Reidel Publishing Company, Inc.
306 Dartmouth Street, Boston,
Mass. 02116, U.S.A.

Printed in The Netherlands by D. Reidel, Dordrecht

TABLE OF CONTENTS

PART A

A SELECTION FROM THE *WISSENSCHAFTSLEHRE*

(Sulzbach 1837, Leipzig 1914–31)

['+A' ('−A') means including (excluding) the *Anmerkung*(*en*)]

Volume One

PART ONE / THEORY OF FUNDAMENTAL TRUTHS

PART TWO / THEORY OF ELEMENTS

Volume Two

§ 525. Explaining a Truth's Objective Ground (−A) 362
§ 530. Proofs by Reduction to Absurdity (Subsection 1) 362
§ 557. How to Prove a Statement Specifying the Composition of an Idea 364
§ 558. How the Proof that a Definition of a Given Proposition Is Correct Must Be Carried out (−A) 366

PART B

EXCERPTS FROM BOLZANO'S CORRESPONDENCE

Letter to J. E. Seidel, 26 January 1833 (Manuscript in Krajské muzeum v Ceských Budějovicích; transcription by Jan Berg) 371
Letter to M. J. Fesl, 8 February 1834 (Manuscript in Literární archív Památníku národního písemnictví v Praze; published in *Wissenschaft und Religion im Vormärz. Der Briefwechsel Bernard Bolzanos mit Michael Josef Fesl* (ed. by E. Winter and W. Zeil), Berlin 1965, p. 58, l. 4 – l. 3 f.b.) 373
Letter to F. Exner, 22 November 1834 (Manuscript in Österreichische Nationalbibliothek Wien; published in *Der Briefwechsel B. Bolzano's mit F. Exner* (ed. by E. Winter), *Bernard Bolzano's Schriften*, vol. 4, Prague 1935, p. 62, l. 32 – p. 67, l. 38) 374
Letter to J. P. Romang, 1 May 1847 (Manuscript in the same archive as Letter to M. J. Fesl (above); published in *Philosophisches Jahrbuch der Görresgesellschaft*, vol. 51, Fulda 1938, p. 50, l. 5f.b. – p. 53, l. 16) 379
Letter to R. Zimmermann, 9 March 1848 (Manuscript in the same archive as Letter to M. J. Fesl (above); transcription by Jan Berg) 381
Letter to F. Příhonský, 10 March 1848 (Manuscript in the same archive as Letter to M. J. Fesl (above); published in E. Winter: *Der Böhmische Vormärz in Briefen B. Bolzanos an F. Přihonský*, Berlin 1956, p. 285, l. 1 – l. 16) 382

BIBLIOGRAPHY 385
A. Works by Bolzano 385
1. Works on Logic, Epistemology and Methodology of Science 385
2. Works on Mathematics 385
B. Works on Bolzano 386

PREFACE

The present selection from the *Wissenschaftslehre* (Sulzbach 1837) of Bernard Bolzano (1781–1848) aims at giving a compact view of his main ideas in logic, semantics, epistemology and the methodology of science. These ideas are analyzed from a modern point of view in the Introduction. Furthermore, excerpts from Bolzano's correspondence are included which yield important remarks on his own work.

The translation of the sections from the *Wissenschaftslehre* are based on a German text, which I have located in the Manuscript Department of the University Library in Prague (signature: 75 B 459). It was one of Bolzano's own copies of his printed work and contains a vast number of corrections made by Bolzano himself, thus representing the final stage of his thought, which has gone unnoticed in previous editions.

The German originals of Bolzano's letters to M. J. Fesl, J. P. Romang, R. Zimmermann and F. Příhonský are in the Literary Archive of the *Památník národního písemnictví* in Prague. The original of the letter to F. Exner belongs to the Manuscript Department of the *Österreichische Nationalbibliothek* in Vienna. The original of the letter to J. E. Seidel is preserved in the Museum of the City of České Budějovice.

J. B.

EDITOR'S INTRODUCTION

Throughout his life Bolzano's interest was divided between ethics and mathematics, between his will to reform the religion of the church and the organization of society and his endeavor to construct a new and philosophically satisfactory foundation for the science of mathematics. Between these two poles, Bolzano's logic was created as an expression of certain fundamental motives generated by his philosophical view of the world.

First, he became more and more aware of the profound distinction between the actual thoughts of human beings and their linguistic expressions on one hand, and the abstract propositions and their parts which exist independently of these thoughts and expressions on the other hand. Second, he imagined a certain fixed deductive order among all true propositions. This motive was intimately associated with his vision of a realm of abstract parts of propositions constituting their logically simple components. Another important motive for Bolzano the moralist and educator was his conviction that our life on earth will be better and happier if more people gain insight into the world of scientific truth.

I. LOGIC AS A THEORY OF SCIENCE

A set of true propositions dealing with a specific domain of objects and containing all consequences concerning the relevant domain we call a *theory*. Such a notion of theory is implicit in Bolzano. His notion contains closure with respect to the relation of *Abfolge* (see Section XI below) and the (indeed obscure) condition of homogeneity. (Bolzano speaks about "truths of a certain kind.") An immediate consequence of Bolzano's distinction between actual thoughts (cf. Section V) and expressions and abstract propositions (cf. Section II) is the difference between a theory and its linguistic representation. *Logic* in Bolzano's sense is a theory of science, a kind of metatheory, the objects of which are the several sciences and their linguistic representations. This theory is set forth in his monumental work *Wissenschaftslehre* (hereafter '*WL*').

Now a *treatise* (*Lehrbuch*) of a theory *T* is, according to Bolzano, a book written with the intention of giving the most intelligible and convincing representation possible of *T* to a selected class of readers (*WL* §1). Hence *B* is a treatise just in case there is a theory *T* and a class of intended readers such that *B* is a treatise of *T* for this class; and *B* is a *scientific treatise* if the involved theory is a science. A theory *T* is a *science* (*Wissenschaft*) in Bolzano's view if and only if *T* is a nonempty set of true propositions and the members of some class of human beings at some time know a non-trivial part of *T* which is a worthwhile object of representation in a treatise. Whether the known part of *T* deserves treatment in a treatise is an ethical question that can be decided objectively, according to Bolzano, on the basis of utilitarian principles. At some places Bolzano conceives of his notion of science as implying an ordered set of true propositions, and this conception would fit his vision of the abstract world of true propositions as ordered by the *Abfolge* relation. Finally, *T* is a *theory of science* (*Wissenschaftslehre*) just in case there is a set *K* of criteria for scientific treatises and *T* is a set of rules necessary and sufficient for the satisfaction of *K*.

A theory of science has to be carefully distinguished from a didactic theory of teaching with treatises. From the definition of a science it follows that a theory of science may itself be a science. Bolzano's *WL* is but one exposition of such a science. But how could a scientific theory of science *T* be composed then, before we have a theory of science that determines how *T* should be composed? This possible charge of circularity Bolzano readily averts with the remark that the composition of a treatise of a theory *T* may be guided by the rules of *T* without there being any organized exposition of these rules (*WL* §2).

Bolzano's very broad conception of logic with its strong emphasis on methodological aspects no doubt accounts for the type of logical results which he arrived at. The details of his theory of science proper are given in the fourth volume of the *WL* and belong to the least interesting aspects of his logic. On the other hand, Bolzano's search for a solid foundation for his theory of science left very worthwhile by-products in logical semantics and axiomatics.

II. PROPOSITIONS AND SENTENCES

Bolzano's notion of abstract non-linguistic *proposition* (*Satz an sich*) is a keystone in his philosophy and can be traced in his writings back to the beginnings of the second decade of the 19th century. His conception of propositions may be characterized by the following assumptions (1)–(13). (Concerning (13), see Section IV.)

In his logic Bolzano utilizes a concept which is an exact counterpart of the modern logical notion of existential quantification. Therefore, he could have stated that:

(1) There exist entities, called 'propositions', which fulfill the following necessary conditions (2) through (13). (Cf. *WL* §§30ff.)

Thus propositions possess the kind of logical existence developed in modern quantification theory. However (*WL* §19):

(2) A proposition does not exist concretely (in space and time).

According to Bolzano, both linguistic and mental entities such as thoughts and judgments (cf. Section V) are concrete. Hence, propositions could not be identified with concrete linguistic or mental occurrences. Furthermore (*WL* §19):

(3) Propositions exist independently of all kinds of mental entities.

Therefore the identification between propositions and mental dispositions sometimes made in medieval nominalism cannot be applied to proposisions in Bolzano's sense.

Now the following two conditions are peculiar to Bolzano:

(4) Not all propositions are simple. A compound proposition is built up recursively from simple parts by means of primitive operations (*WL* §558).

(5) Two propositions are identical if they are generated in the same way from the same simple parts (cf. also (5′) in Section III). A chain of definitions may express the manner in which a complex part of a proposition is built up from simple parts (*WL* §§32, 92, 116, 557, 558).

We may assume that we have a language of a more or less specified structure and that we know what a *sentence* (*shape*) is in this language. Bolzano apparently identified all linguistic entities with concrete occurrences (*WL* §§49, 334) and he did not say anything explicitly about the abstract notion of sentence shape. It is possible, however, to reconstruct a concept closely related to our notion of linguistic shape and based entirely on concepts within Bolzano's philosophy (cf. Section III). From (4) and (5) it then appears that propositions in Bolzano's sense behave somewhat like the closed formulas of a logical calculus. On the other hand, the following assumption would seem to agree with Bolzano's intentions:

(6) No linguistic entity should be a necessary component in propositions.

Hence it would not be possible to identify propositions with sentence shapes or any other objects essentially involving sentence shapes or occurrences.

Furthermore, had Bolzano been faced with the notion of abstract linguistic shape, he would probably have approved (cf. *WL* §410) of the following general assumption concerning sentences (whether shapes or occurrences):

(7) The propositions outnumber the sentences.

For a modern development of a system of non-linguistic Bolzanian propositions it is natural to presuppose that there are nondenumerably many entities of this kind. Consider, for example, a denumerable domain of individuals and the nondenumerable infinity of classes of such individuals, and take any of these classes, say K, and any individual x. We then have the proposition (true or false) that x is a member of K. Admittedly the distinction between the denumerably and superdenumerably infinite was not yet quite clear to Bolzano, although he had one notion of continuum that applied to the set of real but not to the set of rational numbers (cf. Section XII) and had a clear understanding of the fact that isomorphism is a sufficient condition for the identity of powers of infinite sets (cf. Section XIII).

We call a notion of truth for linguistic or non-linguistic semantical entities Y intuitively satisfactory if its definition implies the classical Aristotelian condition that X is true if and only if Y, where 'X' stands for a name or description of the substituend of 'Y'. Now Bolzano would require that:

(8) An intuitively satisfactory notion of truth is definable for propositions (*WL* §19).

As a further axiom for Bolzano's theory of propositions we take:

(9) If a declarative principal sentence (shape or occurrence) expresses some thing, P, then P is a proposition (*WL* §§19, 28).

Now interesting relations hold between propositions and sentences:

(10) If the sentences S_1 and S_2 express the same proposition, then S_1 and S_2 may have different structure (*WL* §127).

(11) If S_1 and S_2 express the same proposition, then S_1 is logically equivalent to S_2.

The latter assumption underlies Bolzano's theory of the reduction of sentences to certain canonical forms (cf. Section IV). The converse does not hold, however, in Bolzano's system (*WL* §32, Note):

(12) If the propositions P and Q are expressed by S_1 and S_2 respectively, and if S_1 is logically equivalent to S_2, then P need not be identical with Q.

III. IDEAS IN THEMSELVES

A proposition in Bolzano's sense is a structure of ideas in themselves. Hence an *idea in itself* (*Vorstellung an sich*) is a part of a proposition which is not itself a proposition (*WL* §48). To be able to "generate" propositions, however, we have to characterize ideas in themselves independently of propositions. This is in fact implicit in Bolzano. He worked extensively with a relation – let us symbolize it by 'Σ' – which corresponds in modern logic to the relation of being a member of the extension of a concept. In terms of this relation Σ, taken as a primitive by Bolzano, certain implicit postulates may be extracted from his writings which concern the existence and general properties of ideas in themselves.

The expression '$x\Sigma A$' will mean, then, that x comes under the idea A. The extension of A may be identified with the set K of all x such that $x\Sigma A$. If K is nonempty, Bolzano calls it the '*Umfang*' of A (*WL* §66). A fundamental postulate schema for the primitive relation Σ is hinted at by Bolzano (*WL* §§48, 69, Note 1):

(a) There is an idea A such that, for all x, $x\Sigma A$ if and only if $\phi(x)$,

where 'ϕ' is (within certain limits) an arbitrary predicate of a presupposed language system. This corresponds to an axiom of abstraction in set theory and asserts that in general a predicate expresses an idea in itself. Two further assumptions are:

(b) In general, an idea A may not be identified with an object x such that $x\Sigma A$ (*WL* §49).

(c) The condition of coextensiveness: $x\Sigma A$ if and only if $x\Sigma B$, for all x, is not sufficient for the identity of A and B (*WL* §§25, 64, 96).

By means of the relation Σ, ideas can be classified as (i) empty (*gegenstand-los*) and (ii) nonempty (*gegenständlich*), where (ii) is further subdivided into (a) singular ideas (*Einzelvorstellungen*) and (b) general ideas (*Gemein-vorstellungen*) (*WL* §68).

Independently of human minds and of linguistic expressions there exist a collection of absolutely simple ideas in themselves. As examples Bolzano mentions the logical constants expressed by the words 'not', 'and','some', 'to have', 'to be', 'ought' (*WL* §78); but he admits being unable to offer a more comprehensive list. He seems to mean that each complex idea A can be analyzed into a sequence S of simple ideas which probably would include certain logical constants. We shall call this sequence S the 'primitive form' of A. The manner in which a complex idea is built up from simple ones may be expressed by a chain of definitions. Now Bolzano implicitly assumed the following *principle of individuation for ideas* (*WL* §§92, 119, 557):

(d) Two ideas are strictly identical just in case they have the same primitive form.

Furthermore, to each Bolzanian proposition P there corresponds uniquely a proposition P^* containing simple ideas only. Such a P^* we call the 'primitive form' of P, and P^* may be found by reducing all complex ideas of P to simple ones by means of chains of definitions. We can now express Bolzano's *generalized principle of individuation for propositions* (cf. (5) in Section II):

(5′) Two propositions are strictly identical just in case they have the same primitive form (*WL* §§32, 558).

If $x\Sigma A$ is the case, then x may be a physical object. It can also be an idea in itself or a linguistic expression ,in which case A is called a '*symbolic*' idea (*WL* §90). When $x\Sigma A$ holds and x is a linguistic expression in Bolzano's sense, A may be a counterpart in Bolzano's theory of our notion of abstract linguistic shape. In particular, if x is a sentence occurrence and if $x\Sigma A$ holds, then in some cases A may be an abstract sentence shape. Whereas a shape usually is taken as a class in modern logical syntax, Bolzano's notion could be described as a corresponding intensional property.

Bolzano elaborated a special theory for a large set of cases where an idea comes under another, second-order idea. For example, according to Bolzano the sentence obtained by inserting an expression for a particular idea A in place of 'X' in

(i) 'X is omniscient'

would, under a reduction to canonical form (cf. Section IV), be an elliptical formulation of a sentence of the following form:

(ii) 'X is something which has omniscience.'

The idea of being something which has omniscience is a so-called concrete idea (*concrete Vorstellung*, *WL* §60). From this idea one can derive a corresponding abstract idea of omniscience, thereby obtaining the sentence form:

(iii) 'X has omniscience,'

and the corresponding proposition:

(iv) [The idea of] A has omniscience.

The idea expressed by the word 'omniscience' is a singular idea of a certain property (*Beschaffenheit*, *WL* §80) of being omniscient which has the relation Σ to the second-order idea of omniscience. This property, in turn, could usually be analyzed into other properties. Now, to every abstract idea b of a property B such that $B\Sigma b$, there is a concrete idea B' with respect to b which is coextensive with B. (It is presupposed that even properties have an extension.) The situation could be schematized as follows: $B' \to b$ (abstraction function), $B \to b$ (relation Σ) and $B' \leftrightarrow B$ (coextensiveness). The symbolic ideas are introduced as a special case of this scheme in which the extension of B' embraces ideas or signs. In Bolzano's theory logical constants such as Something and Nothing fall outside of this scheme. For example: on one hand, the idea of Something cannot be analyzed as the idea of being identical with something that has a certain property, and on the other hand, it is not an abstract idea. Bolzano even considered individual ideas, such as the idea of Socrates, as being neither abstract nor concrete.

IV. THE REDUCTION OF SENTENCES

In the *WL* Bolzano used a partly formalized language embracing an ordinary language extended by constants and variables and certain technical expressions, and he also investigated the relations of his semi-formalized philosophical language to colloquial language (*WL* §§127–146, 169–184). He believed that most sentences of colloquial language were "reducible" to sentences or sets of sentences of certain canonical forms expressed in the philosophical language. When Bolzano speaks of reducibility, he uses phrases such as '*heisst wesentlich nichts anderes als*' (*WL* §171). The criterion of adequacy of such reductions would seem to be that the sentences be synonymous in the sense of expressing the same proposition (*WL* §127).

In Bolzano's theory of reduction, most sentences of ordinary language are reducible to sentences obtained by inserting expressions for particular (singular or general) ideas in themselves for '*A*' and expressions of particular abstract ideas (of second order) having a property as the sole member of their extensions for '*b*' in one of the following two expressions (*WL* §§127, 136):

(i) '[The idea of] *A* has *b*',
(ii) '[The idea of] *A* has lack-of-*b*.'

Here the substituend of '*b*' is also correlated with another, concrete idea *B* (cf. Section III).

In a formalization of Bolzano's philosophical language, the vocabulary would include as predicates both substituends of '*b*' and corresponding instances of 'lack-of-b' in (i) and (ii), respectively. Hence the contradictory of a sentence of the form of (i) is not a sentence of the form of (ii) but another sentence of the form of (i), namely

(iii) 'Prop(i) has falsity,'

where Prop(i) is the proposition expressed by the particular instance of (i) (*WL* §189).

Examples of Bolzano's application of his theory of reduction to ordinary language would be the sentences obtained by inserting a particular expression of an idea for '*A*' in 'There is an *A*' and 'Nothing is an *A*'. These sentences are reducible to the canonical sentences 'The idea of *A* has nonemptiness' and 'The idea of *A* has emptiness' respectively (*WL* §§137, 170). A sentence of the form of 'All *A* are *B*' is reducible to a canonical sentence of the form of (i), where the substituend of '*b*' indicates the abstract idea correlated to the substituend of '*B*' (*WL* §225, Note).

Had Bolzano's theory of reduction been completely developed it might have resulted in the construction of an ideal language for philosophical analysis. In this ideal language sentences of the canonical forms of (i) or (ii) would not play the same role as the atomic sentences of modern quantification theory on the basis of which more complex forms are built up. It seems, on the contrary, that Bolzano intended even the most complicated sentences to have canonical form or to be reducible to a set of such sentences. There is a fundamental vagueness, though, in Bolzano's theory of reduction, for nowhere in his work does he give any systematic rules for the construction of the extremely complicated names that would occur as substituends for the variables '*A*' and '*b*'.

Now the interesting point for the discussion of propositions is Bolzano's implicit assertion that sentences of the canonical form of (i) or (ii) mirror their correlated propositions in the sharpest way. Hence we have to presuppose that:

(13) For all *A* and *b* within the appropriate ranges, the transition from *A* and *b* to the canonical propositions expressed by (i) and (ii) must be explained.

The notion of truth is then defined for propositions corresponding to sentences of the canonical forms of (i) and (ii) substantially as follows (*WL* §24):

(i$^+$) Prop(i) is true if and only if each member of the extension of *A* is a member of the extension of *b*'s extension,

(ii$^+$) Prop(ii) is true if and only if each member of the extension of *A* is a member of the complement of the extension of *b*'s extension,

where *A* and *b* are indicated by the substituends of '*A*' and '*b*' respectively. Bolzano apparently presupposes that the set of sentences of colloquial language and of his own philosophical language could be mapped into the set of propositions, so that an indirect definition of truth for these sentences would be forthcoming.

Bolzano elaborates his theory of reduction by imposing a necessary condition for the truth of sentences of the forms of (i) and (ii). If *A* is nonempty (*gegenständlich*), i.e., if there is at least one member of the extension of *A*, then Prop(i) and Prop(ii) are also said to be nonempty; and if *A* is empty (*gegenstandlos*), i.e., if there is no member of the extension of *A*, then Prop(i) and Prop(ii) are said to be empty (*WL* §146). Now, if Prop(i) is true or Prop(ii) is true, then *A* is nonempty (*WL* §225.4). Hence if *A* is empty, both Prop(i) and Prop(ii) are false (*WL* §234.3). Since the entity indicated by the substituend of '*A*' in (i) and (ii) can be a class concept, Bolzano's condition implies an existential interpretation of classical syllogistic. Curious as it may seem, the very same condition shows that Bolzano was heading substantially for a philosophical language without existence assumptions, where even empty individual domains are permitted.

Thus all propositions are true or false. Furthermore, the following form of the law of excluded middle:

Prop(i) and Prop(iii) are not both false,

is valid even for empty propositions. In the case of certain propositions ascribing properties such as emptiness or simplicity to ideas, the adaptation to the law of excluded middle is made along other lines (*WL* §§196, 225).

V. JUDGMENT AND KNOWLEDGE

Every judgment (*Urteil*) has a proposition P as its "material" and has concrete existence in the mind of a being who takes P to be true (*WL* §§34, 291). A judgment is the acceptance of a proposition. Two judgments are said to be similar only if they have the same proposition as "material" (*WL* §292). However, a proposition should not be considered as a judgment abstracted from its psychological context. There is a difference between the acceptance (*Fürwahrhalten*) of a proposition P and the mere thinking of P. Analogously one could then say that a *thought* has a proposition P as "material" and has concrete existence in the mind of a being who considers P. In his *Logic*, Kant defines the term '*Urteil*' only in this latter way, and is criticized by Bolzano (*WL* §35.6).

Two important subdivisions of judgments are perceptions (*Wahrnehmungen*) and true judgments (*Erkenntnisse*). Perceptions will be defined in (5) below. True judgments are defined essentially as follows (*WL* §36):

(1) U is a *true judgment* if and only if U is a judgment and U is an instance of a true proposition.

If the person X produces and accepts at t the true judgment U, then it often happens that X will be disposed to produce and accept U at later occasions. Hence we can define (*WL* §307):

(2) 'X *knows* U at t' means the same as 'X has produced and accepted U before t and X has the capacity at t to produce and accept the true judgment U.'

Bolzano is now ready to refute the epistemological type of radical skepticism. First he proves the following proposition:

(a) There is at least one true judgment (*WL* §40). He can point out to the skeptic that the fact of the skeptic's doubting implies that he has actual ideas, in other words, that he thinks. Then he must assent to the true judgment: I have ideas. Now Bolzano states that:

(b) Any finite class of true judgments can become a proper subclass of a finite class of true judgments (*WL* §41). One of his prooofs runs roughly as follows. Assume that somebody knows $U_1, \ldots, U_n$. Then he would assent to the true judgment: I know $U_1, \ldots, U_n$; which is a new item of knowledge. From (a) and (b) it follows that the class of true judgments is "potentially infinite" in the sense already described by Aristotle.

Bolzano seems to have viewed man as a machine producing judgments in a specific order starting from a set of instances of conceptual truths (cf. Section VI) and judgments about sense-data. Starting from the basic input, this machine proceeds by direct inferences in the psychological sense. The result of this activity is a sort of living textbook, where the judgments follow each other like the sentences and figures in a treatise. The relation that orders judgments in this way is called 'mediation' (*Vermittlung*). That the sequence $\{U_i\}$ $(i=1,\ldots,n)$ of judgments mediates the judgment W in the system X may be loosely described by saying that the acceptance of $\{U_i\}$ by X together with some other mental activities of X entail the acceptance of W by X (*WL* §300.2).

Between the relation of mediation and other implication relations introduced by Bolzano the following connection holds (*WL* §300.7). Let the judgments U_i $(i=1,\ldots,n)$ and W be instances of the propositions P_i and Q, respectively:

(3) If W is mediated through $\{U_i\}$, then Q is an *Abfolge* from $\{P_i\}$ (cf. Section XI) or there is a sequence of ideas $A_1,\ldots,A_m$ such that either Q is derivable (*ableitbar*) from $\{P_i\}$ with respect to $A_1,\ldots,A_m$ (cf. Section IX) or the probability of Q with respect to $\{P_i\}$ and $A_1,\ldots,A_m$ (cf. Section X) is greater than $\frac{1}{2}$.

Bolzano tries to justify his principle as follows (*WL* §§300, 309): There are two main forms of mediation from $\{U_i\}$ to W in some system, viz., when the specific properties of W are (α) essential, or (β) unessential to the mediation. In case (β), $\{U_i\}$ will contain a logical or an empirical law of the form: If Y then W, such that Y is a subsequence of $\{U_i\}$. Hence this is a case of the second possibility mentioned in (3). Not all cases of mediation can be of kind (β), however, since we would then be forced into an infiinte regress according to Bolzano. But each mediation of the (α) kind is a result of one of the relationships of *Abfolge*, derivability or probability greater that $\frac{1}{2}$; for otherwise there would be a faulty mechanism in our minds. And to prove that the mind is fallible in this sense would require counterexamples where, in spite of all possible caution, a sequence $\{U_i\}$ of true judgments mediates a false judgment W in some systems. The non-existence of such counterexamples is Bolzano's fundamental assumption. As a consequence of this assumption there are only two forms of human fallibility in epistemological mediation: (i) mediations of kind (α) corresponding to the third possibility in (3), and (ii) mediations of kind (β) based on an erroneous law.

Some judgments are the first members with respect to the relation of mediation (*WL* §300):

(4) W is *immediate* in the system X if and only if there is no sequence $\{U_i\}$ such that $\{U_i\}$ mediates W in X.

Some judgments of the following forms are immediate in some system: (a) *I have the experience A*, where A is an actual idea the extension of which can contain an actual idea, a judgment or a feeling, or (b) *This (which I now experience) is A*, where A is an actual idea and not a physical object. Among the immediate judgments there are also some which are instances of conceptual propositions (cf. Section VI). Mediate judgments with a so-called empirical proposition as "material" are called '*Erfahrungsurteile*'. Now Bolzano can give a precise definition of perception:

(5) U is a *perception* if and only if U is an immediate judgment of type (α) or (β) in some system.

Another of Bolzano's fundamental assumptions concerning the relation of mediation is that immediate judgments are infallible (*WL* §309):

(6) If U is immediate in X at t, then X knows U at t.

Doubt about a single immediate judgment would involve doubt about all immediate judgments, since they all arise in the same way. And this would imply doubt about all mediate judgments. Hence there are only the two forms of human fallibility in mediation mentioned above. But of course we cannot always be sure that a judgment U is immediate in X, i.e., the judgment that U is immediate in X is fallible. An odd consequence of (6) is, however, that no judgments containing erroneous laws are immediate. Therefore, we would perhaps come closer to Bolzano's intentions if we excluded judgments containing conceptual propositions from the range of the variable 'U' in (6).

According to Bolzano, there is an "actual" infinity of true propostions (*WL* §§30–33). But in view of our finite capacity of mind, only a "potential" infinity of accepted propositions is possible. Now the question arises whether there are any true propositions which can never be reflected in judgments. In other words: Are there demonstrably unsurpassable limits to our knowledge? In *WL* §314 Bolzano deals with this problem. An attempt to enumerate all unknowable truths would be self-refuting. We might try, then, to indicate certain classes of objects which cannot be referred to in true judgments. Perhaps one could find an idea *A* such that each object of the extension of *A* is unknowable. This would amount to:

(a) No truth of the form: All *A* are *B*, is knowable,

where *B* is an arbitrary idea in itself. But according to Bolzano (a) is of the form: All *A* are *C*, where *C* is distinct from the selected *B*. Hence, a judgment containing (a) would be a counterinstance of (a). Hence there are no limits of this general kind to human knowledge. Bolzano's argument is obviously directed against Kant, who stated the equivalent of (a) with '*A*' denoting the class of supersensual objects.

VI. INTUITION AND CONCEPT

Bolzano defines the *content* (*Inhalt*) of an idea in itself *A* as the set of ideas of the primitive form of *A* (cf. Section III), not in the traditional way as a conjunction of properties. In Bolzano's logic it then follows that the traditional reciprocity between extension and content of ideas does not hold (*WL* §120). Hence there is nothing in the way of considering ideas which are both singular in extension and simple in content even though their linguistic expressions may be complex, as for instance: 'This (which I now experience).' Such ideas Bolzano calls 'intuitions' (*Anschauungen*, *WL* §72):

(1) *A* is an *intuition* if and only if *A* is a simple singular idea in itself.

A further definition of concept (*Begriff*) is now forthcoming (*WL* §73):

(2) *A* is a *concept* just in case *A* is an idea in itself which is not an intuition and no member of the content of *A* is an intuition.

Hence concepts may be simple (empty or general) ideas or complex (empty, singular or general) ideas with no simple singular idea in their content. If a complex idea has an intuition as a member of its content, it is called a 'mixed' idea (*gemischte Vorstellung*, *WL* §73).

Bolzano's intuitions differ from previous similar notions in their abstractness and absolute simplicity (*WL* §76). A singular idea of a physical object cannot be constructed from pure concepts; the complex idea representing the object must originate in an intuition of some physical object (*WL* §74). Bolzano's notion of concept corresponds to the so-called "universal" properties figuring in the modern discussio nof counterfactual conditionals. Bolzano's mixed ideas then correspond to non-universal properties, which are not definable without reference to some specific space-time location.

Bolzano's view of the causal processes involved in most actual intuitions seems to be that a physical object or an actual idea causes a process in the mind which in turn causes the actual intuition. This actual intuition contains an abstract intuition "in itself" with the psychical process as object. Objects not existing in space and time cannot cause intuitions.

The content of a proposition is defined by Bolzano in analogy to the content of an idea (*WL* §123). Propositions are now classified into conceptual (*Begriffssätze*) and empirical (*Anschauungssätze, WL* §133):

(3) A proposition P is a *conceptual proposition* if and only if the content of P embraces concepts only.

(4) P is an *empirical proposition* if and only if P is not a conceptual proposition.

For instance, the theorems of mathematical analysis and Newton's law of gravitation are conceptual truths, even though our knowledge of the latter happens to depend on experience.

An odd consequence of Bolzano's classification as it stands is that an instance of a logical law may be an empirical proposition. For example, the proposition: Bolzano is a philosopher or Bolzano is not a philosopher, where the idea Bolzano does not occur essentially. Perhaps we should reformulate (3) then as follows:

(3′) P is a *conceputal proposition* if and only if the primitive form of P has no essential occurrence of an intuition.

That (3′) may agree better with Bolzano's intentions is likely in view of the fact that in his theory of reduction he reduces the modality of necessity to the notion of conceptual truth (*WL* §182.1). This conjecture is further enhanced by the fact that he proposes the distinction between conceptual and empirical truths as a partial reconstruction of Kant's distinction between knowledge *a priori* and *a posteriori* (*WL* §§133, Note, and 287.9).

An *a priori* [*a posteriori*] truth is one whose truth is [not] seen independently of experience. The notions *a priori* and *a posteriori* can be taken in an objective sense in connection with propositions and in a subjective sense in connection with judgments. In the objective sense, conceptual propositions are *a priori* and empirical propositions are *a posteriori*. Hence, there are both unknown *a priori* and unknown *a posteriori* truths in the objective sense. This sense seems to be indicated by Kant when he declares all mathematical propositions to be judgments *a priori*. However, true judgments containing *a priori* propositions may be *a posteriori* from the subjective point of view. This subjective sense is defined by Bolzano by means of the relation of mediation (*WL* §306):

(5) A judgment W is a *judgment a priori* in the system X iff it holds that, for all sequences $\{U_i\}$ which stand in a chain of mediation to W in X, the terms of $\{U_i\}$ correspond only to conceptual propositions.

Furthermore, W is a judgment *a posteriori* in the system X just in case W is not a judgment *a priori* in X. Elimination of the expression 'conceptual propositions' in (5) by virtue of (3′) may fit Bolzano's intentions best.

VII. THE NOTION OF VARIATION

Bolzano observed that a proposition may change its truth-value upon variation of certain components. For example, the true proposition:

(i) Bolzano is mortal,

can be turned into the falsity:

(ii) Bolzano is omniscient,

by changing the predicate idea. On the other hand, if the idea Bolzano in (i) is changed into other ideas within the (usually explicitly presupposed) range of variation, proposition (i) cannot be turned into a falsity. If the range of variation of the idea Bolzano is extended, however, to include even what Bolzano called 'God', the following truth would be obtained, according to Bolzano:

(iii) God is omniscient.

Thus we see that Bolzano implicitly utilized a replacement operation for propositions, sending P into $P(A/B)$, where the latter proposition is similar to P except that $P(A/B)$ contains the idea B at all places where P contains the idea A. This operation can be extended to a simultaneous replacement in a proposition of the distinct ideas $A_1, \ldots, A_n$ by $B_1, \ldots, B_n$, respectively, thereby sending P into $P(A_1, \ldots, A_n/B_1, \ldots, B_n)$. The replaced ideas $A_1, \ldots, A_n$ have to be distinct, for otherwise there would be no operation of replacement, whereas the replacing ideas $B_1, \ldots, B_n$ need not be distinct.

Each idea occurring in a proposition has its particular range of variation. For example, in the proposition (i) and (ii) above, the ideas Aristotle, Leibniz, Spiro Agnew, etc., belong to the range of variation of the idea Bolzano, whereas ideas such as U.S.A., Socialism or The Number 3 do not belong to it. In doubtful cases the range of variation of an idea in itself may be explicitly indicated in a proposition by a description adjoined to the subject idea. For example, the proposition:

The human being Bolzano is mortal,

is an explicit counterpart to proposition (i) above.

The variation of ideas in themselves is an original and very fruitful conception. Bolzano exploits his innovation with great thoroughness in a logical theory which, on decisive points, anticipates notions expounded (independently by others) in the philosophical discussion of our time, especially in logical semantics and axiomatics.

First of all, the notion of validity (*Allgemeingültigkeit*) of propositions is introduced essentially as follows (*WL* §147):

(1) P is *valid* with respect to the distinct ideas $A_1, \ldots, A_n$ if and only if $P(A_1, \ldots, A_n/B_1, \ldots, B_n)$ is true for all sequences $B_1, \ldots, B_n$, where the idea B_j belongs to the range of variation of A_j.

A proposition being contravalid (*allgemein ungültig*) is defined analogously:

(2) P is *contravalid* with respect to $A_1, \ldots, A_n$ if and only if $P(A_1, \ldots, A_n/B_1, \ldots, B_n)$ is false for all $B_1, \ldots, B_n$.

In (1) and (2) we have to assume that P contians no defined ideas in themselves. Otherwise an unabbreviated proposition could be valid, while a corresponding proposition obtained by definitional abbreviation of certain ideas could be formally nonvalid according to (1). Take P as the proposition: $7+5<7+6$, and Q as: $7+5<13$, obtained from P by the definition: $13 =_{\text{def}} 7+5$. Now $P(7/N)$ is true for all natural number ideas N, whereas $Q(7/8)$ is false, and hence P but not Q is valid with respect to the idea 7. The conception that an unabbreviated proposition could be found, however, for any given conceptual proposition (cf. Section VI) permeates Bolzano's whole philosophy (cf. Sections II–III).

With regard to certain relations among propositions, Bolzano distinguishes the special logical case, where the ideas varied embrace all non-logical ideas of the propositions. This distinction is carried through, e.g., for the notion of analytic proposition, to be dealt with in Section VIII. It is natural to extend the distinction even to Bolzano's fundamental notion of variation. We then define:

(3) P is *logically valid* if and only if P is valid with respect to the sequence of all non-logical ideas of P.

This notion of logical validity corresponds closely to the modern notion of validity in a domain. Bolzano admits, however, that the notion of logical idea is somewhat vague.

VIII. ANALYTIC AND SYNTHETIC PROPOSITIONS

On the basis of his notion of validity Bolzano now makes a logical analysis of the Kantian notions of analytic and synthetic as applied to propositions. First he suggests the following dichotomy (*WL* §148):

(1) P is *analytic* if and only if there is an idea in itself A occurring in P such that P is valid or contravalid with respect to A;

(2) P is *synthetic* if and only if P is not analytic.

There is a disturbing feature in definition (1), though, that Bolzano himself might not have considered serious. Let $F(x)$ be an arbitrary empirical proposition that the individual x has the property F. This proposition could easily be turned into a logically equivalent analytic proposition in the sense of (1) by conjoining the proposition: $G(x)$ or not $G(x)$, which is always true and contains the arbitrary property G vacuously.

Bolzano pays attention, however, to a subclass of the set of analytic propositions, which is much more interesting from a modern point of view, namely the so-called *logically analytic propositions* (*WL* §148.3):

(3) P is *logically analytic* if and only if P is logically valid or logically contravalid.

By analogy we may also introduce the following notion which was never defined by Bolzano:

(4) P is *logically synthetic* if and only if P is not logically analytic.

We now distinguish the following classes of propositions: (I) logically analytic, (II) analytic and logically synthetic, (III) synthetic. This classification is even applicable to the corresponding sentences, and each of these classes can then be further subdivided into: (a) explicit, and (b) implicit sentences. Among Bolzano's own examples of sentences belonging to these classes were: (Ia) '*A* is *A*', 'Everything is either *A* or not *A*'; (Ib) 'Every effect has a cause'; (IIa) 'Even a learned man is fallible'; (IIb) 'Even a learned man is a man'; (IIIa) 'The father of Alexander, king of Macedonia, was king of Macedonia'; (IIIb) 'What I have written I have written'.

Kant's definition of analytic judgments as those in which the predicate is contained in the subject would be contraintuitive if applied to propositions, according to Bolzano, for it would then be both too restrictive by excluding the proposition expressed by the second example of (Ia) and too inclusive by admitting the proposition expressed by the example of (IIIa). The example of (Ib) is a consequence of the definition of the predicate 'effect'. In the proposition expressed by the example of (IIa) the idea in itself corresponding to the predicate 'learned' occurs vacuously, which perhaps was realized by Bolzano (even though he expresses himself obscurely on this point). Finally, the meaning of the example of (IIIb) would be explicitly conveyed by the sentence: 'What I have written I will not change', which is not even analytic in the wider (uninteresting) sense.

In modern logical semantics analyticity has often been considered substantially as a relation holding between a sentence S, a set M of definitions and a language L of specified structure. For example, we may say that S is analytic in L with respect to M if S is a logical consequence of M and if L embraces S and the members of M. Analytic sentences in this sense have a close parallel in Bolzano's true, logically analytic propositions.

IX. CONSISTENCY AND DERIVABILITY

Bolzano also considers logical properties of sets of propositions analogous to those defined in Section VII. He then utilized what is basically a replacement operation on sets of propositions, sending the set $\{P_1, P_2, \ldots\}$ into the set $\{P_1, P_2, \ldots\}(A_1, \ldots, A_n/B_1, \ldots, B_n)$, i.e., $\{P_1(A_1, \ldots, A_n/B_1, \ldots, B_n), P_2(A_1, \ldots, A_n/B_1, \ldots, B_n), \ldots\}$. That a set K of propositions is true means of course that each member of K is true. We may then extend Bolzano's definition of validity as follows:

(1) $\{P_1, P_2, \ldots\}$ is *valid* with respect to $A_1, \ldots, A_n$ if and only if $\{P_1, P_2, \ldots\}(A, \ldots, A_n/B_1, \ldots, B_n)$ is true for all sequences $B_1, \ldots, B_n$.

The important notion of consistency (*Verträglichkeit*) is a generalization of this relation (*WL* §154):

(2) $\{P_1, P_2, \ldots\}$ is *consistent* with respect to $A_1, \ldots, A_n$ if and only if $\{P_1, P_2, \ldots\}(A_1, \ldots, A_n/B_1, \ldots, B_n)$ is true for some sequence $B_1, \ldots, B_n$.

A corresponding relation between sets of propositions could be introduced:

(3) $\{P_i\}$ and $\{Q_i\}(i=1, 2, \ldots)$ are *consistent* with respect to $A_1, \ldots, A_n$ if and only if $\{P_1, Q_1, P_2, Q_2, \ldots\}$ is consistent with respect to $A_1, \ldots, A_n$.

As a special case of consistency in the sense of (3), Bolzano now introduces the relation of derivability (*Ableitbarkeit*) between sets of proportions (*WL* §155):

(4) $\{Q_i\}$ is *derivable* from $\{P_i\}$ with respect to $A_1, \ldots, A_n$ if and only if $\{P_i\}$ and $\{Q_i\}$ are consistent with respect to $A_1, \ldots, A_n$ and $\{Q_i\}(A_1, \ldots, A_n/B_1, \ldots, B_n)$ is true for any sequence $B_1, \ldots, B_n$ that makes $\{P_i\}(A_1, \ldots, A_n/B_1, \ldots, B_n)$ true.

In analogy with the treatment of the notion of validity, Bolzano even distinguishes the special logical case of derivability, which is defined essentially as follows (*WL* §223):

(5) $\{Q_i\}$ is *logically derivable* from $\{P_i\}$ if and only if $\{Q_i\}$ is derivable from $\{P_i\}$ with respect to the sequence of all non-logical ideas in themselves of the propositions P_i and Q_i $(i=1, 2, \ldots)$.

Bolzano's relation of logical derivability is one of his most impressive discoveries and a highly interesting counterpart of the modern notion of logical consequence introduced by Tarski. There are certain differences, however, between the two relations.

First of all, the modern notion of consequence is defined for sentences within formalized languages, whereas Bolzano's derivability holds between abstract non-linguistic propositions expressed in a natural language extended by variables and certain constants. This difference is of vital importance for the study of the relationships between consequence and other logical notions. For example, the equivalence between logical consequence and formal deducibility within elementary predicate logic could not be imagined without a highly developed conception of a formal calculus. Only after logic had become completely formalized by Frege was it possible to formulate a precise notion of syntactic deducibility.

Secondly, Bolzano's condition that the sets $\{P_i\}$ and $\{Q_i\}$ should be consistent with respect to the sequence of all non-logical ideas in themselves of every P_i and Q_i has no counterpart in the definition of modern logical consequence. As a result of this consistency clause, Bolzano's theory is less general and more complicated than the modern theories of consequence. For example, consequence relationships with contradictions of the type: P and not P, as antecedent must be represented in Bolzano's theory by stating that the disjunctive proposition: Q or P, is logically derivable from P, where Q is an arbitrary proposition. Moreover, consequence relationships with contradictions of the type: A is not identical with A, cannot be represented at all in Bolzano's system.

Thirdly, there is a certain incompleteness in Bolzano's semantics that comes to the fore in his notion of derivability. In modern logical semantics we say that a sentence S_2 is a consequence of S_1 if every interpretation over any domain that makes S_1 true also makes S_2 true. Hence we generalize universally over both interpretations and domains. Bolzano's quantification over sequences of ideas in themselves in (4) corresponds rather closely to our quantification over interpretations. Bolzano never made explicit reference to the domain, however, and never thought of combining quantification over domains with generalization over ideas in themselves.

X. DEGREE OF VALIDITY AND PROBABILITY

In connection with his definition of validity Bolzano even introduces the function: degree of validity (*Grad der Gültigkeit*). We write '$g(P, A_1, \ldots, A_n)$' to denote the value of this function for the proposition P upon variation of the ideas $A_1, \ldots, A_n$. To simplify definitions we call Q a 'variant' of P with respect to $A_1. \ldots, A_n$ if Q equals $P(A_1, \ldots, A_n/B_1, \ldots, B_n)$ for some sequence of ideas $B_1, \ldots, B_n$. Now Bolzano's definition of degree of validity can be expressed as follows (*WL* §147):

(1) $g(P, A_1, \ldots, A_n)$ equals the proportion of the number of true variants to the number of all variants of P with respect to $A_1, \ldots, A_n$.

The domain of the logical measure function g is restricted to propostions with a finite number of variants. In order to extend the domain of g, Bolzano adds the further restriction that variants differing only in ideas in themselves with the same extension should be identified. For example, upon variation of the idea 8 and admission of variants with coextensive ideas in the proposition: On the next drawing from this box containing 90 marbles the marble with number 8 will turn up, we get an infinite class of true variants, since each term of the sequence 1, 2, ..., 90 has infinitely many equivalents. (The idea 8, e.g., is coextensive with the ideas 2^3, 3^2-1, etc.)

Within the domain of the function g it obviously holds that P is valid with respect to $A_1, \ldots, A_n$ if and only if $g(P, A_1, \ldots, A_n)=1$. Furthermore, P is contravalid with respect to $A_1, \ldots, A_n$ just in case $g(P, A_1, \ldots, A_n)=0$. Now it is possible to generalize the relation of derivability within the domain of the function g. To say, then, that P is derivable from $\{Q_i\}$ with respect to $A_1, \ldots, A_n$ is tantamount to saying that the conditional proposition: If $\{Q_i\}$ is true then P is true, is valid with respect to $A_1, \ldots, A_n$ – in other words, that the degree of validity of this conditional is 1 with respect to $A_1, \ldots, A_n$. It is natural, then, to consider weaker cases of derivability where the conditional has a lower degree of validity. In this way Bolzano is led to his notion of *probability*.

With '$p(P, \{Q_i\}, A_1, \ldots A_n)$' denoting the probability of the hypothesis P relative to the set of propositions $\{Q_i\}$ and the sequence of ideas in themselves $A_1, \ldots, A_n$, Bolzano's partial definition of the notion of probability can be expressed as follows (*WL* §161):

(2) If $\{Q_i\}(i=1, 2, \ldots, m)$ is consistent with respect to $A_1, \ldots, A_n$, then $p(P, \{Q_i\}, A_1, \ldots, A_n)$ equals the proportion of the number of true variants of $\{P, Q_1, \ldots, Q_m\}$ to the number of true variants of $\{Q_i\}$ with respect to $A_1, \ldots, A_n$.

In analogy with Bolzano's treatment of the notions of validity and derivability, we could even distinguish the special logical case of probability p^* of P relative to $\{Q_i\}$:

(3) $p^*(P, \{Q_i\})$ equals $p(P, \{Q_i\}, A_1, \ldots, A_n)$, where $A_1, \ldots, A_n$ are all the non-logical ideas of P and $\{Q_i\}$.

Bolzano's notion of probability is explicitly conceived as a relation among propositions and their constituents. The notion of logical probability reconstructable within Bolzano's system is the logical relation of an hypothesis to its evidential support. These features and Bolzano's measurement of probability by the technique of variation are strongly reminiscent of Carnap's recent theory of regular confirmation functions. Thus, Bolzano was among the first philosophers who characterized inductive probability.

XI. THE OBJECTIVE HIERARCHY OF PROPOSITIONS

Modern logic distinguishes the three questions: (i) Is the formula ψ a consequence of the formula ϕ? (ii) Is ψ formally deducible from ϕ in the theory T? (iii) Is the sequence f of formulas a deduction in T from ϕ to ψ? The theory of derivability developed within Bolzano's logic of variation yields a method of handling the first question. As was pointed out in Section IX, however, Bolzano had no precise way of dealing with questions of the second kind. Instead he turned to a theory which is intended to answer certain very special questions of the third type transferred to the non-linguistic level of propositions.

Bolzano's *Abfolge* relation involves the notion of being directly deduced in a pre-existing system of true propositions, where the meaning of 'deduced' is not confined to purely logical relations but embraces even direct inferences based on causal and deontic implications. The *Abfolge* is an asymmetric and intransitive relation between sets of true propositions (*WL* §§203, 204, 213). We employ the notation '$\{Q_i\} \to P$' to indicate that $\{Q_i\}$ has the *Abfolge* relation to P. Bolzano then further assumes, inter alia, that $\{Q_i\} \to P$ and $\{U_i\} \to P$ implies the identity of the two antecedent sets (*WL* §206), and that P is never a member of $\{Q_i\}$ if $\{Q_i\} \to P$ holds (*WL* §204).

If $\{Q_i\} \to P$ holds, the members of $\{Q_i\}$ may be conceived as immediate subsidiary truths of P, and for each of these members, in turn, we may look for its immediate subsidiary truths. We can proceed in this manner until we reach what Bolzano calls 'basic truths' (*Grundwahrheiten*), i.e., true propositions P having no set $\{Q_i\}$ such that $\{Q_i\} \to P$(*WL* §214). In this way we may discover a pre-existing unique *Abfolge* hierarchy or "proof-tree" for P (*WL* §§216, 220). The general notion of a subsidiary truth (*Hülfswahrheit*) of P is now forthcoming as a representation of all truths Q distinct from P and included in a proof-tree for P (*WL* §217).

If the proposition: A exists (in space and time), is a subsidiary truth of the proposition that B exists, Bolzano says that A is a cause of B (*WL* §168). Now the causal explanation of the description of the state of a physical object would be an example of an infinite proof-tree, i.e., a proof-tree with at least one infinite branch (*WL* §216). Bolzano seems to require, however, that even infinite branches end in basic truths. For from certain assumptions made by Bolzano it follows that all conceptual propositions (cf. Section VI) have finite objective proof-trees (*WL* §221).

In his logic of variation Bolzano was aware of an inference rule corresponding to Gentzen's "cut rule": from the propositions that P_2 is derivable from P_1 and that P_4 is derivable from P_2 and P_3, we can deduce that P_4 is derivable from P_1 and P_3 (*WL* §155.24). Gentzen's *Hauptsatz* asserts that under certain conditions any proof in elementary predicate logic can be reduced to a normal form by eliminating all cuts. Bolzano is aware of a similar kind of "normal" proof in his logic of variation. This idea is carried through in the theory of the *Abfolge* structure. Here it is shown that the number of simple ideas in themselves involved in the propositions of a branch of a proof-tree will in general increase downwards (*WL* §221). It further follows from the assumptions of the theory that no unnecessary ideas in themselves are introduced into branches of the deductive hierarchy of propositions.

A proposition P can usually be proved in several ways. The proof of P which follows the objective order of *Abfolge* and whose branchtops are basic truths is called the 'demonstration' (*Begründung*) of P. Any other proof of P is called a 'confirmation' (*Gewissmachung*) of P (*WL* §525). A reformulation of the proofs of a non-empirical science by replacement of demonstrations for confirmations is always theoretically possible, the proof-trees of conceptual propositions being finite.

In setting forth his theory of the objective hierarchy of propositions Bolzano was influenced by Aristotle's methodology of the deductive sciences expounded in the Posterior Analytics. Aristotle, however, laid down some very strong requirements which all deductive sciences must satisfy. Bolzano agrees only in so far as he thinks that the ideal of a scientific exposition should manifest the *Abfolge* connection between the propositions involved (*WL* §394). In a strictly scientific exposition, then, the statements are "demonstrated" in accordance with the *Abfolge* hierarchies of the corresponding propositions. But for didactic reasons this ideal should not always be aimed at in the compositions of treatises.

XII. SET AND CONTINUUM

From the time of his early studies forward Bolzano took a profound interest in the philosophy of mathematics. In this and the following two sections I shall discuss some of Bolzano's set- and number-theoretical notions. First, Bolzano defines a *collection* (*Inbegriff*) as a composite object consisting of at least two arbitrary (abstract or concrete) parts or members, where the same member cannot occur repeatedly (*WL* §82). A collection M may be characterized by various combinations of the following properties: (1) if M contains x as a part, it also contains the parts of x as parts, (2) if M contains x, it does not contain any part of x, (3) M is invariant under permutation of its parts, (4) M is not invariant under any permutation of its parts, (5) to each member of M there is an immediate predecessor or successor in M, (6) for all x and y in M there is a z in

M between x and y. Collections are now subdivided into sums, sets, series and continuous series.

A *sum* (*Summe*) is a collection satisfying (1) and (3) (*WL* §84). A *set* (*Menge*) is a collection satisfying (2) and (3) (ibid.). From the definition of collection it follows that a set has at least two members. Another odd consequence is that $\{\{x, y\}, z\}$ is a *Menge* whereas $\{\{x, y\}, x\}$ is not. However, Bolzano also considers homogeneous sets or pluralities (*Vielheiten*) where all members are of the same kind (*Art*) (*WL* §86). The notion of kind is vague indeed, but it might be taken to involve some discrimination of logical types so as to rule out "inhomogeneous" sets like $\{\{x, y\}, z\}$.

A *series* (*Reihe*) is a collection satisfying (2), (4) and (5) (*WL* §85). A *continuous series* (*stetige Reihe*) is a collection satisfying (2), (4) and (6) (*WL* §85, Note). This notion amounts to a densely ordered set in modern mathematical terminology. For example, under Bolzano's definition the set of rational numbers would count as a continuous series. Bolzano also defined a notion of *continuum* (*Continuum*, *WL* §315.7) in such a way that the set of rational numbers is no continuum. In modern terms Bolzano's definition could be reconstructed as follows: In a topological space R, where D is a domain and H is a set of Hausdorff neighbourhoods, the subset K of D is a continuum in Bolzano's sense relative to R if and only if, for every x in K and for any neighborhood N of x in R, there is in the boundary of N a member of K distinct from x. A continuum in the modern sense, viz., a closed and connected set, is a continuum in Bolzano's sense but not conversely. Under Bolzano's definition, e.g., an hyperbola with two branches would constitute one single continuum (cf. *Paradoxien des Unendlichen*, § 48.1).

XIII. INFINITE SETS

Bolzano essentially defines a plurality (*Vielheit*) as *finite* if it is an instance of a natural number (*WL* §87.3). We shall see that Bolzano's notion of natural number is implicitly based on the notion of isomorphism (cf. Section XIV). He describes what we call 'isomorphism' between the sets K_1 and K_2 as follows: Every x in K_1 can be paired with a y in K_2 in such a way that any element of K_1 or K_2 is a member of at least one pair, and no element of K_1 or K_2 is a member of more than one pair (*Paradoxien des Unendlichen*, § 20).

Bolzano now observes that an infinite set can be isomorphic to a proper subset of itself, and he asserts that all infinite sets have this characteristic. We call a set 'reflexive' when it is isomorphic to a proper subset of itself. Hence, Bolzano states in effect that all infinite sets are reflexive. And he is also aware of the fact that all finite sets are non-reflexive (*Paradoxien des Unendlichen*, § 22). Yet he does not take reflexivity as a defining property of infinite sets, as was done later, e.g., by Dedekind.

The statement of equivalence between reflexivity and infiniteness of sets is an important achievement of Bolzano. On the other hand, single cases of reflexivity had been discovered several times before.

In the *Paradoxien des Unendlichen*, and implicitly in *WL* too, Bolzano repudiates the notion of isomorphism as a sufficient condition for the identity of powers of infinite sets. Instead, he subscribes to the doctrine that the whole is greater than its parts beyond the finite case. Accordingly, finite isomorphic sets are equinumerous (*WL* §87.2), whereas the isomorphism of infinite sets does *not* imply their equinumerousness (*WL* §93, Note). As a result, a number of statements follow in the *Paradoxien des Unendlichen* which do not correspond to Cantor's view on this subject. Bolzano's comparison of the powers of infinite sets is indeed impossible to understand, since nowhere does he offer any clear sufficient condition for the equinumerousness of infinite sets.

In *WL* §102, Bolzano considers the series (S_1) 1, 2, 3,..., n,...; (S_2) 1, 4, 9,..., n^2,...; (S_3) 1, 16, 81,..., n^4,.... Obviously, he says, each of these series has infinitely many terms, and the terms of S_m are among the terms of S_{m-1}. He contends that the number of terms of S_m "exceeds infinitely" the number of terms of S_{m-1}. However, in a letter to his pupil Robert Zimmermann on March 9, 1848, that is a few months before his own death, precisely this conclusion is sharply criticized by Bolzano. He has become aware of the fact that S_1, S_2, S_3,... have the same number of terms: 'The *set* of objects represented by n is always exactly the same as before, even though the *objects themselves* represented by n^2 are not precisely the same as those represented by n. The false result was produced only by the unwarranted inference from a *finite* set of numbers, namely those that do not exceed the number N, *to all*.'

Hence, it seems that in the end Bolzano confined the doctrine that the whole is greater than its parts to the finite case and accepted isomorphism as a sufficient condition for the identity of powers of infinite sets. This is a second achievement of major importance in Bolzano's investigation of the infinite.

XIV. NATURAL NUMBERS

In the *WL* there are some incidental remarks revealing that Bolzano considered a natural number as a property characterizing sets of objects (*WL* §§166.4, 174, 188.4, 245) or as a measure of the comprehension (*Weite*) of a set (*WL* §§ 96.9, 97, Note). A development of a theory of natural numbers is to be found in the manuscript *Reine Zahlenlehre* (in the sequel '*RZ*'). Here Bolzano makes a twofold distinction: on the one hand between concrete and abstract numbers and on the other hand between relative (*benannte*) and absolute (*unbenannte*) numbers. These two distinctions overlap. A *concrete number* relative to a selected property *A* (an *A*-number) is a term of a series to be generated in the following way: as the first term select any object coming under *A*, and let each subsequent term be a set derived from its predecessor by adjoining a new object coming under *A*. The *abstract A*-number of a concrete *A*-number *M* is a property (*Beschaffenheit*) the extension of which is the set of all concrete *A*-numbers isomorphic to *M* (*RZ* fol. 1). Zero is not a concrete *A*-number for any *A*; it may be considered as an empty abstract number to any *A* (*RZ* fol. 8). A number relativized to some *A* is also called '*benannte Zahl*'. Abstraction from the kind *A* of a relative number yields the corresponding *absolute number* (*unbenannte Zahl*). Hence, the abstract absolute number of a concrete absolute number *M* is a property the extension of which is the set of all concrete absolute numbers isomorphic to *M*.

Abstract numbers do not exist in space and time and cannot be said to involve an intuition of time, as Kant believed (*RZ* fol. 12). Arithmetic is chiefly concerned with absolute numbers. If relative numbers are used, they are relativized to sets of absolute numbers. The propositions of arithmetic state relations among concrete absolute numbers (*RZ* fol. 9). For example, the proposition $7+5=12$ says that the union of two objects of the abstract absolute numbers 7 and 5, respectively, comes under the abstract absolute number 12. The propositions of arithmetic and *Zahlenlehre* in Bolzano's sense (i.e., mathematical analysis) are *a priori* in the objective sense (cf. Section VI), since all ideas involved are concepts (*RZ* fol. 14). Most propositions of mathematics are (logically) analytic (*WL*

§315.3), and it seems that Bolzano would have declared arithmetic as a system of logically analytic propositions (*WL* §305.6). For example, the proposition $7+5=12$ follows logically from the law of the associativity of addition together with the definitions of the concepts involved.

Bolzano quotes Euclid VII, Def. 2, which states: 'A number is a set composed of units'. In so far as Euclid's definition was intended to delimit the natural numbers, it could be criticized in at least two respects: (a) 1 is a natural number; but if numbers are sets, 1 cannot be a number, since sets have at least two members (*RZ* fol. 4); (b) natural numbers are finite; but there are sets with infinitely many members (*RZ* fol. 5). A finite Euclidean number might be identified with a concrete (relative or absolute) number in Bolzano's sense. Euclid's notion of natural number was still in use among mathematicians of the 18th century, e.g., Wolff and Kästner. However, during the 18th and early 19th centuries, adumbrations of Bolzano's concept formation become visible in Newton, Klügel *et al.* All these works were well known by Bolzano and probably influenced him directly in his construction of natural numbers.

At the end of the 19th century, Frege's definition of natural numbers as classes of concepts with isomorphic classes as extensions appeared in print. And a few years later, independently of Frege, Russell defined natural numbers as classes of isomorphic classes. But half a century before Frege, Bolzano had defined his abstract absolute numbers as properties of isomorphic classes.

In the *WL* there is another approach to the notion of natural number, which is the outcome of the analysis of certain sentences about numbers. That A is a singular idea in itself is defined essentially as follows (*WL* §139.3): A is nonempty (*gegenständlich*) and the idea of a multitude (*Mehrheit*) of A is empty (*gegenstandlos*), in other words, there is at least one object coming under A and there are no x and no y distinct from x such that both come under A. Analogous sentences involving higher numbers are analyzed too (*WL* §243). That exactly two objects come under A is essentially defined as follows: there are an x and a y distinct from x such that both come under A and there are no x, y and z distinct from each other coming under A. This procedure of definition can be continued indefinitely. Now let the statement that there are exactly n objects coming under A be defined by '$\phi(A)$'. Then the abstract number n in Bolzano's sense could be defined as the property of all A such that $\phi(A)$. Bolzano's approach and its obvious consequence are strongly reminiscent of Frege's and Russell's analyses of sentences dealing with natural numbers.

XV. CONCLUSION

The *WL* was intended merely as a prelude to Bolzano's impressively designed work on mathematics. His main ambition was to recreate the whole body of contemporary mathematics in accordance with the vision of an abstract hierarchy of true propositions. For Bolzano this task implied the creation of entirely new foundations for certain branches of mathematics, as may be seen from his highly interesting efforts directed toward basing geometry on topological concepts.

In carrying out this program, most of the means of expression of modern quantification theory were in essence available to Bolzano. He came very close to the modern notions of satisfaction, logical truth, consistency and logical consequence. On the other hand, the formal deductive machinery of quantification theory is practically non-existent in Bolzano's works. This syntactic machinery appears only in Frege, who created the first strict logistic systems at the end of the 19th century. Bolzano's lack of interest in developing particular logical calculi stems most probably from his at times almost excessive preoccupation with the methodological aspects of logic and mathematics and of science in general.

The notion of a calculus in the modern logistic sense was first considered clearly by Leibniz. He dreamed basically of an effectively decidable, interpreted calculus embracing expressions of all "eternal" truths. Bolzano was justifiably critical of this overambitious program, however, and presented instead his own theory of the *Abfolge* structure of non-linguistic propositions, thereby taking his stand aside from the line of development of logic that leads to modern syntactic concept formation. A reason for Bolzano's general lack of interest in questions of logical syntax was no doubt his profoundly intensional, non-linguistic approach to logic.

JAN BERG

PART A

A SELECTION FROM THE *WISSENSCHAFTSLEHRE*

VOLUME ONE

INTRODUCTION

§ 1

What the Author Understands by Theory of Science

(1) When I think of all of the truths that any man knows or has ever known combined into one whole, e.g. all written down in a single book; I could call such an all-inclusive whole the sumtotal of *human* knowledge. No matter how small this amount may be compared with the whole immeasurable realm of all the truths there are, relative to the intellectual capacities of any individual man it would still be very great. Indeed, it would be too great. For surely even the most capable head, under the most favorable circumstances, and with the utmost diligence cannot assimilate – I will not say all, but only whatever is truly worth knowing within the entirety of what the united effort of all human beings has discovered up to the present. We must, therefore, accept some sort of a division. It is impossible by far for each of us to find out everything that may seem worth knowing in any one perspective. And so the one must apply himself to one thing, the other to another, each to whatever, according to our peculiar circumstances, is most necessary to us or most useful. To make the selection of what is worth knowing and the learning of it easier, as well as for many other purposes, it could be useful to divide up the whole domain of human knowledge, or rather that of truth generally, into more discrete parts. And it would be useful to bring the truths that belong to each particular sort, so many of them as are noteworthy, together in books of their own, relating them, as need be, to as many other theses as may contribute to their understanding or proof, so that they achieve the utmost intelligibility and cogency. Permit me, then, to apply the term *science*, to any set of truths of a certain kind so constituted that the notable part of it known to us deserves to be presented in a book of its own in the manner just mentioned. That book, however, or any book written as if someone definitely intended it to set forth all the known facts of a science that are worthy of the reader's attention in such a way

that they can be most easily understood and accepted with conviction should be called a scholarly treatise (*Lehrbuch*) on the science.

So, for example, I shall call the body of all truths that state attributes of space the science of space or *Raumwissenschaft* (Geometry); because those theses constitute a particular species of truths, which incontrovertibly deserves having the part of it that is known to us and worthy of note presented in distinct books and with proofs provided that will give them the greatest possible intelligibility and cogency. Such books themselves I shall call treatises on geometry.

(2) I admit that the meanings I am giving here to both the words science and scholarly treatise are not precisely those in general usage; but I may add that there are no generally accepted meanings for these two words at all and that I will not fail to give these particular definitions a fuller justification below. At present let it suffice to raise just two other meanings of the word *science*, because I will be using them myself in places where there is no danger of misunderstanding. Very many do not understand the word *science* to mean just a sum total of truths of a particular kind, no matter in what order they stand. By science they mean a collection of statements in which the most notable truths of a certain sort are already presented in such an order and combined with certain others, as must be the case when they are written down in a book intended to achieve the greatest ease of comprehension and the firmest conviction. The word occurs in this sense when, for example, we speak of a *really scientific lecture*; by adding the word *scientific*, we only intend to indicate with no uncertainty that the lecture arranges statements in such an order, offers such proofs, in brief has the sort of organization that we require of a scholarly treatise that genuinely serves its purpose. Beyond this, we also sometimes take the word *science* to be equivalent to the word *knowledge*. This interpretation, contrary to the two previous ones, which are called *objective*, could be called a *subjective* interpretation. It occurs, for example, when we say: I have acquired the science of it [ich habe Wissenschaft von dieser Sache]. In this case, *science* obviously means no more than *knowledge*.

(3) Clearly, how we go about the business of dividing up all human knowledge, or rather the entire realm of truth generally, into such parts as I entitled sciences in (1) and how we go about presenting those particular sciences in scholarly treatises of their own are not matters of in-

difference. For even without exaggerating in the least the value of the mere *knowing*, it must be obvious to everyone that among mankind there are countless evils propagated by nothing but ignorance and error; and that we would be incomparably better and happier on this earth if we could, each of us, accumulate just those items of knowledge that are most useful to us in our circumstances. Now if truth in general had been appropriately divided up into particular sciences and if good scholarly treatises on each of them existed and were available in sufficient numbers, the goal of which I speak would of course not yet have been achieved by that alone, but we would nevertheless be significantly closer to achieving it, especially if there were still other adjunct devices and facilities. For now (a) everyone who had only the pertinent prerequisite knowledge would be able to teach himself about any subject of which he needed to be instructed and to learn everything previously known about it. And (b) if everything he found in those scholarly treatises were presented as intelligibly and convincingly as possible, we should rightly expect that even in those parts of human knowledge where passion struggles against the acknowledgement of the superior truth, namely in the areas of religion and morals, doubt and errors would appear much more rarely. Furthermore (c) the more extensive study of certain sciences in more highly perfected scholarly books would produce a much greater proficiency in correct thinking. Finally (d) if the discoveries we have made so far should become more generally known among us, they would surely lead us to many other discoveries. We understand, then, that the benefits of such institutions, instead of diminishing in the course of time, would have to become the more extensive the longer they lasted.

(4) Some thoughtful consideration must make it possible to become aware of the rules we have to proceed by in this business of dividing up truth generally into particular sciences and in composing the scholarly treatises pertinent to each of them. There is also no doubt that the sum total of these rules would deserve to be recognized as a science in its own right; because it would certainly have its uses if we assembled the most noteworthy of these rules in a book of their own and put them there in such an order and provided them with such proofs that anyone could understand them and accept them with complete confidence. I permit myself, then, to give this science, since it is the one that teaches us how to present other sciences (strictly only the scholarly treatises about them),

the name *theory of science*; and so I understand by theory of science the sum total of all those rules by which we must proceed in the division of truth generally into particular sciences and their presentation in their own treatises, if we would proceed in a manner genuinely suited to the purpose. But since it is obvious from the beginning that a science which wishes to teach us how we must present the sciences in scholarly treatises must also instruct us in how to divide the whole realm of truth into particular sciences, in that it will become possible to present a science in a treatise pertinent to it only if we have correctly defined the boundaries of the domain of the science: we could catch the sense of our account of theory of science in more abbreviated terms, as that science which indicates to us how we should present the sciences in scholarly books suited to their purpose.

§ 2

Justification of this Concept and Its Designation

Since I myself just asserted that it is not an indifferent matter how many and which sciences are introduced into the world; it is fitting that I also attempt to give its own justification to the science that I am establishing here under the title of *theory of science*. But since the rules by which one must proceed in such an investigation are themselves supposed to be brought forward initially in the course of this book, I will presently cite only the sort of reasons I can presuppose are known to every one of my readers merely by sound common sense or whatever.

(1) I turn first, therefore, to the mere feeling everyone has and ask whether he would not have to find it strange in fact if we have so many sciences and yet have none that teaches us how we should proceed in their formation and their literary presentation in a scholarly treatise? For that such a science would not be empty of content, that there are of course many rules by which one must proceed in dividing truths generally into particular sciences and in giving an account of each of them: that will be admitted by anyone who has concerned himself with working up scholarly treatises; and even a beginner, without being able to make a definitive statement of one of these rules, will nevertheless have a sense of their existence. It is equally beyond doubt that to assemble them into their own independent and entire unit, to set them forth in a book of their own will advance knowledge among us and through this alone have a

beneficial effect for the improvement of all of the other sciences and treatises about them.

(2) But the doubt may arise in someone's mind whether such a science as I here conceive the theory of science to be is even possible. For since the science, according to the account that has been given of it, is supposed to teach how sciences can be presented in the first place and at the same time is supposed to be a science itself; one might well ask how it can come to be, if we do not know, so far as it does not exist, how a science has to be presented? It is easy to dispose of this reservation. It is possible to proceed according to the rules of the theory of science and so to produce many sciences, among others the theory of science itself, or better stated, their presentations in written form, without being distinctly conscious of these rules. One can have discovered these rules without having at the same time put them in such an order and connection as must be done in presenting them in a scientific treatise. Once these rules are known, one can now further work on every science, including the theory of science itself, and present them in written form, for all this means is bringing truths that are already known to us into such an order and connection as they themselves prescribe.

(3) But if one does not doubt the possibility, one can still doubt the appropriateness of this science. In particular, one can ask whether the territory we obtain by our definition of the science is neither too broad nor too narrow. But it is not possible to support the former, that the explication of our science that has been given would have to include too many or too diverse theories, upon mature consideration. It would be much easier to believe that its compass could usefully be expanded. Some, that is to say, could think it would serve the purpose better to assign to this science the task of indicating not just how genuine scholarly treatises are to be composed, but all other writings that aim at scientific instruction as well. Others, not even satisfied with that, could require that one be concerned not just with how the truths belonging to a science can be presented in written form, but also with how they can be discovered. Still others could finally desire that one state not only the way in which one has to disseminate truths in writing, but also how one has to disseminate them through oral instruction. In that case, we would have to take up in one and the same science, and so in the same scholarly treatise as well, along with the rules to be observed in constituting the sciences

and composing the treatises pertinent to them, all the rules that must be followed in giving a lecture as well. For example, discuss all of the means for arousing and holding attention, discuss the way in which something already understood can be imprinted on the memory, discuss the different ways in which truths must be presented to make them understandable and convincing to just those particular individuals one has before one, taking into account the characteristics that belong to them, and so on. I enter a reminder now that what they would add on to the theory of science is dealt with in another science that already exists, namely *pedagogy* or *didactics*, and that it would be appropriate to keep these two sciences separate. The reason is that the pursuits the two sciences are supposed to guide are of a very different kind and also demand talents that are also different and rarely encountered together. For it is one thing to shape the concept of a new science, to discover the truths that belong to it, to present them in writing and to set them into that order and connection with other theses as must be the case with an appropriately constructed scholarly treatise; and it is something else again to disseminate more widely by means of oral instruction truths that have already been discovered and brought into the proper order and provided with the proper demonstrations. Not everyone who has the capacity to lecture about a science also understands how to write a treatise on it; and, conversely, there are persons who are very well capable of the latter, yet cannot stoop to the former. This has long been remarked, and for just that reason in many countries some citizens (scholars, academicians) are entrusted with the business of a science's written presentation, especially the enlarging of its content, and others (teachers and professors) on the other hand, with the business of oral instruction. Since these two pursuits are actually divided, it is certainly well to give them separate guidance. For the one, it may be given in treatises on the theory of science, and for the other in those on (oral) pedagogy. In my opinion, it would not be equally objectionable to ask of a treatise on the theory of science what was first mentioned, namely that along with instructing us in the art of writing genuine scholarly treatises, it also instruct us in composing other works with scientific content, or that it teach us not only how to present the truths that belong to a science but also how to discover them. But all of this can be accomplished, I believe, without requiring any change in the concept of this science as I have defined it above. For because one cannot present truths before they have been discovered, we are justified, on the definition of theory of science accepted above, in taking up within its scholarly execution the question of how the truths that belong to a science can initially be discovered. And if certain books, even without being genuine scholarly treatises, nevertheless aim at scientific instruction, they must also be composed according to almost exactly the same basic rules as those must be; and this is reason enough to mention them briefly where one is giving instructions for the composition of genuine scholarly works.

(4) If one will now grant me that a science of the sort I have here described under the name of theory of science would serve its purpose, one would probably also have no objection to make to the title that has been proposed. For this (pure German) name expresses the content of such a science as clearly as can be wished. The fact, however, that some scholars, like J. G. Fichte and Bouterweck, have taken and may still take this expression in another sense is probably not of sufficient importance to forbid its use to us in such a natural sense for all the future; besides there are others again, as we shall soon see, who have already preceded me in this understanding of it.

§ 15

Plan for Carrying out Logic According to the Author's Understanding

(1) According to my conception, logic should be a *theory of science*, i.e., it should guide us in how we can divide up the entire domain of truth into particular parts in an appropriate way and cultivate what belongs to each of them and present it in written form.

(2) All of this guidance would obviously be superfluous if we were not clever enough to acquire knowledge in the first instance of a significant quantity of truths that belong to this or to that science. For until we find ourselves not to be in possession of an appreciable store of truths, it is premature to ask which branches of science we should classify them under, in what way, in what order, with what proofs we should present those of them that belong to a certain science we are to cultivate in a scholarly treatise. Now since the former, I mean the discovery of certain truths, could scarcely be a matter of less difficulty than the last cited, or the classification of truths already discovered in particular sciences and

the composition of useful treatises devoted to these sciences; it would certainly be a bad situation if we were provided special guidance only for the latter and not for the former enterprise. So long, therefore, as we do not find it good to provide ourselves with this guidance in an autonomous science, it will be the task of logic to provide it. Consequently, before we begin to teach the rules to be observed in constructing and cultivating the particular sciences, the rules that constitute the essential content of logic, it will be in order first to treat briefly the rules that need to be followed whenever the reflective enterprise has as its end the discovery of certain truths. If I reserve the title of *theory of science proper* for that part of my book in which rules of the first kind appear, the part that offers rules of the second kind can, on the other hand, not inappropriately bear the title of an *art of discovery* or *heuristic.*

(3) But the thought comes easily to mind that not only the rules of the art of discovery, but also those of theory of science proper, not only the rules that should be observed in seeking particular truths, but also those that should be observed in their classification in particular sciences and their written presentation depend in great part on the laws to which the knowledge of truth is bound, if not with all beings, still with us humans. Consequently, in setting out to present those rules in such a way that their correctness and necessity will be evident to my readers, I will first have to premise certain observations on the peculiar nature of the human cognitive capacity. Now because this part will deal with the conditions in which truth can be known, particularly for us humans, permit me to designate it briefly by the term, *theory of knowledge*, or still more definitely, *theory of human knowledge.*

(4) But if the rules of heuristics and the theory of science depend on the laws that are bound up with our human capacity for knowing the truth, there is no doubt that they depend much more on the properties that belong to the propositions and truths in themselves. Without having acquired knowledge of the various relations of derivability and consequentiality that hold between propositions generally; without having heard all about that particular manner of connection that obtains between truths when they are related to one another as premises and conclusions; without having any acquaintance with the different kinds of propositions and with the different kinds of ideas as those components into which propositions are directly analyzed; we are surely in no position to define

the rules as to how new truths are to come to be known from truths that are given, how the truth of a proposition before us is to be tested, to be judged; whether it belongs to this or that science; in what order and connection with other propositions it has to be adduced in a scholarly treatise if its truth is to be quite evident to everyone, and so on. It will therefore be necessary for me to devote a lot of attention in my discussion to propositions and truths in themselves; it will be required to first treat ideas, as the constituents of propositions, then propositions themselves, then true propositions, and finally inferences, or propositions that express an inferential relation. I shall call this part of my book the *theory of elements*, because here I shall be discussing approximately those topics which modern treatises on logic commonly handle under the title of theory of elements.

(5) Since it would still not be impossible that some of my readers should doubt even whether there are truths in themselves at all, or at least whether we humans possess a capacity for knowing such objective truths, it will not be superfluous to demonstrate this before anything else, i.e., to show that there are truths as such and that we humans do have the capacity to know at least some of them. In order to give this part of my book its own name, too, I choose – because the observations offered here can and even must constitute the beginning of any course of instruction where one cannot count on finding readers who are already acquainted with them or who are not in some other way sufficiently assured against ever falling into a state of all-encompassing doubt – the name *theory of fundamental truths*.

Consequently, all of the discourse below will be divided into these five parts, which are, to be sure, not equal in extent:

First Part. *Theory of Fundamental Truths*, including the proof that there are truths in themselves and that we humans also have the capacity to know them.

Second Part. *Theory of Elements*, or the theory of ideas, propositions, true propositions and inferences in and of themselves.

Third Part. *Theory of Knowledge*, or concerning the conditions underlying the possibility of knowing the truth, particularly among us humans.

Fourth Part. *Art of Discovery*, or rules to be observed in the enterprise of thought when it is aimed at discovering the truth.

Fifth Part. *Theory of Science Proper*, or rules that must be observed in dividing up the domain of truth generally into particular sciences and in presenting those sciences in specialized scholarly treatises.

A more detailed justification of this plan as well as the account of the subdivisions into which each of these parts should be analyzed still further will be presented in what follows.

PART ONE

Theory of Fundamental Truths

CHAPTER ONE

ON THE EXISTENCE OF TRUTHS IN THEMSELVES

§ 19

What the Author Understands by a Proposition in Itself

In order to indicate as clearly as possible to my readers what I mean by a *proposition in itself (Satz an sich)*, I shall begin by explaining first what I call an *assertion* or a *proposition expressed in words*. I use this term to designate a *verbal statement* (most often consisting of several, but at times of just a single word) if it is an instrument of asserting or maintaining something, if it is therefore always either true or false, one of the two, in the ordinary sense of these words, if it (as one can also say) must be either correct or incorrect. Thus, for example, I call the following series of words, "God is omnipotent," an assertion; for these words are used to state something, and in this particular case something true. But I equally call the following series of words an assertion, "A square is round;" for this combination of words too is used to declare or state something, although something false and incorrect. On the other hand, the following combinations of words, "the omnipotent God," "a round square," would for me not be called propositions, for they may well *represent* something, but do not declare or state anything, so that for just this reason we can not say, strictly speaking, either that they contain something true nor that they contain something false. If what I understand by asserted propositions or assertions is now known, I go on further to note that there are also propositions that are not presented in words, but which someone is merely thinking, and I call these propositions in thought. But just as in the term, "an asserted proposition," I am obviously distinguishing the proposition itself from its assertion; so in the term, "a proposition in thought," I am also distinguishing the proposition itself from the thought of it. Now what I call a *proposition in itself* is precisely what one necessarily has to represent by the word, proposition, in order to be able to make this distinction along with me. It is what one must think of as a proposition if one is still able to ask whether anyone has or has not asserted it, has or has not thought it. This is also what I understand by just the word, proposition, when, for the sake of brevity, I use it without the addition of *in itself.* In other words, then: I understand by a *proposition in itself* any statement that something is or is not, indifferently whether this statement is true or false, whether or not anyone has put it into words, whether or not it has even been thought. If an example is called for, in which the word proposition occurs with the meaning fixed here, I offer the following, alongside which many others could be set: "God, as omniscient, knows not only all true, but also all false propositions, not only those that some created being holds to be true, or of which he only forms an idea, but also those that no one holds to be true, or even has an idea of, or ever will have an idea of." So as to fix all the more firmly in the reader's mind the concept which has, I hope, already been made understandable by the foregoing, and also to make his confidence that he has correctly understood me still more certain, the following remarks may serve. (a) If we choose to represent as a proposition in itself what I am calling for here, we may no longer think of what its *original* meaning signifies, and so not of something *propounded,* which would have the consequence of presupposing the existence of a being by whom it is propounded.* Such sensuous connotations, which cling to the original meaning of a word, must be set aside in thought. Indeed this is also true of the technical terms that occur in so many other sciences. Thus, for example, in mathematics, when concerned with the concept of a square root, we may not think of a root, known to the botanist nor of a geometric square. We do not have to think of a proposition in itself as something propounded by someone; no more may we confuse it with an idea present in the consciousness of a thinking being, nor with an affirmation or *judgment.*

Of course it is true that every proposition is conceived or thought by God, even if by no other being, and, if it is true, it is also recognized to be true. Consequently, in the divine understanding, it occurs either

* The German text concerns the relation between *Satz*, here translated as *proposition* and *etwas Gesetztes*, for which *something propounded* has been used. 'Proposition' and 'propounded' do share a common root, but the relation is not nearly so obvious as it is in the German. Furthermore, the more concrete connotations of both *setzen* and *proponere*, in the sense of to put or to place, have been completely lost in the English derivatives of the Latin verb. The sentence that follows is therefore hardly pertinent to the English expressions used in translation. *(Translator's note.)*

as a mere idea or even as a judgment. But the proposition is nevertheless something different from an idea or a judgment. (c) For this reason, one may not attribute being (existence or actuality) to propositions in themselves. It is only the proposition asserted or thought, i.e., only the *thought* of a proposition, likewise the *judgment* containing a certain proposition that has existence in the mind of the being who thinks the thought or makes the judgment. But the proposition in itself, that makes up the content of the thought or judgment, is not any existing thing; so that it would be equally improper to say that a proposition has eternal existence as that it came to be at a certain moment and then at another moment ceased to be. (d) Finally it is obvious that although a *proposition in itself* is as such neither thought nor judgment, it can still concern thoughts and judgments, i.e., contain the concept of a thought or judgment in some one of its component parts. Indeed, the very proposition I previously offered as an example of a proposition in itself shows this to be so.

Note: If, after what has been said so far, one already knows well enough what one has to be thinking of or not thinking of as a proposition, the following question about one proposed example may nevertheless create some confusion. This example occurs in Savonarola's *Compendio aureo totius Logicae*, Lips. L. X. No. 18 under the heading; *Insolubile* propositum *(h.e. propositio se ipsam destruens) nec est concedendum nec negandum: Hoc est falsum, posito quod per subjectum demonstretur ipsamet propositio*. I.e. *This* (namely what I am saying right now) *is false*. – The question arises whether this combination of words deserves to be called a *proposition* and, in that case, whether this proposition is true or false? – Of such combinations of words, Savonarola says that one may neither affirm them or deny them. "Et si dicitur, omnis propositio est vera vel falsa: dicendum est, quod non sunt propositiones. Nam diffinitio propositionis, quae est oratio vera vel falsa, non competit eis in veritate. Habent tamen *figuram* propositionum. Sicut *homo mortuus* habet figuram et similitudinem hominis, non tamen est homo: ita et hae dicuntur *propositiones destruentes se ipsas, vel insolubiles*, non tamen propositiones simpliciter." – What say the readers? We are supposed to believe that Savonarola is right and precisely for the reason that the subject of a proposition can never be the proposition itself, no more than the part can constitute the whole. That notwithstanding, I venture to confess myself of the opposite opinion and I believe that common sense will decide in my favor. For what linguistic scholar would hesitate to call the words, "What I am now saying is false," a sentence that gives its meaning in full? – But as regards the objection that this proposition would have to be its own subject at the same time, which seems as preposterous as claiming that the part of a whole makes up the whole itself; this objection is resolved, I believe, by the distinction between a *proposition as such* and the mere idea of it. It is not the proposition itself, as proposition, that constitutes the subject idea in that proposition, but only the idea of it. That there is a basis for this distinction is demonstrated by the fact that not just here, but everywhere, we must differentiate between the thing itself and the concept of it, if we want to avoid confounding ourselves in the grossest absurdities. But if I declare the above statement to be a complete proposition, then I must also decide to call this proposition true or false, one of the two. As one will surmise, I do the latter and say that the proposition: "What I am asserting right now is false" is itself a false proposition, for it is equivalent to the following: "What I am asserting right now I declare to be false, and I am not asserting it." And of course that is untrue! In no way does it follow from this that I would have to assert the following: "What I am asserting right now is true." Savonarola seems to have believed this, and only because this proposition struck him as being almost as absurd as the first one, he preferred to take refuge in the claim that the two word-combinations were not propositions at all. I say, on the contrary, that the only special characteristic of the proposition, "What I am asserting right now is false," is that its contradictory can not be formed quite in the same way as with so many other propositions (whose subject idea has only a single object), namely, not by placing the negative 'not' before its predicate, 'false'. This will not do in the case of our proposition because the consequence of a change in its predicate is a change in its subject as well. For the latter, or the concept expressed by the words "What I am asserting right now" will be a different one when I say "What I am asserting right now is false" from what it is when I say "What I am asserting right now is not false." – There is a similar phenomenon with all propositions in which the subject or predicate presents a reference to the proposition itself or even to any of its component parts. Thus, for example, the two propositions following, which seem to be verbally contradictory, are both true: "The next to the last word in the statement

I am making right now is an article" and "The next to the last word in the statement I am making right now is no article." The following two, on the other hand, are both false: "The number of words that make up the proposition I am expressing right now is seventeen" and "The number of words that make up the proposition I am expressing right now is not seventeen." For the latter sentence really consists of seventeen words, because it has one more (namely, the word *not*) than the former, and so on. The contradictory of the proposition: "What I am asserting right now is – or I declare it to be – false" is not, therefore, the proposition, "What I am asserting right now is true", but "What I am asserting right now, I do assert." – But enough of these subtleties!

§ 21

That Others Have Already Made Use of this Concept

If the concept of a proposition in itself has so many claims to be incorporated into logic. it is to be suspected that it has not been entirely ignored previously. That has actually not been the case. But although the necessity for introducing it has only rarely been explicitly asserted, I find nevertheless that on certain occasions most logicians have at least had obscure intimations of this concept.

(1) With good reason, I will offer the conjecture below that the concept of a *truth in itself* was not unknown even to the ancient Greeks, so it may be supposed that at least on occasion they also attached the aforementioned concept to the word *proposition* (πρότασις, ἀπόφανσις, λόγος, ἀποφαντικὸς), for a truth in itself is also a proposition in itself. But the fact that they declared propositions in general to be a kind of *discourse* does not justify us at all in concluding that they had regarded only propositions expressed in words as genuine propositions. It could be that only the sensible character of speech prevented them from expressing themselves about this concept quite as abstractly as they actually wanted to be understood by their readers.* I will not engage in dispute on the matter, however, but I will gladly defer to the superior insight of those who know the ancients more deeply.

(2) I am more certain, however, that on more than one occasion

* In this connection, see the distinction in Aristotle's *Posterior Analytic*, L. I. c. 10 between *outer and inner speech*, λόγος ἔξω and λόγος ἔσω or ἐν τῇ ψνχῇ.

logicians have had the concept of a proposition in itself at least obscurely in their minds. For it is clear that what the logicians understand by what they call a proposition or judgment is not always the account indicated by the initial explanation, namely, nothing but a phenomenon occurring in the mind of a thinking being, much less something expressed in words, whenever they investigate the question as to whether there are two completely identical ideas. As I already mentioned in Section 16, No. 2, they answer this question in the negative and the reason they give is that what one might incline to regard as two ideas in this case is really only one and the same idea present in *thought* two times. Here, therefore, they undeniably make a distinction between the *idea in itself* and the *thought* of which it is the material. But anyone who distinguishes the idea in itself from the thought of it must also distinguish the proposition in itself from its appearance in the mind. A second circumstance in which it is still more manifest that propositions are not always understood to be propositions thought is provided in the theory of the syllogism when the question comes up whether there is a difference in the arguments when the order of the premises is reversed. Almost all logicians give a negative answer to this particular question. They demonstrate by this that they are not regarding the propositions in the argument as thoughts, but as propositions in themselves. Since what every syllogism presents is fundamentally only a proposition, namely a proposition of the form, "From truths *A* and *B*, truth *C* follows," and since furthermore the ideas of the two truths *A* and *B* appear in this proposition as mere members of a sum and so entirely without any rank order, it is true of course that once we regard the entire argument, and so its individual parts as well, the propositions *A*, *B* and *C*, as propositions in themselves we cannot speak of reversing them. But if we are speaking of propositions thought, then there is a temporal sequence among them, and Prof. Krug is correct in maintaining that with respect to the impression that stating an argument makes on the mind of a thinking being, the order in which its individual parts follow one another is not an entirely indifferent matter.

(3) Yet the concept of a proposition in itself did not just loom obscurely in the minds of almost all logicians. Some of them recognized it and expressed it distinctly. The following examples of this, which may hardly be the only ones, have occurred to me: Leibniz explicitly notes in the *Dial. de connexione inter verba et res (Oeuvres philosophiques*

publiées par Raspe) that propositions do not all have to be thought, and he uses the two expressions, *propositio* and *cogitatio possibilis*, as synonymous. All that presupposes that what he meant by propositions was propositions in themselves. – Mehmel says in his *Analytische Denklehre* (P. 48): "The judgment considered objectively, that is, in abstraction from the mind of which it is an act, is called a proposition. There must necessarily be as many kinds of propositions as there are kinds of acts of propounding that belong to the mind." – Here there is not only the *concept* that was established above, but the same *word* has also been chosen to designate it. In Prof. Herbart's *Lehrbuch zur Einleitung in die Philosophie*, section 52 he says: "With a judgment, thinking is only the means, the vehicle, as it were, of bringing the concepts together; it depends on *them* whether they will be fitting to each other or not. Here, too, therefore, logic must be kept isolated from any admixture of the psychological." – While this statement is not drawn up as clearly as the preceding one, we can nevertheless see well enough that Herbart does not choose to regard the judgment as a phenomenon in the mind, but as something objective, and so just as I regard the proposition in itself. Prof. Metz also says in his *Handbuch der Logik*, Section 12, entirely in agreement with me: "Since something is *propounded* by any judgment, namely a certain relationship of the given ideas to the unity of consciousness; each, *in abstraction from the act of the mind it is*, is called a *proposition* (positio, thesis)" – and in this connection he refers to the passage just cited from Mehmel's *Denklehre*. Hr. Gerlach (*Grundriss der Logik*, Section 67) writes: "A judgment can be defined both subjectively and objectively. Subjectively it is the *consciousness* of the relationship of two ideas; objectively, however, it is a proposition *(propositio, thema)* in which the relationship of two ideas is determined." Correct as this is, the following addendum is all the more striking: "The former is the condition of the latter." I say precisely the opposite: we would not have propositions *in thought* if there were no propositions *in themselves*.

§ 24

Various Meanings of the Words: True and Truth

Doubtless it is something to be criticized when one and the same word has a number of meanings that are easily confused with each other.

Actually, however, this criticism only rarely strikes at the meanings that have been attached to a word, not just by some scholar but by common linguistic usage. For even if it is the case with many words that common linguistic usage ascribes to them several meanings, they are nevertheless almost always so constituted that the circumstances prevailing in each case permit us to pick out without difficulty with which meaning the word is to be taken. This also holds of the various meanings that common linguistic usage ascribes to the words, *true* and *truth*; we do not have to be at all concerned that they will produce confusion. At most, confusion could arise from a few meanings that have been recommended only by scholars. Happily, however, they have not yet made them prevalent. For just this reason, I shall not deal with these until further on; but here enumerate only those meanings of the words *true* and *truth* known to *ordinary* linguistic usage.

(1) Without doubt the *first* and the *most characteristic* is that one understands by truth a certain property that can belong to *propositions*, equally whether they are asserted by anyone or not, indeed whether anyone even has an idea of them or not; a property, namely, by virtue of which they describe something as it *is*. This, for example, is the meaning in which one takes the word *truth* when one says that, "of the three propositions: There never were any winged serpents; they are extinct; there are still some of them, it is necessarily the case that one of them has truth, although we do not know which one."

(2) The *second* meaning arises from this one, when we apply to the *proposition itself*, which has the property of truth, the name, a *truth*. This is what we are doing, for example, in the expression that occurs so often: '*the knowledge of truths*'. In this case it is obvious that what we understand by *truths* are the propositions that are true. In this sense, the *false* stands in contrast to the *true*, or what comes to the same thing, the truth. Consequently, we say, for example, "of the three propositions mentioned above, one of them is certainly a truth, but the other two must be false." Which of these two senses of the word truth is intended in any present case, we apprehend very easily from the context, which indicates whether one is speaking of a mere property of propositions or of the propositions themselves.

(3) A third sense of the word truth arises when we ascribe the property, which really belongs only to the proposition in itself, to any *judgment*

that contains this proposition, and so call judgments that contain a true proposition true judgments, or simply *truths*. Of course we could dispense with this sense of the word *true* and very conveniently call judgments that contain a true proposition in itself *correct* judgments. In the meantime, it is nevertheless quite obvious that the expression, a true judgment, is not liable to any ambiguity; and whether one understands by the word *truth* a mere property of propositions, or a proposition in itself, or a judgment, the context will show clearly enough. So, for example, it is very apparent that in the following statement, "Even truth can at times be harmful," one understands by the word 'truth' the knowledge of it, i.e., a judgment that includes a true proposition. Truth in this sense of the word is contrasted to error. So, for example, we say, "If the truth sometimes does harm, then there must be useful errors."

(4) In a fourth meaning, we are sometimes given to call all of a *set of many truths* in one of the senses described above, i.e., a set of propositions or judgments that are true, *truth* in the singular number. Thus, Jesus says that he has come into the world to witness to the truth (John 18.37); where it is obvious that by truth a certain set of many propositions is understood.

(5) Finally, not the noun truth, but the adjective true is taken in yet a *fifth* sense, in which it is ascribed not merely to propositions and judgments, but also to anything when we want to say that it is really what it is supposed to be according to the designation we are giving it. In this sense, we are wont also to call the *true* the *genuine*, *real*, its opposite, however, the *false, spurious, illusory*. Thus we say, for example, "That is the true God," when we want to say that this is a being which not only seems to be God, but really is. Sometimes we hear the expression that sounds entirely self-contradictory: "true lie," which means nothing other than a statement that not merely seems to be a lie, but really is one. It is easy to understand how this sense of the word true arose by way of mere abridgement; in that, instead of speaking at such length as in the following statement, for example, "The proposition that this being is God not only seems to be true, but he really is God," we would use the brief expression, "This being is the true God." Confusion can not easily arise from this sense either, for the mere fact that here the adjective true is applied to things, which are neither propositions nor judgments, lets us recognize the presence of this case, and the example just

given shows us how we have to clarify the abbreviated statement form.

§ 25

What the Author Understands by Truths in Themselves

What understand by the expression, *truths in themselves*, which for the sake of variety I shall sometimes replace with the expression, *objective truths*, is only what we already understand by the word, *truths*, when we take it in the second of the senses just advanced (i.e., in the concrete objective sense), which may actually be the most usual sense.

Thus, to repeat, I understand a truth in itself to be any proposition whatsoever which states something to be as it is, leaving it undetermined whether this proposition has or has not been thought by anyone. Be that one way or the other, to me the proposition should bear the title of a truth in itself if only what it asserts is as it asserts it to be. In other words, if only what it ascribes to the object it concerns really belongs to it. For example, a number can be specified for the mass of blossoms on a certain tree standing in a particular place last spring, even if no one knows what it is. And so I call a proposition that specifies this number an objective truth, even if no one knows it, and so on. So that my readers will not be left with the least doubt about such an important concept as this one is, even though they have fully understood me, let the following further remarks, which really contain only certain plain and obvious theorems about truths in themselves, be set forth.

(a) All truths in themselves are a species of propositions in themselves.

(b) They have no real existence, i.e., they are not the sort of thing that would exist on sime place or at some time or in any other way as something actual. *Known* truths or even truths only *thought* do have real existence at a definite time in the mind of the being that knows them or thinks them, namely an existence as certain thoughts which began at one point in time and ceased at another. But one can impute no existence to the truths themselves, which are the *material* of these thoughts, that is to the truths in themselves. If all the same we do occasionally ascribe to some truths in themselves, for example the truths of religion, or moral, mathematical or metaphysical truths the predicate of eternity, as when we say, "it remains eternally true that vice brings unhappiness, or that the straight line is the shortest between two points," and the like; all we wish to say thereby is that these are propositions that express a fixed (eternal) enduring relationship, while other propositions, e.g., "the bushel of grain costs three dollars" or "it is snowing" and the like, assert only a transitory relationship (occurring at a certain time and also it may well be in a certain place). For this reason, in order to be true, they also require the addition of such a specification as to time (and often place as well), "It is snowing today, in this place."

(c) From God's omniscience it follows, to be sure, that every truth, even if it should be known to no other being, will not only be thought, but to Him, the all-knowing, it will be known, and will be permanently represented in his intellect. Therefore there is really not a single truth that is not known to anyone at all. But this does not prevent us from speaking of truths in themselves, in the concept of which it is not at all presupposed that they have to be thought by anyone. For if being thought does not lie within the concept of such truths, all the same it can follow from another fact (namely, God's omniscience) that they must be known at least by God himself, if by no one else. The situation of the concept of a truth in itself in this respect is like that of many other concepts (really all of them), for which we must distinguish between what constitutes their content, or what we have to think in order to have thought them, from what belongs to their object as a mere property (and which we do not need to think in order to have thought only of the concept itself). Thus the thought of a line which is the shortest line between its two end points is certainly a different thought from that of a line in which every segment is similar to the others, and as we distinguish between these two thoughts, so the concepts in themselves which we are thinking of when we think these thoughts are differentiated. The concept of the line which is the shortest one between its end points is therefore quite a different concept from the concept of the line in which every segment is similar to the others. All the same, it is beyond doubt that a line falling under the first concept, i.e., a line which is the shortest one between its end points, at the same time possesses the property which the second concept denotes, i.e., that each of its segments is similar to the other, and this holds true vice versa as well. We see from these examples that one may not dispute the difference between two concepts merely because they are correlative. Therefore, although all truths in themselves are at the same time known truths (namely known by God), nevertheless the concept of a truth in itself is very definitely to be distinguished from that of a known truth or (as we also say) a cognition. It follows that the logician must be free to speak of truths in themselves with the same full right with which (to offer a second example) the geometer speaks of spaces in themselves (i.e., of mere possibilities of certain locations) without thinking of them filled by matter, although it might be demonstrated perhaps on metaphysical grounds that there is not and could not be any empty space.

(d) If I stated above that a truth in itself "is a proposition that says something as it is," the words used are not to be taken in their original, not even in the ordinary sense, but rather in a certain higher and more abstract sense. In which sense becomes clear (I believe) from the clause added, "that I wish to leave it undetermined whether such a proposition has actually been thought by anyone and expressed in words, or not." Most of my readers, therefore, will already have thought of the further remarks I propose to make about each one of the above words by themselves. The word *proposition* used, by way of its derivation from the verb, to propound, readily calls to mind an action, a something that has been propounded by someone (and so produced or altered in some way). In fact there must be no such thought in the case of truths in themselves. For they are not propounded by anyone, not even by the divine intellect. A thing is not true because God knows it, but on the contrary God knows it because it is so. So, for example, God does not exist because He conceives Himself to exist, but only because there is a God does this God conceive Himself existing. And likewise, God is not omnipotent, wise, holy, etc., because he represents Himself to be so, but the other way around, the thinks Himself omnipotent, etc., because He actually is, and the like. The verb, to say, is to be taken in an equally loose sense, as everyone will understand by himself. For to be sure, no truth is capable of saying (speaking) in a literal sense. It would be easier to overlook the fact that the way of putting it, that a truth says "something as it really is" may also be understood only loosely. This is because not all truths say something which actually is (i.e., has existence); in particular, all those that concern objects which themselves have no actuality, e.g., other truths, or their constituents, the ideas in themselves. The proposition, a truth is not something that exists, certainly does not say anything that exists, yet it is a truth.

§ 26

Differentiation of this Concept from Some that Are Related to It

It will serve a still sharper grasp of the concept I have specified here if I also bring out the *difference* between it and some *related* concepts which for that very reason could easily be confused with it.

(1) In the first place, one must distinguish, as I have already said several times, the concept of a *truth in itself* from the concept of a *known truth*. However it may be that (as I have already granted) every truth is at the same time a known truth (at least a truth known by God), the *concept* of a truth in itself nevertheless always continues to be distinct from the concept of a known truth. The latter is formed by combining the former with the concept of a judgment; known truth or knowledge is a judgment which is true.

(2) We further distinguish the concept of truth from that of *certainty*. Truth in itself is a property that belongs to propositions in that they can be divided into true and false. Certainty, on the other hand, is a property that relates to judgments in that only judgments can be divided into certain and uncertain.

(3) Let us not further confuse the concept of truth in itself with that of reality. There are surely truths that *refer* to something real, i.e., assert properties of something real, but nevertheless the truth is never this reality itself; rather, as I have already said (25b), no single truth, as such has reality or existence.

(4) Finally let us not confuse the concept of a truth in itself with either the concept of *conceivability*, i.e., the possibility of a thought, nor with that of knowability, i.e., the possibility of knowledge. Conceivable is obviously a wider concept than true, for everything which is true must be conceivable, yet not everything conceivable must be true. Knowability is on the other hand a concept we can not call either wider or narrower than the concept of truth. All the same it is to be distinguished from the latter, because (as a closer look shows) it includes this among its constituents. For if I have correctly observed the usage of the word, know, knowledge is used only of true and never of false propositions; and the expression, "knowledge of the truth," still more 'true knowledge' is consequently really a pleonasm, because we do not call mistaken views knowledge at all. What is directly above knowledge and error, or the

genus of which knowledge and error are both species, we call judgment, also view, or opinion, if the connotation of uncertainty is removed from the latter words. Now the concept of knowledge and that of knowability can both be derived from the concept of judgment. Knowledge is namely (as I already said in (1)) a judgment that is true. Knowability of an object, however, is the possibility of making a judgment about it that is true. If these definitions are correct, then there is no doubt that the concept of knowability already includes that of truth as one of its constituents, and is therefore essentially different from it. The necessity of a distinction between these two concepts is also confirmed by the fact that truth, as everyone grants, admits of no degrees, no more or less, while knowledge can assume infinitely many degrees (namely in reliability as well as forcefulness). But if knowledge has a quantitative aspect, we must concede that the possibility of knowledge, i.e., knowability also has a degree (in the same respects at least).

§ 30

The Meaning of the Claim that there Are Truths in Themselves

The expression, *there are*, of which I make use in claiming that there are truths in themselves, requires an explanation of its own, so that it may not be misunderstood by anyone. For according to its real and customary meaning, in which it occurs, for example, in the proposition, there are angels, we want this word to signify being or existence of a thing. It can not be taken in this sense here, however, because, as I have recalled several times, truths in themselves have no existence. So what is meant when we say that there are such truths? Nothing, I answer, but that certain propositions have the character of truths in themselves.

But nothing definite should be said here in the beginning about the *number* of these propositions, so that we could regard our claim as proven if we were to have shown that there is only a single truth, or what comes to the same thing, that the claim that there is no truth whatsoever is false.

Let us take this remark together with the one immediately preceding: it finally becomes evident that the meaning of the claim that we wish to demonstrate here can be expressed with the utmost clarity in this way: "The proposition that no proposition is true, is itself not true," or somewhat more briefly. "That no proposition is true is itself not true."

§ 31

Proof that there Is At Least One Truth in Itself

Not even a person of the most limited insight can fail to grasp the proof of the clear statement to which we have just reduced this claim (§ 30). That no proposition is true contradicts itself, because it too is a proposition and in intending to declare it to be true we must at the same time declare it to be false. If every proposition were false, then this proposition itself, that every proposition is false, would also be false. And so not every proposition is false, but there are also true propositions; there are also truths, at least one.

Note: The argument by which I prove here that there is at least one truth was already well known to the ancients. Aristotle (Metaphysics 1. IV, Ch. 8) criticizes the self-contradiction of the proposition that nothing is true. Sextus Empiricus (adv. Log. L. II. 55) sets our argument out in great detail:
Και δή τοὺς μὲν πάντα λέγοντάς ψευδή, ἐδεὺξαμεν προσθεν (namely Book I. 390, 398; but not so convincingly there) περιτρεπομένος εὶ γὰρ πάντα ἐσι ψευδῆ, ψευδος ἔσαι καὶ τὸ πάντ' ἐσὶ ψευδῆ, ἐν πάντων ὑπαρχον. Ψεύδους δὲ ὄντος τω, παντ' ἐσὶ ψευδῆ, τὸ ἀντι κεί μενον αὐτω ἀληθὲς ἔσαι, τό, οὐ πάντ' ἐσὶ ψευδῆ (Only instead of πάντα, 'all', the more definite πᾶσαι ἀπὸφανσεις, 'all propositions', should have been used.) In Lambert's *Organon*, Book I, §§ 258, 262, we find a similar claim proved in a similar way. Bouterweck (*Idee einer Apodiktik*, Book I, PP. 375 and 378) also had the same idea in mind. – This argument can take many other forms, moreover. For example, one does not need to use just the very proposition that everything is false in order to prove that something is true; one can choose any other proposition whatsoever, *A* is *B*, and note that – in so far as it is false, the claim that it is false will be a true claim. If necessary, we could also just call the attention of those who hold that nothing is true to the fact that it is nevertheless true at least that there *are* propositions. And anyone who doubted that need only be reminded that surely the words: there are no propositions, themselves involve a proposition, and so on. But I do not believe that this argument would be superior in clarity.

§ 32

Proof that there Are a Number of Truths, Indeed an Infinite Number

(1) From the preceding section it is clear that there is at least one truth, because the contrary claim contradicts itself. But perhaps there is only one single objective truth, only this very one, that there is one truth? – So as to resolve this doubt, I will now show that there are a *number* of truths, indeed an *infinite number.*

(2) If someone claims that there is only a single truth; then let me be permitted to designate it, whatever it may say, by A is B; and I will now prove that there is at least a *second* truth besides this one. For anyone who assumes the opposite must also grant that the assertion, "There is no other truth besides A is B," is true. But this assertion is obviously distinct from the assertion, A is B, itself; for it is made up of entirely different parts. This assertion, consequently, if it were true, would immediately be a second truth. Therefore it is not true that there is only one truth, but there are at least two of them.

(3) But it can be proved in the very same way that two truths could also not be the only ones. For whatever these two may say, it is still obvious that the assertion, "Nothing is true but the two propositions, A is B and C is D," is a proposition entirely distinct from the two propositions, A is B and C is D. Therefore, if this proposition were true, it would immediately constitute a new truth, thus a third one, and we should have falsely presupposed that there were only two truths.

(4) We can see for ourselves that this sort of argument can be extended further and further, from which it follows that there must be an *infinite number* of truths, since to assume any *finite* number of them entails a contradiction. Let us suppose, namely, that someone chooses to admit only n truths: then these truths, whatever they may say (even if one of them consists, as the case may be, in the very assertion that there are only n truths) can be presented by the following n formulae: A is B, C is D... Y is Z. Now since our opponent wants us to accept nothing at all as true outside of these n propositions, he is maintaining something that we can clothe in the following form: "Outside of the propositions: A is B, C is D... Y is Z, there are no further true propositions." This form makes it evident, however, that this proposition has entirely different constituents, and is thus an entirely different proposition from any

of the n propositions: A is B, C is D... Y is Z. Now since our opponent holds this proposition to be true, nevertheless, he himself refutes the assertion that there are only n truths, since it would be the $(n+1)$st proposition which is true.

Note: So far as I know, this way (which so easily presents itself) of proving that there are a number of truths, indeed an infinite number, has not been used previously. Apparently it has seemed sufficient to force the sceptic only to admit that there is at least *one* truth. Moreover, this form of argument can also be modified. In particular, one can also proceed, avoiding the apagogic form, as follows: If the proposition, A is B, is true, then it cannot be denied that the assertion: "The proposition that A is B, is true," is also a true proposition. And the latter, so far as its constituents are concerned, is already different from the proposition, A is B, itself. So it is a second truth, distinct from it. – Likewise, from any true proposition of the form A is B we can derive the proposition: therefore, some B are A, and thus present a new truth distinct from the one given, and so on.

CHAPTER TWO

ON THE POSSIBILITY OF KNOWING THE TRUTH

§ 34

What the Author Understands by a Judgment

Since I want to show in this chapter that we possess some cognitions, I shall have to define the concept I attach to the word *cognition.* But since this concept, from my point of view, includes that of a *judgment*, it is the latter I must clarify first.

(1) I do say, however, that I too am using the word *judgment* quite in the same sense as we take it in ordinary linguistic usage, e.g., in the following proposition: God's judgments, unlike those of men, are not fallible.

(2) To anyone who does not find this sufficient, let the following be said: There lies in the concepts signified by the words asserting, deciding, saying, believing, holding for true, and others like them a certain common component, which is bound up in each of them with a particular subsidiary concept as well. Now let us take away these subsidiary concepts and think merely of what the meanings of those words have in common: then we are thinking of what I call *judging.*

(3) Consideration of the following propositions will provide, finally, a third means by which the concept I attach to the word judgment can be discerned in the most precise way: (a) Every judgment includes a proposition, which is either in accord with the truth or not in accord; in the first case the judgment is called a correct one, and in the second an incorrect one. (b) Every judgment is an entity (i.e., something that has existence). (c) Nevertheless, the judgment has no existence of its own, but exists only in the mind (or, if one prefers to say, the soul) of a certain being who for that very reason is said to be *the one judging.* (d) There is an essential distinction to be made between the actually *judging* about a proposition and merely *thinking* about it or conceiving of it. E.g., right now I am thinking of the proposition that there are pigmy tribes; but I am merely thinking of this and not asserting it, i.e., I am not making

such a judgment. (e) In God's infinite intellect, *every true proposition* is also present as an *actual judgment*. False propositions, on the other hand, also appear in God's intellect, but not as judgments God makes, but only as *ideas* of objects of which He is judging. (f) The *judging* of which we *humans* are conscious is an act of our mind that follows upon a prior mere *consideration of ideas*, and is dependent upon it. The act of judging depends only *indirectly* on our will, namely only insofar as we have a certain arbitrary influence on the consideration of ideas. (g) But we make each of our judgments with a certain degree of force, sometimes more, sometimes less, depending on the nature of the prior consideration, which I call the *confidence* with which are we judging. (h) No more than it lies directly in our power to construct a judgment so or otherwise (e.g., affirming or denying) does the degree of *confidence* with which we judge lie in our direct control. (i) If one proposition seems just as probable to us as its opposite, then we can neither judge that it is true nor judge that it is false, but we are in doubt. *Being in doubt* about a proposition thus means representing the proposition to ourselves, but from a lack of sufficient reason. asserting neither it nor its opposite,* Doubting is as necessary with some ideas as judging is necessary with others. (k) Yet at times we say: "You *should* not be in doubt, you *can* believe this, you *may* – indeed you even *should* believe it, you may depend upon it with complete confidence," and the like. These are improper forms of speech, the sense of which is only that if one were to pay due attention to certain ideas, he would make such a judgment and with that degree of confidence. – Doubtless these propositions will suffice to make the concept I attach to the word *judgment* clear to every one of my readers.

§ 35

Examination of Other Definitions of this Concept

(5) A considerable number of philosophers define the concept of judgment by using in their definition one of the words: *perception, observation,*

* Anyone who is in doubt, so far as he is, is not yet judging at all. But anyone who expresses the judgment, I am in doubt, is really making a judgment.

consciousness, insight, cognition, or a similar word. Along with what I cited in § 23.10, I shall mention the following. Hollmann (*Logik*, §§ 18 and 291): "Judicium appelatur actus intellectus, quo id, quod ad rem aliquam vel pertinere, vel non pertinere, vel plane eidem repugnare deprehendimus, de eadem vel affirmamus vel negamus." Reimarus (*Vernunftlehre*, § 115): "Judgment is the knowledge or insight of the agreement or non-agreement or the contradiction of two concepts." Kant (*Critique of Pure Reason*, p. 93): "Judgment is the indirect *knowledge* of an object," or (in the Preface to the Metaphysical Elements of Natural Science) "an act through which the given ideas first become knowledge of an object." Herr Schulze in the definition already cited in (1), as well as the one that appears in the 4th edition: "Judgment is a particular kind of cognition, namely by way of thinking of the relations ideas of all kinds have to certain concepts and also to the judgments already formed by way of that thinking." Likewise Gerlach, Fries, Calker, and others. However different they may be in their particulars, these definitions seem to me to have an error in common, that the words that are emphasized in them: perception, observation, consciousness, insight, cognition, etc., all have to be construed as already incorporating the concept to be defined, judgment, and what's more, tied up with the subsidiary concept, entirely inappropriate in this instance, that the judgment is *true* and *correct*. The point is that in my opinion one can not say that someone *perceives* or *observes* something, becomes *conscious* of it, has *insight into it* or *knows* it, unless he is making a judgment. Yet it is true that we do not at all regard the expression, 'perceiving a thing' or 'observing' 'becoming conscious of it' as synonymous with the expression, 'making a judgment about it.' But this is not because the *perceiving* or *observing* or *becoming conscious* is not already a judgment in and of itself, but only because the judgments we call a perceiving, etc., have a different content from those we call judgments about the objects perceived. The former are judgments that have the being making the judgment himself as their subject; in the latter, however, the subject is the object which was perceived. If we should say of someone that he *perceived* the rose standing in front of him, he must have made the judgment, I see a rose (granted, even, that he had not expressed it in words). But since the subject of this judgment is not the rose but himself, we are quite correct in saying that he has not yet made any judgment about the rose, but has merely *perceived* it.* It is still more obvious that the words, *cognition*, *insight*, and the like, include the concept of *judgment*. But that all these words, taken in the sense linguistic usage characteristically assigns to them, also carry with them the subsidiary concept of the *correctness* of the judgment that is made, becomes clear from the fac- that we by no means say of someone who is merely falsely *imagining* that this or that relation obtains between objects *A* and *B* that he *perceives* or *observes* that relation between them, or is *conscious* of it, or has *insight* into it, or *knows* it, and the like. From this it is apparent that the above definitions not only commit the fallacy of a circle, but are too narrow, for at best they are appropriate only to *true* judgments.

§ 36

What Would the Author Understand by a Cognition?

In order to make plain which concept I attach to the word *cognition*, I remark first of all that I am here taking all of the expressions: knowing something, having knowledge or cognition of something, being aware of it, understanding, perceiving, etc. in one and the same sense, in that I am abstracting from the various subsidiary ideas that accompany each of them and taking note only of the significance they have in common. Consequently, I understand by the word cognition any judgment that includes a true proposition, or (which comes to the same thing) is correct or in accord with the truth. According to me, therefore, every cognition is a judgment, but I do not turn it around and call every judgment a cognition, because there are also incorrect judgments, which I do not call cognitions, but mistakes. Every cognition presupposes a being who forms judgments and at least some correct judgments among them.

§ 40

How It Can Be Proved that We Know At Least One Truth

(1) The highest degree that doubt can attain arises when someone doubts

* In my opinion, it is so generally the case that every perception is a judgment that we must even grant to animals, so far as we ascribe to them the *capacity for perception*, a *capacity for judging* also (although not equally a capacity to *recall* this judgment once again, i.e., to be able to make the judgment about themselves that they made it, and therefore no *clear consciousness*).

not merely this or that particular truth (e.g., the existence of a physical world and the like) but even doubts whether there is anything whatsoever that is true or whether even a single one of these truths can be known. I call this sort of a doubter a perfect or complete sceptic; and in the present section it is to be shown how we can persuade such a person that he does know at least one truth.

(2) Such a consummate doubter, during the space of time in which he is such, can not make even a single judgment, whatever its content might be. For any judgment is nothing but a proposition which the person making the judgment himself holds to be a *true* proposition, with more or less confidence; thus, (with more or less confidence) he believes that he knows at least one truth. He is consequently not to be called a complete doubter. Putting it the other way around, in order to cure a doubter in some measure, all we need do is to bring it about, in whatever way, that he makes *some* judgment, the truth of which is even to him so incontrovertibly evident that – if we call his attention the next moment to how he knows at least one truth, in this judgment, he would try in vain to make himself doubtful of it once again.

(3) There may be various means leading to this goal. For since any answer to a question put forth, even if it consists in a mere, I know not!, includes a judgment, fundamentally any question we can hope the doubter will answer in such a way that its truth is incontrovertibly evident to himself is adapted to our purpose. Now since it is obvious that when he is doubting, the doubter can be no more conscious of anything than he is of his own mental state of doubting, we could well believe that it would be most to the purpose to choose the truth that he is doubting as the one we attempt to bring him to acknowledge; all the more since doubters are for the most part very willing to make this admission. On closer examination, however, it is apparent that this admission is inconvenient in one respect. Anyone who says that everything (i.e. every proposition) would be doubtful to him is expressing a judgment which, true as it may be when understood from the moment immediately preceding, when it is applied to the moment at which he makes it, it appears to be an incorrect judgment. For if everything (i.e. every proposition) were really doubtful to him, for that very reason he would not venture even to maintain the proposition that he is in such a state and so he would not have expressed the judgment that he takes everything to be doubtful.

Now if we should discover to him the mistake he is making here, by showing him how he only falsely believes he doubts everything, while he nevertheless does not doubt this one proposition; we would be giving him a new proof of the uncertainty of his own judgments and thereby we would only heighten the mistrust with which he confronts any judgment that wants to impose itself upon him.*

(4) It would seem better to me, therefore, if we attempted to elicit from the doubter not the admission that he doubts *everything*, but instead only this, that he doubts this or that *particular* proposition. Only it must be, to be sure, a question of a kind that our doubter can not and will not answer with anything but "I know not," not only in his present frame of mind [as a doubter] but later also, when his confidence in his own insight has grown again. Such a question might be, for example, whether there are men on the moon, or houses like the ones we build. But even such a question would be threatened by having the doubter subsequently withdraw the answer he had given at the beginning, and modify it by expressing himself as follows: "I know not whether I know not; it seems to me that I know not." Although this latter is not essentially different from what was said at first, it does diverge from it apparently and after a fashion; and the worst of it is that this means of altering the answer he gave at one time would always be at the doubter's disposal. As he just declared that he had said too much when he said before he did not know how to answer the question, in that he should have said it *appeared* to him that he did not know; so he can now again declare that even this is still saying too much, and that what he should have said was merely that it appeared to him *that* it appeared to him that he did not know how to answer that question, and so on. By continually modifying his answer in this way, the doubter hinders us from presenting any of the judgments he has just made as a truth which, on his own admission, he does see. Now should there not be a truth that is still more evident than not knowing some matter of fact, a truth of which it must be obvious to the doubter that in the very moment when he wanted to deny it or doubt it, he would be confirming it? – I believe there is such a truth, and it is this; *that he has ideas.* The fact that someone has ideas he

* In Hegel's *Religionsphilosophie* (*Works*, Vol. 2, p. 71), to be sure, it is said "If the doubter doubts doubt itself, then doubt disappears." – But I must confess that I do not understand this at all. How is the doubt supposed to disappear by way of doubt itself?

demonstrates precisely by the admission that he is doubting. And if he repudiates this admission later on and modifies it so as to say that it only *appears* to him that he is doubting, the truth that he has ideas loses not a bit of its certainty thereby. In my opinion, the best means for curing a complete doubter is therefore this one, that either without preparation or after a conversation such as has been intimated above we put before him the question: whether it is not true at least *that he has ideas*? – He may give the answer that suits this question or he may give no answer: all the same he will feel inwardly certain that of course this is true. He really does have ideas, and among others even ideas of entire propositions, because otherwise it would be impossible for him to doubt whether these propositions are true or not true, and so on. Now if this is the way he feels, we have already won. For if on one occasion he does not doubt that it is *true* that he has ideas, then neither will he doubt that he knows this truth, and consequently that there is at least one truth known by him.

§ 41

How It Can Be Proved that We Are Capable of Knowing an Indefinitely Large Number of Truths

Once someone has granted that he knows one truth, it is easy, in my opinion, to demonstrate to him that he also is capable of knowing a *number of truths* (not at the same moment, but in definite periods of time following one on the other).

If he should first make the mental admission to us that he knows *one* truth: then let us – that truth, whatever it may state, designated by A – ask him "whether the proposition that he knows the truth A is a true proposition?" He will necessarily give an affirmative answer to this question and in so doing he must admit that this proposition is not only true in itself but that he also *knows* its truth himself. Let us now call his attention to the fact that this proposition, or the judgment: I know truth A, is distinct from proposition A itself: thus he will already know that there is still another truth besides A, which he has just known.

It is obvious of itself how this form of argument can be extended and used to prove to anyone who first of all sees that he knows one truth that he also knows three, four, and more truths.

Generally, it can be shown in this way that the number of cognitions a man may be able to recollect is never so large that he could not increase it if permitted further consideration. For if he brings all of these cognitions into his consciousness and now forms the judgment: "I know all of these propositions," then this judgment itself is a new truth, not included among the previous propositions, one he has just now come to know, which consequently increases the sum-total of his cognitions by one.

Note: The thoughtful reader will see without any reminder from me that the proof I went through here can be carried out in many other ways. For example, it is well known that from any truth of the form *A* is *B*, a new truth of the form, some *B* are *A*, can be derived. So we shall easily be able to bring anyone who sees that he does know one truth, namely, all *A* are *B*, to see that he also knows still a second truth: some *B* are *A*.

PART TWO

Theory of Elements

§ 46

Purpose, Content and Sections of this Part

Since logic is supposed to instruct us in how we can divide the entire domain of truth into various particular sciences and present them in scholarly treatises, it is necessary that it first make us acquainted with certain properties that belong to truths or even to propositions in general. Good order demands that the properties belonging to *propositions* generally be handled prior to those which are found only in *true* propositions. But since every proposition consists of certain even simpler parts, namely of mere *ideas*, it will be to the purpose first to speak of mere ideas before we deal with the characteristics of propositions. Furthermore, it will be apparent even on considering the attributes of propositions in general, still more so on considering true propositions or truths, that among the latter there is an entire class which has as its essential feature the fact that some assert the relation of mere derivability between given propositions and some assert a genuine ground and consequence relationship; and that knowledge of these truths, which are commonly called inferences, is of great importance for logical purposes. For this reason, it will be appropriate to provide, following the theory of truths, a section of their own for the most useful kinds of inference. And so the part we are just now beginning will fall into four subdivisions, which I will call chapters:

First Chapter: Theory of Ideas in Themselves

Second Chapter: Theory of Propositions in Themselves

Third Chapter: Theory of True Propositions.

Fourth Chapter: Theory of Inferences.

As long as there is no independent science that sets the objects here identified apart and discusses them, we may count it to the credit of our science if in dealing with these objects it also takes up one thing or another that has no great usefulness for logical purposes, but still has something noteworthy about it in another respect.

The title, theory of elements, which I have selected for this part, suits the studies that are supposed to be presented in it insofar as they concern the individual components (elements) out of which, as they are combined, the presentation of a science in a scholarly treatise comes into being. For everything in a scholarly treatise is compounded from ideas, propositions, and from true propositions and inferences especially.

CHAPTER ONE

ON IDEAS IN THEMSELVES

§ 48

What the Author Understands by Ideas in Themselves and by Ideas Possessed

(1) The fact is that in the foregoing I have already used the word *idea* often, at times in paragraphs supposed to be understandable even to novices. But on such occasions either I took the word in a sense I could presuppose was already known from common usage, or, if on occasion it was encountered in a sense peculiar to myself, this occurred in such a context and together with such other propositions that one should nevertheless see approximately what it is that I understand by it. This approximate understanding can not suffice for the future, however. On the contrary, this concept's importance and its difficulty both demand that my reader be brought to the most precise understanding of it possible, by a discussion specifically intended for that purpose. Grasping the concept of an *idea in itself* can be rendered much easier, however, if we set it alongside that of an *idea* in the *ordinary* sense, which I also call an idea *possessed* or a *subjective* idea. Therefore, the more certain it is that both have equal claims to be taken up in the presentation of logic, the more willingly we choose to discuss them both together.

(2) To anyone who has understood correctly what I call a proposition in itself, what I mean by an *idea in itself*, or at times simply an *idea*, also an *objective* idea I can most easily explain by saying that for me it is anything that can occur as a constituent in a proposition, but does not constitute any proposition standing all by itself. Thus, for example, an entire proposition is expressed by combining the following words: Caius has cleverness. But something is expressed by the word Caius alone, which, as we have just seen, can provide a constituent of a proposition although by itself it does not constitute any entire proposition. This something, then, I call an idea. In the same way, I also call what the

word *has* signifies and, finally, what the word *cleverness* denotes in that proposition, ideas.

(3) Were I to have someone before me who was not yet acquainted with the concept of a proposition in itself then I would attempt to bring that of an idea in itself home to him by deriving it from the concept this word signifies in ordinary usuage. Everyone knows, or we can at least easily explain to anyone, what is called an idea in the usual sense. Namely, whenever we see, hear or feel anything, or perceive anything, by way of whatever outer or inner sense; whenever we only imagine or think of something, but without making any judgment about it at all or asserting anything about it – in every case it can be said that we have an idea of something. So idea in this sense is the general term for phenomena in our mind, among which we designate particular species by the terms: seeing, hearing, feeling, perceiving, imagining, thinking, and the like, insofar as there are no judgments or assertions. So what I see when someone is holding a rose in front of me is an idea, namely the idea of a red color. But what I smell on coming closer to this object is also an idea, namely that of the particular scent we call the scent of roses, etc. In this sense every idea presupposes a living being as the subject in which it occurs; consequently I call it *subjective* or *thought*. The subjective idea is thus something real; at the particular time at which it is present, it has a real existence in the mind of the subject for whom it is present. As such, it also produces all sorts of *effects*. This is not true of the *objective* idea or *idea in itself* which belongs to every subjective idea. I mean by it something not to be sought in the realm of actuality, something that makes up the direct and immediate *material* of the subjective idea. This objective idea requires no *subject* to whom it is present, but would have being – to be sure not as something existent, but nevertheless as a certain *something* – even if no single thinking being should apprehend it. And it is not multiplied when one or two or three or more beings think of it, as the subjective idea related to it then exists in plural number. Therefore the title *objective*. The objective idea that a word signifies is, insofar as the word is not ambiguous, for just this reason a unique one. Of the subjective ideas that the word arouses, however, there are innumerably many and with every moment their number increases with the word's use. We are accustomed, however, to *equate* all of the subjective ideas that have a single objective idea as their material, so far as we pay no

attention to the differences they may have in liveliness, etc. For example, the subjective ideas that arise in my readers' minds on looking at the next word – *nothing* – may all be apparently alike, but there are nevertheless many of them. The objective idea this word signifies is, on the other hand, a unique one. But with the word *bat*, there are two objective ideas it signifies, differentiated in Latin by the words *clava* and *vespertilio*.

Finally, there can be objective ideas that – with the exception of God – are not taken into the consciousness of a single thinking being. The number of grapes that ripened in Italy last summer is an idea in itself, even if there should be no one who is actually thinking of this number, etc.

§ 49

Differentiation of the Concept of an Idea in Itself from Some Related Concepts

In order to omit nothing which can make it easier for my readers to grasp the really difficult concept of an idea in itself, I must call attention to how it differs from some other concepts that are related to it.

(1) When in (3) of the preceding section, for lack of a better word, I used the expression that an idea in itself is the *material* of what is called an idea in the usual or subjective sense, this could almost be construed as if what I understood by the idea in itself is nothing but the *object* to which an idea (thought of) *refers*. But this is not what I mean. On the contrary, I wish to distinguish the object to which an idea refers or (as one can call it in more abbreviated terms) the *object of an idea* very sharply not only from an idea thought of but also from the idea in itself on which it is based, in such a way that when an idea thought of has one, or no or several objects, I require one or no or several objects, indeed the same ones, to be ascribed to the pertinent objective idea. By the object of an idea I understand that (sometimes existent, sometimes non-existent) something we are accustomed to saying the idea *represents* or is an idea *of*. It is easiest to understand what the object belonging to an idea is supposed to be if it is a real (existing) object. Everyone will certainly understand me if I say that Socrates, Plato *et al.* were the objects to which the idea, Greek philosopher, refers. But from this example it is also clear how much reason one has to distinguish the object of an idea, taken by itself, not only from the idea thought of but also from the pertinent object-

ive idea. For the latter, as I have already noted in the preceding section, is never something existent. The object to which an idea refers, however, can of course be something existent, as in the present example (Socrates, Plato, etc.) Besides, there is the further distinction that sometimes there are several objects referred to by one and the same idea, as is the case with the idea just adduced: Greek philosopher. It is not quite so easy to distinguish the objective idea and its object (if it has one) when the latter is not an existent. Yet scarcely anyone would deny that in the very same sense in which we can say that the idea, philosopher, refers to the objects, Socrates, Plato, etc., so too does the idea, principle, refer to the things that are called the Pythagorean theorem, the principle of the lever, the principle of the parallelogram of forces, etc. The single difference is that Socrates, Plato are existents, whereas the principles named here are, being propositions in themselves, not existents. Finally, there are also ideas that have no object at all, such as the ideas; nothing, $\sqrt{-1}$, etc. Nevertheless, in the case of the word *nothing* what we are thinking is certainly an idea (a subjective one). Consequently there must also be an objective idea corresponding to this subjective idea. But surely an object for these ideas to have is not to be conceived of. If, then, an idea in itself has several or none or only a single, but existing object, the distinction between it and its object is easy enough to see. The temptation to take the objective idea and its object for one and the same can be greatest when a subjective idea has only a single object, and a non-existent one as well: the idea, the highest moral law, for example. Nevertheless, the difference will be perceived even in this case if I set down the reminder that the *material* on which this subjective idea is based must be an *idea*, while the *object* to which this idea (both the subjective and the objective) refers is a proposition.

(2) Even less permissible than taking the object to which it refers for the idea itself (an idea in itself) is taking the word introduced as its sign for the idea. A word is always some sensory object (present at a specific time and place), a combination of sounds or of letters and the like, for example; but an idea in itself is, as we have said, nothing that exists. There are even ideas, not just objective but subjective ideas (thoughts) also for which we have no words at all. On the other hand, we often have several words that signify only one and the same idea, for example the words, triangular and threecornered. Difference enough so as not to confuse word and idea.

(3) It is to be noted finally that the word *idea* is often used, not only in the language of ordinary life but in textbooks of logic, in such a broad sense that even whole propositions and judgments are comprehended under it. This is the case, for example, whenever one speaks of true or false ideas, for not ideas in themselves, but only propositions or judgments can be true or false. At times we do this even when we are making an apparent distinction between judgment and idea, as in the following statement: "The harsh judgments Caius expresses of me are a result of the ideas he has been given of me." By the ideas Caius had been given of me I do not understand in this case anything other than certain judgments of which *I* am the object. As I have already remarked in §48 (2), the word *idea* is not supposed to be taken in such a broad sense here. Therefore, we may never take ideas in themselves to be propositions, but always take them to be only (real or possible) parts of such propositions. Nevertheless, this will not prevent us from admitting ideas that include an entire proposition, even several propositions, as parts. For complete propositions can also be combined with certain other ideas in such a way that nothing is asserted by the whole which arises from this combination unless something further is added to it. Such a whole will accordingly not yet bear the name of a proposition, but may only be called a mere idea. Thus, for example, the words, "God is omnipotent," certainly express a complete proposition, and this proposition also occurs in the following combination of words: "The knowledge of the truth that God is omnipotent." All the same, what is expressed by this latter combination of words is no longer a complete proposition, but can only become one by yet another addition, for example, if we say "The knowledge of the truth that God is omnipotent gives us much comfort." Thus what the words, "The knowledge of the truth that God is omnipotent," express all by themselves is to be called a mere idea, although one such that it includes within it a whole proposition as a component part.

§ 50

Justification of this Concept

I can suspect from the outset that many will find the concept of an idea in itself that has just been set up hard to take in. I can conceive of the charge being laid against me that it is very strange, indeed improper, to speak of ideas that no one is thinking of. Nevertheless, I believe I can maintain not only the *denotative character* (reality) of this concept but also the *necessity* of introducing it into logic.

(1) What I understand by the *denotative character* of an idea in itself is nothing but the fact that there are objects that fall under it. In this connection I wish to see the *there are* explicated in quite the same way as in the form of expression, that *there are* truths (§ 30). What moved me to define an idea in itself precisely as I have presented it above is just the concept I have formed for myself of propositions and truths in themselves. In particular, it would seem to me incontrovertible that any proposition, no matter, how simple, is composed of definite parts. These definite parts, such as subject and predicate, do not first emerge (as seems to be some people's opinion) only with the verbal expression of a proposition. These parts are already contained in the proposition in itself and if they were not there they could never enter into its expression. Furthermore, it seems obvious to me that the components constituting a proposition in itself, which is nothing thought, could not themselves be thoughts and so could not be ideas thought, but only ideas such as I have described objective ideas to be. So if one feels oneself forced to admit that there are propositions in themselves, i.e. propositions from which thought propositions only arise by way of their being apprehended in the mind, then, I believe, one must also admit ideas in themselves, such that thought ideas or thoughts only come to be by way of their being apprehended in the mind of a thinking being.

(2) But if the concept of an idea in itself has denotation, its noteworthy character will already warrant its being set forth in logic. And if the reasons I have used above (§§ 15 and 16) to prove that it is proper for logic to consider propositions and truths in themselves – as distinct from mere thought propositions and known truths – are not thoroughly incorrect and refutable, then it will be necessary to speak of ideas in themselves in their own right as well, distinguished from thought ideas. For without differentiating the former from the latter one can not properly comprehend the attributes of propositions and truths in themselves either.

(3) Moreover, I would not even deny that it sounds peculiar to speak of ideas that are present to no one. But what is peculiar arises, in my opinion, from the absence of any quite suitable term. For of course it is

true that there is something inappropriate about the word *idea*, because we are used to conceiving of an idea always in terms of a certain modification in the mind of a thinking being. This term is therefore very well suited for subjective (thought) ideas, not quite for what I call objective ideas. To my knowledge, however, there is no other word in our language that would be more appropriate, unless it might be the word concept, with which it is much easier than it is with the word *idea* to think of something that exists nowhere, not even in the mind of a thinking being. But it has become the custom to understand by the word *concept* only a particular class of ideas (namely, those that are not intuitions of sense) that are worthy of attention. So if we wanted to use this word for ideas generally, we would once again lack a name for that class.

§ 51

That this Concept Is Already Encountered in Others

Since, as we saw in §§ 21 and 27, there were several philosophers who recognized the concept of a truth in itself and also the still wider concept of a proposition in itself clearly enough, it could be expected from the outset that the concept I am designating here with the expression, an idea in itself, would not have remained completely unnoticed. For since it is generally taught that any thought truth is composed of several parts, namely particular ideas, how might one not have noticed that the truth in itself (the material of what one is thinking of in thinking a thought truth) would have to contain parts corresponding to the particular parts that constituted the thought truth, i.e. ideas in themselves? But because the concept of these parts is still more abstract than that of the entire proposition, it is not surprising that only infrequently were they explicitly mentioned.

(1) I do not venture to determine whether what *Pythagoras* taught of his *numbers*, *Plato* of his *Ideas*, *Stilpo* and later on the *nominalists* in their conflict with the *realists* of their *universals* was based upon the more or less clearly apprehended concept of an idea as such.* But we have already become acquainted (§ 21 (2)) with a situation in which it became obvious to not merely some, but almost all logicians that we have to distinguish

* It would seem to me that the nominalists had correctly noted that a concept is nothing existent, the realists, that it is not a mere name.

the idea in the ordinary sense of this word, i.e. the thought idea, from its real material or the idea in itself. This came up with the question whether there were several ideas (or *concepts*, as usually said at this point) identical with each other. This was commonly denied, and just for the reason, added explicitly, that identical concepts were really only "one and the same concept thought of several times." Here there was obviously a distinction drawn between the concept in itself and the thought of it. What, then, could have been understood by the former other than what I call an idea in itself? – Even though these very logicians, where they should have clearly defined what they call concepts (or ideas), commonly define them only as a kind of thought (phenomena in the mind), we may not immediately draw the conclusion from that that they never had any notion of the concept of an idea in itself. It could come about by a sort of inconsistency, excused by both the poverty of language and the intent to be more easily intelligible, that they described concepts as thoughts and nevertheless sometimes understood thereby, as we can see from their further discussion, something somewhat more abstract, namely the possible material of such a thought, an idea in itself. So that one can judge the better whether I am correct in such a presumption, I will cite some of them in their own words: In Maimon's *Logik, Abschnitt* 3, § 6, we read: "So-called identical concepts would be merely equivalent expressions for the same concept." And in Kiesewetter's *W. A. d. Logik*, p. 115: "Even if I conceive of the concept man by way of the attributes animal and rational a thousand times, there still remains only one and the same concept man." – How could Kiesewetter have said this if he had not, here at least, understood by a concept something quite different from a thought? For if I thought of a concept a thousand times, of course I had not one thought – but a thousand. Prof. Krug writes (*Logik*, § 37) that "two concepts that were really identical would basically constitute only only object of thought." What Herr Krug calls object of thought here is exactly the same as what I call an idea in itself; for (§ 25) he also carefully distinguishes the two ways of regarding a concept, as object of thought and act of thought. The same scholar writes (§ 35, Note 4): "In itself every concept is precise, but not every concept is thought by us in its precision." Here too the thought of a concept has been distinguished from the concept in itself as I would wish it to be distinguished everywhere. Even in the *Handbuch der Philosophie*

(Vol. 1, § 142) Krug distinguishes between real and unreal, i.e. unthought concepts. How could there be the latter, if nothing other than thoughts were to be understood by concepts? Herr Metz (*Logik*, 82) says: "Two identical concepts are only one concept thought twice or designated differently." So Herr Metz distinguishes, like the designation (the word), the thinking of a concept from the concept itself. And so here at least he can not have conceived of the latter as a mere thought.

§ 54

Ideas in Themselves Have No Existence

After we have become sufficiently familiar with the concept of an idea in itself, we go on to consider their *properties*, first of all those that are *intrinsic*. We can make a reasonable beginning by mentioning some that all of them have in common, following that with such as belong only to particular and especially note-worthy kinds of ideas, or, what means the same thing, by which some of them are distinguished from others. In order to follow some order in this connection too, I will speak first of kinds of ideas which can be conceived without specifying any of their component parts except *generically*. Then [I will consider] some which require for their conception that some of their parts be *individually* specified.

Now one property that belongs to *every* idea in itself whatsoever is that they have no *actual* existence. Although I have already noted this in § 48, it still seems appropriate to bring this property up for discussion again here, for there it was only mentioned in passing. No one will doubt the truth of this assertion who has acquired from linguistic usage knowledge of the concept we attach to the words: *existence*, *actuality*, also *actual existence*. Anyone who understands us when we say that God has an actual existence, that the world is also something actual, but that a round square is nothing existent will allow without contradiction that ideas in themselves belong to the class of those things that have no actuality. *Thought* ideas, i.e. thoughts, do possess an existence in the mind of whoever thinks them. And insofar as all ideas are apprehended in God's infinite understanding, there is not a single idea as such to which there does not correspond an idea thought of, and so actual, and eternally actual in the divine understanding. But we must not confuse these thought ideas with the ideas in themselves, which are only their material. The latter have no existence.

§ 55

Ideas in Themselves Are neither True nor False

A second property that belongs to all ideas is that neither truth nor falsehood can be ascribed to them.

Only entire *propositions* are true or false, but we understand by ideas parts of propositions that are still not themselves propositions. Consequently, neither truth nor falsehood can be ascribed to them. If common linguistic usage nevertheless speaks of *true* and *false* ideas, this comes about only in one of the following two cases:

(a) Insofar as these ideas are regarded as parts of certain propositions; we *are* accustomed to calling an idea a true or correct one if it is related by a proposition of the form, This is *A*, to an object which it actually represents, i.e. if the proposition, This is *A*, is itself true. And in the contrary case, if this proposition is false, we call the idea false or incorrect. For example, we say that the idea, "of a being which has creative power" is a true or correct idea of God. Contrariwise, the idea "of a being which can only affect existing substances, but can not give existence for the first time to a substance itself" is a false or incorrect one. It is obvious that here we are not ascribing the designations true or correct, false or incorrect to the ideas in and for themselves, but only to their application to certain objects, which happens by way of propositions, and so to the propositions. For this reason, we do not call even the ideas mentioned in the example true or false *simpliciter*, but only true or false ideas of God. Thus they are declared to be true or false only insofar as they are supposed to be ideas of God, i.e. that they can fill the place of *A* in a proposition of the form, God is *A*. This linguistic usage contradicts our claim that ideas in and for themselves have neither truth nor falsity so little, then, that instead it substantiates it.

(b) Still, there is a case in which we speak of true or false ideas without first considering them as applying to a specific object. Sometimes we also call an idea true, if we merely wish to say that this idea not only has the form and appearance of an idea that represents an object, but that there actually is an object represented by it. In the contrary case, we call an idea false when it only has the figure of an idea of an object, without actually having any object, indeed without even being able to have one. For example, we say that the idea of a body bounded by four identical surfaces (tetrahedron) is a true idea, while the idea of a body bounded by five identical surfaces is a false idea. All we wish to say by this is that the first idea of course has objects corresponding to it, which is not true of the second one. Here, then, the words true and false are not being taken in the strict sense, but in the derivative sense described in § 24 (5). Neither, then, does this linguistic usage offer any objection to the above claim.

§ 56

Parts and Content of an Idea in Itself

A very remarkable property which belongs to most, if not all ideas in themselves is their being composed of parts. Our consciousness informs us that we distinguish in almost every idea thought certain parts, from the combining of which it comes to be. The idea signified by the expression, earthling, may provide us an example. For certainly what we are thinking of with this expression, and what we are supposed to be thinking of is exactly the same as what we are thinking of with the words: "A creature that lives on the earth." But in this latter expression the several words of which it is composed make it unmistakable that the idea they are all needed to designate is also composed of several parts. It is certain that in the idea, earthling, there is present the idea of a creature and the thought that this creature lives on earth. But if the idea *thought of* is composed of several parts, which we can clearly distinguish in consciousness, then there is no doubt that the idea *in itself* which constitutes the material of this idea thought of, must also be composed of at least as many parts. So ideas in themselves are also composed of parts. Now the sum of the parts of which a given idea in itself consists we are also accustomed to calling in one word its content. So every composite idea incontrovertibly has a content.

Since we understand by this content only the *sum* of the components of which the idea consists, but not the way these parts are bound up together, an idea is not completely defined merely by giving its content. On the contrary, sometimes two and more different ideas can arise from one and the same given content. Thus the two ideas, a learned son of an unlearned father and an unlearned son of a learned father, obviously have the same content, yet they are very different. The same holds of the ideas 3^5 and 5^3, and many others.

§ 58

Closer Examination of the Most Notable Ways in which Ideas Are Compounded

(1) In the first place, we can distinguish between the more proximate and the more remote parts of any idea that consists of more than two parts. Taken in and for itself, we can of course distinguish between more proximate and more remote parts in any whole that consists of more than two parts. For if we first divide it into a number of parts less than the number of all of its parts, then several of these parts or perhaps even all of them will still be composed of parts themselves, and so we can call the parts constructed first the most proximate parts and the parts of these parts the more remote parts of the whole. But if it would be entirely arbitrary into how many and what kind of parts we first divide the whole and if the most proximate parts we arrive at in this way do not differ from the more remote ones in any circumstance that is worth noting, it would be purposeless to wish to draw any distinction by using these designations. So we must first show that with ideas the more proximate and the more remote parts can be differentiated in a way that has a real use. Now this would be the case if we called those parts of an idea that language has words to designate, by citing which the idea can easily be represented again, its *most proximate* parts; and called those into which the former can be further divided its *more remote* parts. In this sense, for example, the most proximate parts of the idea, earthling, are those expressed by the particular words in the following expression: "creature that lives on the earth." The more remote, however, would be those into which the ideas, creature, live, etc. can be resolved. It is easy to see that this distinction will be useful if we are concerned with the way and manner in which one can make others aware of what kind of an idea one understands by a certain word.

(2) It could, however, be a still more notable difference between the parts of an idea that some of them are themselves ideas while others are entire propositions. Let us consider in particular, to give an example, the idea, earthling. It would seem to me that one of its parts, expressed

by the word *creature*, taken all by itself, is an idea. The remaining part, on the other hand, "that this creature lives on the earth," would seem to me an entire proposition, which nevertheless is combined with the idea creature in such a way that the thought arising out of this combination (the thought of a creature which lives on the earth) asserts nothing and consequently does not yield us any proposition, but only a mere idea. That there are parts of ideas which are ideas themselves has always been accepted and so requires no further justification. For to the objection that if an idea consisted of several parts which were themselves ideas, it would therefore not be one idea, but a sum of several ideas, the answer is easily found. This no more follows than it can be inferred from the fact that the particular parts of a machine are already machines themselves, it does not well deserve in some aspect to be called a single machine. Just as it is only one thing with respect to the result which it alone produces, and not its individual parts as such; so is the idea that is the product of combining several ideas only one with respect to the objects represented by it, or the place it can take in a proposition, and the like. It is more controversial whether whole propositions can also appear as parts of ideas. Of course linguists will all agree with us when we say that the words above, "that live on the earth," are the expression of an entire proposition (although not an independent one). For they do, too, and point out to us in those words all of the parts required for a proposition, a primary ending (i.e. a subject), a verb (i.e. a copula), etc. But the following example, already cited in § 49, is a still more obvious demonstration that mere ideas can also encompass entire propostions. No one will deny that the thought the words, "the knowledge of the truth that God is omnipotent," express is a mere idea. Nevertheless it is plain to see that an entire proposition appears in this thought as a component part, namely the truth that God is omnipotent.

(3) Now if it is correct that the parts of an idea can be ideas sometimes and sometimes entire propositions, then there is still a second way of understanding the distinction between the most proximate and the more remote parts of an idea. Those parts namely, which are parts of a certain idea only because they are parts of a proposition included in it can quite appropriately be called its more remote parts, all others its most proximate parts. On this definition of the concept, the most proximate parts of the idea, earthling, would be the idea, creature, and the proposition,

"that lives on the earth." For these parts are not included in the idea named merely in so far as they occur in one of its parts that is a proposition. The idea of earth and the idea of living, on the other hand, would be identifiable as remote parts.

(4) If we look at the relation the parts of an idea bear to each other, we will note that some of them are attached to each other directly, others only indirectly. It is self-evident that every part of an idea must be connected with every other part, if not directly, at least indirectly. For they are both parts of one and the same whole, namely the one idea in which they occur. But it is equally certain that there must be parts that are directly attached to each other. For wherever there is an indirect connection, there must also be at least some direct one, since the former can only come about by way of the latter. Thus, in order to give a few examples of direct connection, in the compound idea, 'nothing' (= not something), the two ideas, not and something, are directly connected. So, too, it would seem to me, are the concepts of having and of obligation in the compound idea of having an obligation, directly connected. We would have an example of a connection that is merely indirect, on the other hand, in the concepts of man and honesty, which hang together in the idea of a man who possesses honesty only by virtue of the concepts, who and possession.

(5) From this example we see right away that in compound ideas there are occasionally individual parts, namely ideas, by which the connection between certain other parts is mediated. The relative pronoun, who, just mentioned, is such a component, which combines the idea, man, with the proposition, who possesses honesty, into a single idea. But the concept of possessing is also such a component, by which that relative pronoun is united with the idea, honesty, into the proposition, who possesses honesty. In that way it is also indirectly connected with the other parts of the whole idea.

(6) In what follows, having to do with the theory of propositions, we will discuss in greater detail ideas which, like the concept of possessing, unite two other ideas into a whole proposition. Here, then, only a few more words about the nature of the connection mediated by the relative pronoun, who. It appears to be an essential feature of this connection that the members connected, as in the example before us, are an idea and a proposition. The former has such an importance place in the idea

that arises from its combination with the proposition following that not inappropriately it can be called its principal part. Finally, as far as the proposition which the relative pronoun attaches to that idea is concerned, it is obvious that this pronoun itself provides an important component of it. Still, without presupposing prior knowledge of the theory of the various parts into which every proposition decomposes, we can not define the position assumed by the relative pronoun in such a proposition in greater detail. Here, then, it will suffice merely to remark that it is not always the same. For we can be convinced of this truth, even without having any more exact knowledge of those parts, by considering sufficient examples. The different endings with which the relative pronoun appears in the following ideas: "a person *who* is tall," "a girl *whose* eyes are blue," "a man *whom* no one trusts," already betray the fact that the concept of this pronoun has a very different relationship to the other parts of these propositions.

(7) From the foregoing it is clear that the components of which an idea is constituted often appear in a certain *sequence* which is not at all arbitrary. By changing it another idea is created, as we have already noted in § 56. It is self-evident, however, that we must not think of this order of the parts of an idea, so often essential, as a temporal succession. For an idea in itself is not anything actual and consequently we can not say of its parts that they exist alongside each other at the same time nor that they succeed one another at different times. It is another matter with the idea *thought of*; this is something actual and with us humans it is true that its parts follow one on the other in time. We will begin to think of the one that is first in the objective idea somewhat earlier than the one that is second, and so on.

(8) However, if I maintain that the parts present in a compound idea sometimes occur in a definite order and sequence, by virtue of which we can call one of them the first, another the second; I do not therefore maintain that this is the case with *all* of them. Could there not be parts for which the order is arbitrary, or rather, that occur in the idea in itself in no order or sequence whatsoever, but as the members of a sum? Such is actually the case, if I am not mistaken. If, to introduce just one example, we think of the idea of an A which has the properties b, b', b''..., we are, to be sure, because of our particular nature, unable to think of the properties b, b', b''... all at the same time. We think of them one after

the other, and so in a certain temporal succession, which can be this at one time or that at another. Nevertheless, we feel clearly that this temporal succession does not belong to the idea in itself, that nothing about it changes, whether we think of the attributes b, b', b''... in the order just given or in another, e.g., b'', b', b.... In the objective idea, then, the individual ideas of the attributes b, b', b''... are components that do not occur in any distinct order, say, the one in the 4th place, the other in the 5th, etc. Instead, all of them stand together in one place (namely the last).

§ 60

Concrete and Abstract Ideas

The idea of *something which has (the property) b* also belongs to the kind of ideas considered in the preceding section. The place of the principal idea is occupied by the most general idea of all, namely that of *something* or an object in general. I call such an idea a *concrete idea* or simply a *concretum*. The idea of b that occurs here, as the idea of a mere property, appropriately bears the title of an *attributive idea*. Considered as a constituent of the concrete idea, it may be called its abstractum or the idea *abstracted* from it. Thus, for example, I call the idea of an animal, i.e. the idea of something that has the property of animality, a concrete idea. But I call the mere attributive idea of animality itself the abstractum from that concreto. For whatever attributive idea b you please, as is easily judged, we can find a concretum that includes it, for which it will accordingly constitute the pertinent abstractum. It is the idea: something, which has (the property) b. I permit myself to designate the latter idea briefly by B. – It will be obvious to anyone that the answer to the question as to whether a given idea is concrete or abstract or neither of the two concerns a mere *intrinsic* property of it. For we can determine from the given idea itself, without comparing it with any others, whether it falls under the form, something which has (the property) b or is the idea of a mere property. But whether an idea, B, is the concretum that belongs to b, and consequently b the abstractum that belongs to B; that, of course, is a question we are able to answer only upon consideration of both the ideas of B and b. Or, which comes to the same thing, the answer to this buestion expresses a *relationship* between the two ideas. There are obviously also ideas that are neither abstracta nor concreta. For example, the idea, 'something', is not itself to be called a concrete idea, for it is not of the form, "something which has (the property) *b*." No more, however, can it be called abstract, because it is not the idea of a property. The like holds of the ideas, nothing, this *A*, Socrates, and many others that neither designate properties nor are of the form I have just laid down for every concrete idea. Abstract ideas can also be simple, but a concrete idea is always compound, for it consists of the idea of something and the proposition, "which has *b*." But although the concreta are always composed of the abstracta that belong to them, their linguistic expression (e.g. animal) is usually the shorter one – and that of the abstractum (animality) is customarily compounded from the one for the concretum. This comes about because when languages are devised the concreta are usually designated by their own names earlier than the abstracta are and even at present we find occasion to speak of the former much more often than of the latter. One last comment to be made is that we can not always clearly determine from a linguistic expression whether the idea attached to it is an abstract or a concrete idea. Often one and the same word is used here to designate the abstractum, there to designate the concretum, so that we must guess from the context in which it appears in which sense it is being taken. Thus the idea the word, virtue, really signifies is the idea of a property, therefore an abstractum. Often, however, we take this word in a concrete sense as well and understand by it something (namely an act or behavior*) that has virtue, as in the proposition: "Virtue may at times go unrecognized, but it never goes unrewarded," and the like.

§ 61

There Must also Be Simple Ideas

As the word itself indicates, I understand by a *simple* idea one that includes no parts at all, whether they be mere ideas themselves or whole propositions. That there are such ideas I believe I am able to demonstrate in the following way. Of every object, no matter how complex, it holds true that it must contain parts that are not also complex, but absolutely simple. If the number of parts that constitute a whole is finite, the truth

* Literally, according to the text, a being, but the point cannot be made in the same way in English. *(Translator's note.)*

of this claim is self-evident. For in this case after a finite number of divisions, of bisections, for example, we must always arrive at parts that are not further divisible and are therefore simple. But there can also be wholes that contain within themselves an infinite number of parts. We have an example of that in any spatial extension, any line, surface, or solid. With such objects, we never arrive at parts that are no longer complex, but simple, no matter how often the division is repeated, if it always produces only a finite number of parts, as is the case on continued bisection, trisection and the like, for example. That makes it appear as if such a whole consisted of no simple parts at all. I maintain, however, that even this whole must have parts that are simple. Complexity is a property which obviously can not subsist without the parts that produce it (i.e. that contain its ground or condition) being present. If these parts are themselves complex, what they account for is only a complexity of a *particular* sort (namely, of parts constituted in such and such a manner), but not the *complexity* which *generally* characterizes the whole. In order to account for this, then, and to account for it sufficiently, as a condition that requires no further condition, there must be parts that are no longer complex, but simple.* Thus lines, surfaces and solids, for example, also contain parts that can not be divided further, but are simple, namely points. Points, however, are of course not homogeneous with the whole they constitute precisely because they produce it only in infinite number. Therefore they are not ordinarily called parts by geometers, who use the word *part* in a narrower sense, of homogeneous parts only. – Every idea, then, no matter how complex it might be, even if it contained, if it is possible at all, infinitely many parts must nevertheless have parts that do not admit of any further division. For that very reason they can not be propositions, because every proposition, as such, is still complex. Consequently, since the parts of an idea can certainly not be anything but either propositions or ideas, they must be *ideas*. And so it is demonstrated that there must be simple ideas. Now since according to § 56 there are certainly *complex* ideas as well, there is no doubt that we can base a legitimate *classification* of ideas on this distinction between them, namely whether they are simple or complex. For we can guess that this difference is of great importance from the preceding discussion and it will become still clearer in what follows.

* This was Hegel's judgement, too (*Log*. Vol. I, p. 142).

§ 63

Are the Parts of an Idea the Same as the Ideas of the Parts of Its Object?

There has been frequent use of the expression, that if the *idea of an object* is correct, i.e. if it really is an idea of it, not merely judged to be, it must have a certain *correspondence* to that object. This expression's obscurity gave some an excuse for thinking of the correspondence that must hold between an idea and its object as a kind of similarity in their composition, and assuming accordingly that the parts of which the idea consists could not be anything but the ideas of the parts of which the object consists. Thus Abicht's *Logik* says on p. 362: "The *concept* of an object must have as many parts of ideas as there are particular details of the *object* of the concept." And on p. 363: "A complex concept is recognized to be complete if there are reasons to see that its object has *such* parts and only *so many* of them to display." One of the consequences of this opinion was that it was often held that a completely simple object could only be comprehended by means of a completely simple idea, and the like.

This view seems incorrect to me. In the first place, because (in my opinion, at least) there are also ideas that have no object at all, e.g. the idea of nothing or that of a round square, among others. With such ideas it is obvious that one can not pass off the parts of which they consist as ideas of the parts of their object. So one must at least choose to restrict the above claim to ideas that have an object. But if it is true that whole propositions often appear among the parts of which ideas are composed (§ 58 (2)), then one will once again not be able to say that every part of an idea is the idea of a part contained in its object. But one will concede willingly that in such a case it is not the entire proposition, but only an idea present in it that points to a part to be found in the object. This is really sometimes the case. Thus, the idea of a rightangled triangle, i.e. of a triangle that has a right angle, presents in the proposition, "that has a right angle," the idea of a right angle, which in fact points to a part present in the right-angled triangle. Something similar holds of the ideas, a mountainous country, a book with engravings, and many others. But that this is not always the case, or that not every one of the parts of a proposition occurring in an idea presents an idea that refers to a part existing in the object, we can see from ideas such as the following: "A country that has no mountains," "a book without engravings," and the like. For these obviously do not, by way of the ideas of mountains, engravings, present in them, refer to parts the object understood by them has, but rather to parts it lacks. This is even more indisputably evident with ideas like: "the eye of the man," "the gable of the house," and the like. Who could deny that the idea of a man is present in the first and the idea of a house in the second? So if the view we are contesting were correct, the whole man would have to be a part of his eye, the whole house a part of its gable, and so on. Finally, there are also objects which, being completely simple, have no parts at all, while their idea is plainly composed of several parts. A spiritual substance is a completely simple object; all the same, the concept of it is composed of several parts. So we will have to give up the notion that every single part of which an idea is composed refers to a corresponding part in its object. But we could still say that even if not every component into which an idea can be analyzed disclosed a distinct part in its object, the reverse is nevertheless true. Each one of the latter must be indicated by one, or several, constituents of the idea. But we can see in a moment that this can not always be expected, at least not of those parts in an object it does not have to possess in order to be an object that falls under the given idea. So surely no one will expect that there should be present in the idea of a flower, which this rose tree falls under, constituents that reveal how many roses, buds and leaves precisely this rose tree has. But may it be a different matter with parts an object must necessarily have in order to be called an object of a given idea? If there is reason to believe (as seems to be the opinion of many) that the idea of every property an object must have insofar as it is supposed to be the object of a certain idea is present in this idea as one of its components; then there is no more arguing the point that for every part that necessarily belongs to an object in order for it to fall under a certain idea, there must be present in the latter some particular part that represents it (an idea that refers to it). For that a thing consists of such and such parts is one of its properties. But since I hold that opinion to be incorrect, for reasons which will be developed in the sections immediately following, I have no reason at all to believe that the idea of an object must be composed of the ideas of all those parts it must have in order to belong under the idea.

§ 64

Are the Parts of an Idea the Same as the Ideas of Its Object's Properties?

The *correspondence* between an idea and its object mentioned at the beginning of the preceding section, which some thought of as a kind of similarity in composition, others believed was rather to be found in the fact that the idea of an object had to contain within it as its parts the ideas of all of the object's properties. Just as every object is, as it were, nothing but the sum-total of all its properties, so, it was believed, the idea corresponding to it would also have to be nothing but the sum-total of all the ideas of these properties. That this would not hold, however, of those properties of an object it does not *have* to possess in order to be the object of a certain idea will be granted us without objection. This already involves an admission that the claim just asserted must be expressed more precisely in the following way: Every idea of an object is a mere sum-total of the ideas of all those properties it is necessary for it to have as an object of the idea. Is the proposition *now* true? It is surely not to be denied that many properties of an object really are and must be conceived together in the idea we form of it. For example, the equality of all the sides is a property of the equilateral triangle and of course the idea of it is present as one component in the concept thereof. For we do understand by an equilateral triangle nothing else but "a triangle, of which all the sides are equal." The consideration of such examples ties in with the fact that it is very easy to understand how we can assert the necessity that a given object has a certain property if this property has already been conceived to be a component of the idea of it. This is probably the principal reason the opinion of which we have just been speaking has won so many adherents. For most logicians do seem to believe that the idea of any object is composed of nothing besides the mere ideas of its properties. All the same, I venture to contradict this opinion, which prevails almost universally. I maintain not only that there are various parts of an idea that do not express properties of the corresponding object at all, but that in every object there are also properties, which – even though they belong to it of necessity insofar as it is supposed to fall under a certain idea as object – are by no means conceived of as among the idea's components.

(1) The first part of this claim I hope will be conceded to me without difficulty as soon as one has considered the following fact. In order – so far as this is possible – to compound an idea of an object out of mere ideas of its properties b, b', b''..., it is still requisite that there be some other ideas as well, which serve to connect them. In order to represent the object that has properties b, b', b''... in itself, one must form the idea "of a something which has (the properties) b, b', b''...". But present in this idea, besides the ideas of the properties b, b', b'', ..., there are many other ideas as well, in particular the idea of *something*, the idea of the relative pronoun, *which*, and the idea of *having*. Besides, the ideas of many properties of an object are composed of a number of other ideas that do not represent properties of this object at all. Thus the idea of equal sides, which manifests a property of the equilateral triangle, and one which does appear as a component in the concept of it, is itself composed of the ideas of a side and equality. Accordingly, these ideas are present in the idea of an equilateral triangle as (remote) components. Nevertheless, no one will say that these ideas express properties of the equilateral triangle itself. An equilateral triangle is neither a kind of side nor does it have the property of equality. This latter property is to be found only in the relationships in which the lengths of its sides stand to each other.

(2) Now in order to prove the second part of my claim, I could (a) appeal to a great many examples in which it is obvious that we ascribe a property to a certain object as necessarily following from the concept of it, although we are not in the least aware of actually having thought of this property within that concept, indeed without its even having been known to us previously. Thus, for example, we can discover after some reflection that every square has the property that its side is related to its diagonal as $1:\sqrt{2}$, although we are not at all aware that the idea of this property is already there in our concept of a square as one of its components. (b) A second proof can be derived from the existence of so-called *equivalent ideas*. There are undeniably ideas that we differentiate from each other with the utmost clarity even though we see at the same time that they have the same objects and that the properties which can be deduced from the one can also be deduced from the other. For example, the two ideas, equilateral triangle and equiangular triangle, are of this sort. Who would not look on these two ideas as different? And

all the same they certainly have the same objects and the particular properties that follow from the one can also be derived from the other. The like holds of the two ideas: "a space which includes all points equidistant from a single point" and "a surface which for a given dimension contains the greatest physical content," two ideas which are so different that the one is scarcely reminiscent of the other. Yet every mathematician knows that they have the same object. Now if the idea of every property of an object had to be present as a component of the concept of it, then all such equivalent ideas, because they have the same objects, would also have to have the same components, and consequently be *one and the same idea.* (c) I find a third and still more decisive proof for my claim in the following circumstance. There are ideas from which follow not only very many, but actually an *infinite* number of properties for their object. The idea of $\sqrt{2}$ is such a case, for example. For it represents a quantity which, as is well known, is composed of infinitely many parts, $1+\frac{4}{10}+\frac{1}{100}+\frac{4}{1000}+\frac{2}{10000}\ldots$. The character of each of these parts (what numerator belongs with each denominator) is determined by the idea of $\sqrt{2}$. Now since we can look on the properties of these particular parts at the same time as properties of the quantity itself, we have in $\sqrt{2}$ the example of an idea that informs us of an infinite number of properties of its object, all entirely different from one another. For obviously the determination of each of the fractions $\frac{4}{10}$, $\frac{1}{100}$, $\frac{4}{1000}$, $\frac{1}{10000}\ldots$ furnishes its own property of the whole quantity (differing not merely in our sense, but objectively). So if it were true that every property of an object which follows necessarily from the idea of it must be thought of as a component of it, then the idea of $\sqrt{2}$ and any idea like it would have to be composed of an infinite number of different parts. In order to be able to say, then, that we are *conceiving* of such an idea, we would have to think, at least obscurely, of the infinitely many parts of which it consists. We would consequently have to be in a position, with our finite power of thought, to comprehend an infinite number of ideas at the same time. (d) Nevertheless, much as I have already stressed these three reasons, I will not regard them as decisive. For we can not conclude with certainty, merely from the fact that we are not conscious of one idea as a component of another, that it is not actually a component of it. So the doubt can always be raised against proof (a) that the attributes derived from the concept of an object only after long deliberation, which of course we had not clearly conceived, could still have been obscurely represented. Against (b) it can be said that the distinction we draw between so-called equivalent ideas may only consist in the fact that we have a distinct conception of these components of the one and those components of the other and conceive indistinctly of the rest of them. Finally, to weaken (c) one could reply that it is no obvious contradiction for a finite power of thought to have an infinite number of ideas in a finite period of time if, as is the case here, it is not required that it be distinctly conscious of these ideas. I have introduced these three arguments, therefore, not so much in order that the reader should feel himself moved by them to definitely commit himself as to the incorrectness of the view that is disputed here, but only so that he may become more fully acquainted with the consequences the adoption of this view has. In my opinion, the only decisive arguments are the ones to follow now. First, let me recall that here I have to prove my thesis, although it also applies to ideas thought of, only of ideas in themselves, since they are all that I am concerned with here. Obviously, however, those ideas that happen to make their appearance along with the thinking of an idea in itself are by no means to be counted to that idea. So the ideas of the sounds of the letters t and l that may be present in my mind when I am thinking of the objective idea of a triangle certainly do not belong to that concept. Just from the fact that a certain property occurs to us involuntarily, that we have the idea of it whenever we have the idea of a certain object, i.e. that our subjective idea of the object is always accompanied by the idea of that property, it does not yet follow that the *objective* idea of it includes the idea of that property as one of its components. Even if it were the case, for example, that the property of equiangularity occurred to us spontaneously in connection with the concept of an equilateral triangle, it would by no means follow from this fact alone that the objective concept of an equilateral triangle also includes within itself the concept of equiangularity as a component. This presupposes, it now becomes easy to show, that there are properties which necessarily belong to the object of an idea yet without being present as components of it. For that such a case arises with the concept, 'equilateral triangle', and with countless others as well, becomes evident from the following observation. As is well known, the property of equiangularity belongs to all equilateral-triangles, yet everyone must concede that the concept of this equiangularity does not lie within the concept of an equilateral triangle taken in and of itself. For this concept arises when the concept of triangle is attached to the proposition: "which is equilateral." Now it is obvious that the concept of equiangularity is present neither in the concept, 'triangle', nor in the proposition, "which is equilateral." So it is certainly not present in the whole composed of nothing but these two parts either. In so far as it is true that in *thinking* of this concept, with every occurrence of the *thought* idea of an equiangular triangle the idea of equiangularity puts in an appearance, this has nothing to do with the idea in itself. It surely consists of no other parts but what we give it. Otherwise one would have to say that it is impossible in itself to connect the concept, triangle, and the proposition, which is equilateral, together without adding on a number of other parts besides, among others such as contain the concept of equiangularity. But this would certainly be very false, since the construction of ideas in themselves is something quite arbitrary, so much so that we can even unite components that represent contradictory properties. For example, the combination of the following ideas: "a triangle which is equilateral and also not equilateral," presents a concept, only one that has no object. Just as the concept of an equilateral triangle emerges when we attach the proposition, "which is equilateral," to the concept of a triangle, the concept of an equilateral tetragon arises when we add the proposition, "which is equilateral," to the concept of a tetragon. If the first concept were supposed to include the concept of equiangularity within it, this would have to hold of the second as well. But no one will say of this latter concept that it includes the concept of equiangularity among its parts, since it is well known that it is not true at all that every equilateral tetragon is also equiangular. (e) According to § 61, there must be simple ideas. Now if *a* is an idea of this sort, the proposition, "the idea *a* is simple," holds true. And the concept of simplicity expresses a property that *necessarily* belongs to the object the concept, "the idea *a*," has, namely to *a*. So if the supposition I am disputing were correct, the concept of simplicity would have to occur as a component of the concept just named. But it does not; for surely this concept occurs neither in the concept of an idea nor in the idea *a* itself. (f) Even simple concepts, perhaps with some exceptions, must represent something. Now however this something is constituted, it must have the property, "that it is something." Now if every property of an object that necessarily belongs to it as soon as it is supposed to be the object of a certain idea had also to be present in the idea itself as a component, all simple ideas would have to contain the idea, something. But in order to remain simple, they could contain no other besides this one. As a consequence, all simple ideas (with the exception of some that do not refer to any object at all) would be equivalent to each other, or to put it better, there would be only a single simple idea, namely the idea of something. Who does not see that this is absurd? – (g) But if one acknowledges that there are a *number* of simple ideas, distinct from each other (as one must), then it can be asserted with respect to each of them that this *a* is not the other *b*. But since this assertion already expresses a property of the idea *a*, it would have to a component in the idea of it, i.e., in the idea: "the idea *a*." But this is absurd, however. (h) Finally, suppose one were justified in concluding, merely because an object has property *b*, that the idea *A* which refers exclusively to it included the idea of *b* as a part. Then *A* would have to be of the form, "X, which has *b*," and one would have to admit either that the attribute *b* belongs to *every* object that falls under *X* or that it belongs only to some of them. In the first case, one would have to grant that *X* itself also contained the attribute *b* and was consequently of the form, *Y* which has *b*. What would follow for the idea A is that it contained the component *b* two times, and if one chose to go on asking the same question and giving the same answer, infinitely many times. There would still always be a putative idea which, omitting even an infinitely repeated component *b*, otherwise contains all of the other parts of *A*. Of this idea, which I will designate by *Z*, just the same thing would have to be true, which I previously asserted of *X*, namely that property *b* either belongs to *all* of the objects falling under Z or surely to *some* of them only. Let us assume the former: then *Z* itself is an example of an idea of which all of its objects have a certain property without this property being signified by any one of its parts. In the second case, it can be maintained at least that the idea of a *Z* which has the attribute *b* is a *denotative* (*gegenständliche*) idea, i.e. that it has an object. But this claim asserts a property of the idea mentioned here which belongs to it necessarily, and yet which does not appear as a component in the idea of it, i.e. in the idea, "an idea of a *Z* which has property *b*." For the thought that this idea has denotation (*Gegenständlichkeit*) certainly lies neither in the concept of an idea generally (for there are also imaginary ideas), nor would it be of any use if this thought were actually included in the ideas *Z* or *b*. For from the fact that the ideas *Z* and *b* are denotative taken in themselves it does not follow that the idea of an object that has these properties *Z* and *b* combined is denotative. So we see in each case that there are properties which follow from the idea of an object although they are not represented by any of its components.

§ 66

The Concept of the Extension of an Idea

I have already mentioned several times that if not all ideas, nevertheless most of them refer to a certain *something*, very definitely to be distinguished from the ideas themselves, which I call their *object*, taking this word in its broadest sense. There was already an explanation given in § 49 of the concept I attach to this expression. But in order to avoid every possible misunderstanding here, it will be useful to add this to what has been said: The form of expression, "A certain idea refers to an object," which I always take in the sense that this object is one the idea mentioned represents, is being used in an entirely different sense when we are accustomed to saying of a proposition of the form, "*X* has (the attribute) *b*," that *by this proposition* we *refer* the idea *b* to the object *X*. In this case the word *refer* means nothing different from *assert*. The idea *b*, or rather the property it indicates, is merely being *ascribed* to the object represented by *X* by means of this proposition. Now if it is true (as for example, the proposition: God has omniscience), then of course we can say that the object represented by *X* (by the concept, God) is *represented* at the same time, not to be sure by the abstractum *b* itself, but by its concretum, i.e. by the idea of "something which has the property *b*." But if we have a false proposition before us (for example the proposition: Man has omniscience), then it is not true at all that every object contained under *X* (namely a man) is at the same time an object which the idea, "something which has *b* (something which has omniscience)," also includes, despite the fact that in the thought proposition we are *ascribing*, i.e. *referring* to it the property *b* suggested by this idea.

(2) I call ideas that have one or more objects *denotative* or *non-empty ideas*; those, on the other hand, that have no object corresponding to them, *objectless* or *empty*. Once one knows of an idea that it represents

some objects, then one can still ask which ones it represents and how many? Anyone who answers this question, or gives us an account of what objects a certain idea refers to is giving us an account of the *domain*, the *extension* or the *sphere* of the idea. What I understand by these expressions is that property of an idea by virtue of which it represents just these and only these objects and no others. So only with ideas that have objects do I find a domain as well, and this is just as much the case when they have only a single object as when they have a number of objects. We are told of this domain of ideas when we are given a detailed account (by the mediation of other ideas, of course) of the one or more objects they have. So, for example, we are given an account of the domain of the idea, man, if we are given an account, in detail, of all of the beings that are represented by this idea. We say of every single one of these objects, and of every collection of several of them which is not yet the sum of them all that it is a *part* of the given idea's domain, or *belongs* to its domain, or *falls under it*, is *contained in it*, *subordinated to it*, or *can be subsumed under it*, or is *comprehended* or *included by it*. So we say, for example, that Julius Caesar belongs to the domain of the idea, man, the duty of truthfulness to the domain of the idea of duty generally, and the like.

(3) In order to specify the domain of an idea that has several objects completely, one must, according to the explanation just given (2), not only specify the set of objects that belong to it but also give an account of which particular objects they are. Like every set, the set of these objects has a certain *quantity*, which we call the *breadth of the domain*. With many ideas the set of objects that belong to it is infinite. For example, the ideas, line and angle, include infinitely many objects they refer to, since it is well-known that there are infinitely many lines and angles. Now since the infinite, in just the respect in which it is infinite, does not admit of any determinate account, the breadth of the domain of such ideas can never be specified completely in and of itself. It can only be specified in comparison with others and in some aspect, for example by saying that the domain of idea *A* has the same breadth as that of another idea *B*, or that its domain is a part of the latter's, or *vice versa*, and the like. These kinds of specifications, since they depend on the relationship of one idea to another, will be discussed in the following section.

(4) What has been said so far has already given us sufficient reason to see that specifying an idea's domain, or answering the question as to whether an idea is denotative or not and if so, what objects it has and how many, is not always something that can be done by considering the ideas in and for themselves. It requires taking account of a great many other things and often requires knowledge of accidental circumstances. For that the idea, "heirs to the empire of Genghis Khan," has precisely four objects, for example, history must teach us. Nevertheless, the extension of an idea does not belong among its relations, but among its intrinsic properties. For if we want to determine the extension of a given idea, what we have to attend to is not just any object we please, discovering now this extension and now that as we choose now these objects and now those. No, we must take note of which objects and how many there are within the totality of things generally that it represents. Now since there is only one such set, there is also only a single extension for any given idea. Consequently this extension is not counted among its relations. Also the fact that a given idea has an extension and which one it has is something quite unalterable, not at all something that can increase or diminish with time. For example, the idea man, if we call every being endowed with reason and sensation which at any time has lived or will live on earth by that name, has its determinate extension, in which there has been no change nor ever will be. The idea, "a man now living," also changes its objects only if we change the meaning of *now*, i.e. change one of its components and so change the idea itself.

(5) The *pictorial* terms we use in calling the concept of the domain of an idea its *extension*, its *sphere*, words with meanings that are in accord with terms for *extending over*, *including*, and *breadth*, already betray the fact that we give a material sense to this concept by comparing it with a *space*. What we think of in conceiving of the domain of an idea is the compass, or rather the size of some spatial extension. And we conceive of the particular objects that belong to the idea's domain as particular parts of this space. The expression, *sphere*, indicates that we often think of the domain of an idea as a physical space, indeed shaped liked a ball. But it is by no means necessary to stick to this idea. We could just as well use the picture of a surface, in fact of a mere line. Actually, we shall see from what follows that representing it by a surface, specific-

ally a plane surface, deserves to be preferred over other representations.

§ 67

There Are also Objectless Ideas

True as it is that most ideas have objects, even an infinite number of them, I still maintain that there are also ideas which I have called *objectless* in the above, i.e. ideas which have no objects and consequently have no extension at all either. This seems to me to be most indisputably the case with the concept signified by the word, *nothing*. It seems to me absurd to say that even this concept has an object, i.e. a something it represents. If anyone objects to saying this is absurd and chooses to find on the contrary that it is absurd to claim that there is an idea with no object at all, and so representing nothing, this only happens because he understands by ideas merely thought ideas, i.e. thoughts, and takes the *material* these have (the idea in itself) to be their *object*. We may very well say that the thought of nothing has a material, namely the objective concept of nothing. But that this too has a certain object falling under it is a claim that is difficult to justify. The same holds of the ideas, a round square, a green virtue, and the like. Of course we are thinking of something in using these expressions and must be thinking of something, but what it is is not the *object* of these ideas, but the idea *in itself*. With these examples, moreover, it is self-evident that no object could correspond to them, because they ascribe contradictory attributes to the same thing. There could, however, be ideas that are objectless not just because they ascribe contradictory properties to their object, but for some other reason. Thus the ideas of a golden mountain on a vine now blooming may have no object, although there is nothing contradictory in them.

Note: For the most part, the theory of the extension of ideas is set out in such a way as to presuppose that every idea must have some extension. I admit that appearances favor this opinion. For in the first place *linguistic usage* seems to demand that we presuppose for every idea an object represented by it, and that accordingly we not count what the word, nothing, and others like it express as ideas at all. Nevertheless, there are propositions, even true propositions in which the sort of parts

I have described here as objectless ideas do undeniably occur. And we have seen in § 50 that there is reason to include all the parts of a proposition that are not propositions themselves under a common term. The most anyone could complain of, therefore, is that the term, *idea*, does not suit this purpose. But perhaps even this objection has no basis, and for any term we chose to use for the concept of an idea our imagination would add on the auxiliary concept of an object belonging to it, just because *most* ideas have such an object. But an objection of yet another sort could still be raised. It could be said that the ideas I have regarded as objectless also have an extension, since they are sometimes compared with respect to their extension, and one of them regarded as *broader*, the other as *narrower*. So there will be no objection to saying that someone who demonstrates the impossibility of round polygons is doing *more* than someone else who is only demonstrating the impossibility of round squares, since the latter can be inferred from the former, but not *vice versa*. But this inference only seems to be valid insofar as we presuppose that the concept of a round polygon is broader than that of a round square. I too grant that the impossibility of a round square can be inferred from the impossibility of a round polygon generally, but I deny that the minor premise, "Round squares are a species of round polygons," is a prerequisite for this inference, and that we must first admit that both concepts have an extension in order to be able to draw that conclusion. The assertion that no round polygons are possible at all already follows from the proposition, that no polygon is round (or every polygon is something which is not round). But the conclusion that no square is round is also derivable from that major premise, by way of the minor premise that all squares (not just round squares, of which there are none) are polygons. That round squares are impossible can consequently also be derived in like manner. Moreover, in § 108 we shall become acquainted with a sense in which the relation of sub-ordination can be applied even to objectless ideas.

§ 68

There Are also Ideas that Have Only a Finite Set of Objects, and Singular Ideas as Well

To my knowledge no one has ever disputed the fact that there are ideas which refer to an *infinite set* of objects (§ 66 (3)). But that there are also ideas which have only a *single* object or a *finite* set of objects has not been universally acknowledged. Let us then prove both by way of some very obvious examples.

(1) Must we not admit that the ideas we connect with the expressions, the philosopher Socrates, the city of Athens, the fixed star Sirius, and many others, and moreover all ideas of the form, *this A*, in its *more precise* sense (§ 59) represent only one single object? – The reply might be made that with each of these expressions we are of course only *thinking* of one object, but that the idea we form of this object would in fact fit other objects as well, or that even if there are not several objects in *reality* which have all of the properties we are thinking of in our idea, there are in the realm of *possibility* at least. But my reply to this is: What we are *thinking of* under an expression is the idea we attach to it. However it may be that the *words*, the philosopher Socrates, etc., can be interpreted in such a way that they fit still another object, the idea we attach to them at present nevertheless has only one object, just because we are *thinking* of only one. Now consider the further point that there may be still other objects that have all of the properties thought of in our idea, in the realm of possibility, if not in actuality. As far as that is concerned, we may not forget that our idea of the philosopher Socrates requires the reality of its object (about 2000 years ago). From that it follows that something which did not exist in reality at that time could not be an object of this idea. The like holds true of all ideas that require, by their nature, that their object should be something real (at a specific time or even at all times). We may never say of such ideas that they comprehend more objects than there are *real* things constituted as they describe them. For merely possible things, which have no reality (at a specific time), precisely because they lack reality, do not belong among their objects. – Other examples of ideas that have only a single object are the ideas of the highest moral law, the Pythagorean theorem, and similar ideas. Their object (as we see) is a mere proposition in itself. Thus it is something which neither has existence nor can assume it, i.e. it is neither to be counted among real things nor among possible things. We shall recall a whole class of ideas that have and can have only a single object when we mention ideas of the form: "*the totality of things that have* (*the property*) *b*," e.g., the whole world, the entire human race, the sum of all truths, and the like. With such ideas it is already a matter of

their form that, insofar as they are denotative (insofar as there are actually things with property *b*, and several of them), there can be only a single object (always composed of parts, often of infinitely many) represented by them. For the sum-total of *all* the objects having a certain property *b* can still be only one thing. – Some logicians have called ideas of this sort, ideas which have only a single object, *individual* ideas. But since every idea is as such only one idea and consequently can also be called individual, it will be more correct to call them *singular ideas*, following Krug (*Fund.* § 79, Note 2). Every other idea which represents several objects (at least two) will then, in contrast to such singular ideas, be called a *common idea*, also a *general* idea.

(2) No one can any longer dispute the fact that there are among these some that comprehend only a *finite* set of objects, once the existence of singular ideas has been granted. If, for example, *A*, *B*, and *C* are singular ideas, then the idea, one of the things *A*, *B*, will obviously have only two objects. The idea, one of the things *A*, *B*, *C* will obviously have only three; and the idea, one of the things *A*, *B*, *C*, ... will generally have as many objects as there are distinct things here assembed. Another example that belongs here is provided by the idea: "a geometric theorem known to Euclid," for this idea can surely have only a finite number of objects, since no human intellect comprehends an infinite number of truths. It is just as obvious that the idea, "a whole number between 1 and 10," has no more and no less than eight objects, and so on.

§ 70

Real and Imaginary Ideas

Just as a redundancy can occur in composite ideas, in that their content includes components which assert properties of the object they refer to that are already consequences of other components, the opposite situation can also occur. Components are included in an idea which ascribe to the object it could represent because of them properties that contradict those implied by the rest of the properties. We can not say of such an idea that it in fact represents an object. All we can say is that its individual parts and the way in which they are combined are similar to what is the case with ideas that have an object, but that they lack one, because the properties concerned contradict each other and are consequently never encountered together in reality. The compound idea, "a triangle which is four-sided," offers an example. For this idea's principal part, 'triangle', refers to a certain object, and the auxiliary part, "which is four-sided," is attached to the principal part exactly as if it were supposed to mention a property this object has. Something which belongs under the idea, 'triangle', can never have the attribute, 'being four-sided', however. Consequently, there is really no single object that would be represented by the whole idea, "a triangle which is four-sided." Another example of such an idea, in which the contradiction is less apparent, would be the idea, "a body bounded by five identical sides," for that such a body is impossible becomes clear to us only upon relatively longer reflection. There are, then, as we see from these examples, ideas that despite being constituted in their particular parts and the way they are combined just like ideas that refer to an object nevertheless have no object, just because the properties they are supposed to ascribe to it contradict each other. Or to put it still more clearly: There are ideas of the form, "an A which is B and P at the same time," where B and P are such that the two propositions, "Every B is M" and "Every P is not M" are both true. Now if it is necessary to take note of redundant ideas, it is easy enough to guess that ideas like those with which we have just become acquainted deserve still greater attention. They could (I believe) not inappropriately bear the title, *self-contradictory* ideas. They have usually been called *empty* ideas sometimes, sometimes *impossible* ideas, sometimes *imaginary* ideas. All the rest have been given the name of *possible*, *actual* or also *real* ideas. I shall retain these designations, then, no matter how little they are to the purpose, because of the civic rights they have already achieved. And so I shall now give warning of certain misunderstandings that could be occasioned by an incorrect interpretation of them. (a) We may not refer the expression, *empty*, to the content of these ideas and thus we may not hold them to be empty of content. For all ideas which like the present ones, belong to the class of compound ideas have a content, i.e. certain constituents. If they are called empty, then, it is only in connection with their extension, i.e. the sum of objects they refer to or which are represented by them, namely because there is no object at all that would be represented by an idea of this sort. In this connection we are nevertheless not to forget that contradictory ideas are not at all the only ideas that can be called *empty* in this sense of the word. For there are various other ideas which, althought they are not contradictory, still have no object, like the idea of nothing, for example. The distinctive feature of contradictory ideas consists only in the fact that they have no object because they ascribe contradictory properties to the object they are supposed to refer to. (b) The adjectives, *impossible*, *possible*, *real* and *necessary*, are still easier to misunderstand when they are applied to mere ideas. Ideas in themselves have and can have no existence. Therefore one should never ascribe to them either necessity (which always presupposes existence), or reality, or mere possibility, or even impossibility, if what is understood by the last is not the mere denial of a possibility, but (as is usually the case) it is imagined that the thing to which it is ascribed could very well exist if only its existence were not hindered by the existence of certain other things. That the term *impossible* was applied to self-contradictory ideas probably came about only because it was either thought that it was not possible to cite an object to which they referred or because even merely thinking of such an idea was held to be something impossible. Now as far as the first reason is concerned: to be sure it is completely correct; nevertheless the remark made under (a) also holds true here, namely that the impossibility of finding an object for such a contradictory idea is by no means an attribute that belongs to this sort of idea exclusively. But it seems completely incorrect to me to say that the thought of it, i.e. the subjective idea of it, is impossible. For we really have such ideas, whenever we hear the verbal combinations, a round square, a regular pentahedron, and others like them expressed. Or we would have to say that what we think of with such combinations of words is either nothing at all or no more than with the entirely meaningless word Abracadabra. But the fact that we put forward propositions like "there can not be a regular pentagon, a negative square" as truths and that gaining insight into these truths requires us to consider in each of them its own object, the one for the former entirely different from the one for the latter, already proves to us that this is not so. (c) The expressions, *imaginary* and *real*, must also be used with caution if we are not to be misled by them. Ideas in themselves (as has already been said many times) are not thoughts, much less images. They can therefore never be called merely *imaginary*, i.e. merely *imagined*. If all the same we apply such a designation to self-contradictory ideas, this is to indicate that they can merely become the material of thought (of imagination), that there could never be encountered outside our thought an existing object that corresponds to them. In contrast to them, the other ideas have been called *real*, supposedly because it was thought that there was always some actual (real) object corresponding to them. But this (as I have reiterated several times) is incorrect, because there are also ideas that refer to nothing real whatsoever, yet contain no contradiction, and consequently can not be called imaginary, either.

Note 4: Even those logicians who accepted the distinction between real and imaginary ideas have not all understood it in the same way. Leibniz (*New Essays*, L. II, Chap. 30, § 5) accused Locke (it seems to me not unjustly) of vacillating in his understanding of this distinction, in that sometimes he required for the reality of an idea that an object corresponding to it be merely possible, sometimes that it be actual. Leibniz improved on this mistake by giving his definition, that the absence of any corresponding object makes an idea imaginary (or chimerical) only if the idea explicitly presupposes its reality. This agrees completely with my views also, and therefore I call the concept of a regular decachiliagon real even if there is no actual object that would be such a decachiliagon. For that concept, like all spatial concepts, does not include the idea of reality, but only the idea of the mere possibility of an object so constituted. On the other hand I would call the idea, "Alexander, the father of Philip," chimerical, because there is no actual object that would correspond to it althought it presupposes that there should be one. For the same reason, I find myself forced to charge someone – not one *qui parle en hiver de roses et d'oeillets* – but one who constructs the idea of roses that bloom only in winter, with imagining a chimerical idea, for this idea includes the concept of existence. In my opinion it would generally be useful to distinguish between two kinds of imaginary ideas: namely those in which the contradiction arising from the assumption of objects corresponding to them follows from purely conceptual truths and those for which it follows in some other way. But this distinction will only become quite clear when I have explained in § 132 what I mean by pure conceptual propositions, so a few examples may serve here. The concept of an equiangular yet right-angled triangle would be an example of the first kind, for the impossibility of such a triangle is apparent from mere concepts (from mere a priori truths). An example of the second kind would be the idea of Alexander as father of Philip, for the two contradictory properties this idea would confer on its object follow from it only upon introducing the empirical fact that Alexander was not Philip's father, but his son. Those who distinguished between merely *empty* and *imaginary* ideas also seem to have had their eye on this distinction, and counted among the former the idea of a 997 year old man. Jakob (Log. § 154) divides ideas into *ideal* and *real*, depending on whether the object the intellect is thinking of by means of them is merely possible or also actual. In his distinction he seems to presuppose that every idea has either an actual or a possible object. I believe, on the contrary, that there are ideas which have no object at all (to which class, for example, the imaginary ideas belong) and others which to be sure have an object, but only such that it makes no claim to either actuality or possibility. The idea designated by the words, "a mathematical truth," seems to me to be of this kind, for this idea has many objects because there are many mathematical truths. But since none of them, as a truth in itself, has existence or even possibility, one can not say that the objects of this idea are *actual* or *possible*.

§ 71

Two Consequences

(1) According to the definition in the preceding section, the concept of an imaginary idea requires that it be compound. For this very reason, then, ideas that are *simple* always belong to [the class of] *real* ideas, even if they are not denotative. (§ 66.)

(2) Not every idea that includes an imaginary component has to be imaginary itself; there are ideas of this kind that are not only real but even denotative. Thus, to give just a single example (and after it has been introduced, one will surely be reminded of many others), the idea, "the mathematician who first applied the concept of $\sqrt{-1}$," certainly contains the imaginary idea, $\sqrt{-1}$, as one of its parts, yet it is undeniably a denotative idea.

Note: Thesis (1) had already been stated by others, e.g. by Locke (*Essay*, B. 2, Chap. 30). Leibniz (Nouv. Ess. *ib.*) also agreed, as did Lambert (*N. O. Dianoiol.* § 654) and others.

§ 72

What the Author Understands by Intuitions

Everyone will grant that of the many kinds of ideas with which we have become acquainted in the foregoing the most noteworthy, with respect to their content, are the *simple* ideas. With respect to extension, however, the most noteworthy are those that represent only one *single* object. How much more worthy of note would those ideas have to be that combined both of these characteristics, i.e. ideas that were simple and yet at the same time had one single object! The only question is whether there are ideas of that kind. When we reflect on the fact that enlarging an idea's content, i.e. bringing in some new specifications and so making it still more composite, is ordinarily the only way of making it narrower, we feel ourselves disposed to doubt whether any idea could be simple through and through and be so narrow in extension that it has only a single object. For example, the idea of a watch becomes significantly narrower in extension as soon as we add the characteristic that it should be suited for carrying in the pocket, so constructing the idea of a pocketwatch. Its extension becomes still narrower when we make the further addition that this watch is supposed to have a golden case, or create the idea: gold pocketwatch, and so on. The prevailing impression, then, is that in order to obtain an idea that has the smallest extension, i.e. denotes only a single object, we must include a very large number of specifications in its content; and consequently that such an idea could never be simple. But suppose I were to prove that even among the ideas we human beings possess, and so among the class of *subjective* ideas there are very many which are entirely simple yet are genuine singular ideas? From this it would incontrovertibly follow, since there is an *idea in itself* corresponding to every subjective idea, that there are also simple singular ideas among the objective ideas. I believe I can succeed in proving what I have said here in the following way. Every time we turn our attention to the modification produced in our mind by some physical object brought before our senses, e.g. a rose, the *first*, *immediate* effect of this act of attention is that there arises in us an *idea* of that modification. Now this idea is a *denotative* one; its object is, namely, the modification that is going on in our mind at the time. It is nothing but that, so it is a unique object, and therefore we can say that this idea is a *singular idea*. To be sure, many other ideas, among them some that are not singular ideas, are created on this occasion, by the mind's ongoing activity. Whole judgments are also included, particularly concerning the modification occurring in us, in that we say, for example: This (that I am just now seeing) is the sensation or idea of red; this (that I am now smelling) is a pleasing fragrance; this (that I now feel when a thorn pricks my finger-tips) is a painful sensation, etc. The ideas of red, pleasant fragrance, pain, etc. in these judgments have a number of objects, of course. But the subject-ideas present here, which we designate by the word *this*, are certainly genuine singular ideas. (§ 68.) For what we understand by the *this* is precisely this unique modification going on in us right then and not any other one that might be taking place somewhere, even if like this one it should be our own. The further point is no less certain that these ideas are all *simple*. For if they were composed of parts, they would not be the *first* and *immediate* effect that arises from the observation of the modification going on in our mind at the time. No, the particular parts which constitute that compound idea would have been created earlier and more directly. From the fact that we use several words to *refer* to these subject-ideas, this (that I am now seeing), this (that I am now smelling), etc., we are not to conclude that they are compound themselves, however. Reminders of this have already been made in §§ 59 and 69. It has been demonstrated, then, that with every case of observing a modification just occurring in our mind ideas arise in us that despite their complete simplicity have only a single object, namely the observed modification itself, to which they are related as the first and immediate effect to its cause. All that we have still to be concerned with is an appropriate name for this sort of idea, then. I believe that in Germany the word *Anschauung* (intuition) has been used if not in exactly the same sense yet in a very similar one since its introduction into logical usage by Kant. Now since I know of no other word that would be more fitting, I beg permission to use the word *Anschauung* (intuition) not only for subjective ideas of the sort described, but also for the objective ideas corresponding to them. I will, then, call every simple singular idea an *intuition*, a *subjective* one if the idea itself is subjective, an *objective* one if it is objective. Beginners would be reminded that the original meaning of the word [*Anschauung*] should not be permitted to mislead them into thinking exclusively of ideas that are conveyed to us by the sense of sight. We call ideas of every other sense, indeed even ideas that come to us from no external sense at all, intuitions also, if only they are simple and have a unique object.

§ 73

What Is It that the Author Calls Concepts and Mixed Ideas?

(1) Anyone who grants me the existence and the noteworthy character of ideas such as the *intuitions* described in the preceding section will not dispute the fact that ideas which are not intuitions and do not contain any intuition as a part are also sufficiently worthy of note to deserve a title of their own. So I call them *concepts*, because I am of the opinion that this word has also been used in a very similar sense, ever since concepts and intuitions began to be distinguished. Thus, for example, I call the idea of *something* a mere concept, for this idea is no intuition, since it has not one but infinitely many objects. Nor does it include any intuition as a component, for it is no compound at all. Likewise, I call the idea of *God* a mere concept, for this idea is not an intuition either, since although it has only a single object it is nevertheless not simple: I understand by God that being which has no cause of its reality. Furthermore, this idea includes no intuition as a component, because none of the ideas of which it is constituted is a singular idea, and so on.

(2) If a compound idea includes intuitions among its parts, I will call it, even supposing that all of its parts are intuitions (if that should be possible), a *mixed* idea. Accordingly the idea, "the rose that is producing this fragrance," is a mixed idea, for the idea of "this fragrance" which it includes as a part is an intuition.

(3) As the part of a mixed idea I regarded as the *primary* part (or the principal part, § 58) is sometimes an intuition, sometimes a concept, sometimes I call the whole idea itself a mixed *intuition*, sometimes a mixed *concept*. Thus I shall call the redundant idea of *this*, which is a color, a mixed intuition, because the principal part of that idea, this, is an intuition. On the other hand, the idea, "the truths contained in this book," I call a mixed concept, because the principal part of this idea, truths, is a concept. For a more precise distinction, or in contrast with such mixed intuitions and concepts, I call the rest of them *pure* intuitions and *pure* concepts.

§ 75

Some Remarks on the Difference between the Ways in which Intuitions and Concepts Are Designated

It will place the difference between intuitions and concepts in a still clearer light if we bring forward something about the different ways in which these ideas are *designated* in language.

(1) So far as *intuitions* are concerned: I note that we are unable to evoke for a second time a single intuition we have once had, if what this is supposed to mean is that these two subjective ideas correspond to one and the same objective idea. Every subjective intuition has its own object, namely that modification outside of us or within us which is the immediate cause of its arousal. But such a cause is never present but one time, since a change even in the very same subject at another time is already a second modification. It follows, then, that every two subjective intuitions have two different objects to which they refer. Consequently, they must also necessarily pertain to two objective intuitions that are distinct from each other. The color, the smell, the pain I am perceiving right now may be ever so like one I perceived at some other time. Nevertheless, it is always a different one, and therefore the objective idea that refers exclusively to the one can never have the other as its object. But if it is impossible for two subjective intuitions to pertain to a single objective idea even in the very same man, it is so much the less possible to evoke in *another* man a subjective intuition that would correspond to the same objective idea as an intuition present in *ourselves*. So if what we mean by the communication of an idea to someone else is the evocation of a subjective idea in him that pertains to the same objective idea as our own does, then we must maintain that *intuitions can not be communicated at all*. It is another matter with *pure concepts*, which we can communicate to each other by all sorts of means, among them mere words. Thus, for example, everyone who understands English will connect the words, and, nothing, one, two, three, etc., with ideas which correspond to the same concepts in themselves.

(2) We say nevertheless that we do *communicate* with others about our intuitions. This only means that we make them acquanted with their various *properties*. Thus we are particularly concerned, when our intuitions are aroused by the stimulation of our sense organs by an external

object, to make that external object known to others. If it is an object that lasts, that comes before us frequently, and if it is sufficiently important to us, it will be given a mark constructed just for it, a *proper name*. Thus proper names generally designate only *mixed* ideas of the form: "the object which is the cause of my having such and such intuitions at a given time." This is true not only of proper names such as signify an external object affecting our senses. It is also true of proper names that signify objects which have long since ceased to affect our senses, for example, Socrates. If someone asked, but what kind of intuitions enter into our idea in such a case, I would reply that we think of Socrates as "the philosopher named Socrates who lived in Greece so many hundreds of years ago." Now there are certain intuitions occurring here, at least the sounds that compose the *name* Socrates, if no others. But proper names are used to indicate only a small number of objects. We specify the great majority of the others by describing a relation exclusively their own in which they stand to certain other things already known to us. Time and space relationships have the greatest use here and for the most part even an approximate specification of them is sufficient, such as words like *now*, *a little while ago*, *soon*, *here* or *there*, and the like involve. Sometimes, to be more certain, we add the *kind* or *class* the thing we intend belongs to. For example, if I were merely to say, *this here*, pointing to a rose bush standing right in front of me, you would not know whether I meant the whole rose bush, or only this particular rose on it, or only this petal, or whatever. But I remedy this indefiniteness if by adding a common noun I specify the class of things the intended object belongs to, and that class of which there are no other members in the place and time indicated. And so, instead of saying, this here, I say only this leaf, this color, and the like.

(3) It is easy to see how we could use this device, which is commonly applied only for designating *mixed* ideas, to designate a *pure intuition*, when that is required for intellectual purposes. We need only say that the specifications of time and place in our expression, as well as the specification of the kind or class to which our object belongs, are added merely for the purpose of making our idea *known* to others, by no means so that he should regard these specifications as parts of this idea. The other person not only learns from this statement that our idea is a singular idea, i.e. an intuition, but he will also be able to judge what

object has produced it in us and what other properties it may have. But that it is impossible for us to arouse in someone else the *same* intuition we have was already asserted in No. 1).

(4) The following fact also deserves to be mentioned here. There are words in all languages that we use ambiguously, namely so that sometimes they mean a pure *concept*, sometimes a *mixed* idea. What is worst is that we often leap from one of these meanings to the other without being clearly aware that we are doing so. The case of which I speak arises primarily in connection with the names we apply to certain kinds of *natural objects* (particularly the most subordinate species), for example in connection with the words, man, lion, gold, and so on. In the one sense we understand words of that kind to mean nothing but things that possess such and such properties, capable of being represented merely by *pure concepts*, and of which we have either been or could be informed. So we are prepared to acknowledge any object to be a thing with this name just as soon as it possesses these properties, no matter how different it may be in other respects from the things we have previously called by this term. We are taking the word, man, in such a sense, for example, when we decide to understand it to mean nothing but a being that unites a rational soul and an organic body, and we are consequently prepared to call even the inhabitants of the moon men in case it could be proved to us that they are rational creatures with an organic body, no matter how many differences there might be otherwise between them and us with respect to both intellectual powers and bodily form. So it is obvious that the idea our words designate on such an interpretation is a *pure concept*. The idea of 'man' would remain so even if we were to restrict it still more narrowly, indeed if we finally said that would call men only such rational beings endowed with senses which were like those encountered on earth in all of their properties that could be represented by pure concepts. For since there is only a determinate number, even though a large one, of properties common to all rational beings endowed with senses that live on the earth, an idea could be put together from the ideas of these attributes, i.e. from mere concepts, which would represent what we mean by man, without any intuition being included. It will be quite a different matter if we decide that the term man is not supposed to mean beings of that *sort* (as previously), but rather precisely *those* rational beings endowed with senses that are encountered on earth, and

not others, no matter how much like them they may be. Now the idea designated by this term is no longer a pure, but a *mixed* concept, which includes an intuition. For even after removing all the components from this idea that signify certain properties, capable of being represented by a pure concept, of the beings it is supposed to apply to, i.e. all those that are mere concepts, there still remains in the requirement that they should be beings that live on earth the intuition which is incorporated in the term *earth*. And just as we now have in the word, man, an example of an expression which seems to designate a pure concept, yet is sometimes used in such a way that it includes a real intuition, so there is the contrasting case of words that seem to designate a mixed idea and all the same are sometimes taken in such a way that they express only a pure concept. Such is the situation with the words, gold, silver, oxygen and other inorganic materials.* Our scientists are not at all disinclined to apply the same names to all the materials in the universe if they only have the same intrinsic properties that we find in these materials on earth. But we know extremely little of these intrinsic properties except from the effects of this material on some other material, ultimately on our own sense organs and ourselves (i.e. on our faculties of sensation and imagery), and so from mere relations to certain objects given only by way of *intuition*. Therefore, we commonly express them only as such relationships, and describe gold, for example, as a physical object that produces the idea of a yellow color in our visual organs, will appear to be green in thin sheets, is 19 times heavier than water, and so on. Now if we take this description, as it is, to be the explication of our concept of gold, then the idea we have of that material is of course mixed with a great many intuitions and so it is an impure concept. But we can look at the matter in still another way. We can regard the effects gold has on our sense-organs and on other objects given only through intuition as only the signs of certain properties of the gold definable by mere concepts. We can look on the words, that it looks yellow to us, for example, only as the expression of a certain intrinsic property in the gold which is the reason it produces the idea of yellow in an organ like our eye, and so on. If we understand it all this way, then the intuitions that occur in our verbal expression of the concept of gold could belong not to the

* Locke has already remarked on the double meaning of words like these. (Essay. B. 4, Chap. 6, §8.)

content of this concept but merely to the means by which we *identify* the unknown intrinsic properties of the gold, definable by mere concepts, which are supposed to be combined to form the concept. An *intuition* does not cease to be pure because in order to identify it we include one or another *concept* in the expression of it. No more does a *concept* cease to be pure merely because, in order to specify some of its parts, we resort to certain intuitions. Indeed, if we give it precise consideration, this occurs with *all* concepts the components of which we specify to others by means of words. For what we basically demand is always that our listener should think of those concepts that arise in his mind in connection with the expression of certain words. But these words (the sounds) themselves he becomes acquainted with only by means of intuition. The distinction would be greatest when we are able, by a word we know the meaning of, to bring before our mind the very concept it signifies. On the other hand, if someone defines the components of which the concept of gold is made up merely by telling us that they were the concepts of those intrinsic properties of gold by virtue of which it produces the idea of yellow in our visual organs, and the like, we still do not learn from that what sort of concepts they are. We can say, then, that on such an explication the concept the word, gold, signifies is basically a pure concept, but one that is not fully known to us. It is somewhat like the value of x in an equation as long as we have not yet solved it.

§ 78

Differences among Concepts with Respect to Content and Extension

If a question is raised about the status of the two kinds of ideas with which we are now acquainted with respect to content and extension, it follows directly from the definition given of intuitions (§ 72) that every pure intuition must be *simple* with respect to its content and a mere *singular idea* with respect to its extension. There is no difference among intuitions in this regard, then. Among *concepts*, however, all of the *distinctions* with which we have become acquainted in connection with *ideas generally*, both with respect to content and extension, can have a place, according to the exposition in § 73.

(1) That there can be *compound* concepts requires no special proof,

but that there are also concepts that are *simple* we do have to prove. Anyone who wanted to deny this would have to maintain that all compound ideas, and all propositions as well, arise from nothing but piling one mere intuition upon another, which is absurd. For several ideas of the form, 'this', 'this', etc. (where every one of these demonstrative pronouns refers to some real object) never come to be united into a *single* idea just by virtue of the fact that we think of them one right after the other or even simultaneously. Their union requires still another idea, e.g. the idea of the word, *and*. Who could believe for even a moment, furthermore, that the idea the word *not* refers to, or the ideas of *having* or of *obligation* and a hundred like them arise from a mere combination of various intuitions? There must, then, be simple ideas that do not belong to the class of intuitions, i.e. there must also be simple concepts.

(2) With respect to *extension*, we learned in § 66 that there are ideas that have an *infinite* number of objects, others for which the number of these objects is *finite*, still others that have only a *single* object, and finally such as have *no* object whatsoever. The examples that were introduced above could prove to anyone who chose to read through them over again now that *concepts* (pure concepts) belong to each of these classes. All the same, let us have some new examples here. "A created substance," "one of the four cardinal virtues," "the community of all the virtuous," (or the so-called kingdom of God), "a means of altering the past," – are four pure concepts. The first comprehends an infinite number of objects, the second only four objects, the third has only a single object and the fourth is entirely objectless.

Note 1: Doubtless it would be very useful for logic to prepare a list of all the simple concepts there are or at least all there are within the total sphere of human knowledge. Such a list would be still more valuable if we could become convinced of its completeness just by inspecting it. This task seems to me so difficult, however, and the attempts to solve it that have been made so far appear to me so unsuccessful that I have no inclination to venture a new one. If someone just wanted to have it proved that there is not only one, but a *number* of simple concepts, that would be easy to accomplish. If there were a single simple concept, all of the other pure concepts would have to be merely combinations of it with itself, and consequently the difference between them would depend solely on the greater or lesser number of repetitions of one and the same component. But the truth is that there are only two single concepts that are capable of being combined with themselves without requiring another concept to combine them. The one is the concept of affirmation (yes), which in any number of repetitions produces a concept equivalent to itself (yes, yes). The other is the concept of denial (not), which combined with itself 2 or 4 or $2n$ times offers a concept equivalent to *yes* and in every combination of an odd number of times a concept equivalent to itself. Since only two kinds of idea arise by way of these two concepts, in whatever number of repetitions you please, it is therefore clear that there must be many other simple concepts to be able to explain the possibility of such a great number and variety of concepts as we find in our consciousness.

Note 2: The opinions of logicians are very much divided over the existence of simple concepts. Among those who maintained that they exist, Locke has a high place. For although he spoke of ideas in general, without adding whether he meant mere intuitions or the sort of ideas I have called concepts, there is no doubt that he did assume that some of of the latter were simple. Thus be adduces the ideas of unity, motion and others, which can surely only be counted as concepts, as simple ideas. Now although, as Leibniz accused him of being, he might have been very much much mistaken in these examples, in that the concepts he offered can be analyzed, it is sufficient for us to know that he was convinced that there must be simple concepts because there are complex concepts. Kant, although in his Logic he passes over silently the question of the existence of simple concepts, appears to have presupposed that there are and must be such. For in the *Critique of Pure Reason* he writes on p. 40: "No concept can be conceived to contain an infinite number of ideas." Now I would not agree with this claim, to be sure, since I believe that among concepts in themselves there can be some that are composed of infinitely many parts, but it seems to me obvious that the claim implies the necessity of accepting simple concepts. For if there were no simple concepts, presupposing that we can call the component ideas which constitute the complex concept concepts themselves, every given concept would have to be analyzed into an infinite number of parts.

§ 80

Ideas of Qualities and Relations

No one will doubt that there are ideas of mere *properties*. But since there are very many different kinds of properties, we also distinguish several types of ideas that refer to them. It is all the more necessary to bring them into logic, since the largest and most important part of our knowledge is concerned with the properties of things only. Or rather, since every truth and consequently every piece of knowledge can be regarded as depicting a property of certain objects. At present, where we are considering only such distinctions as can be apprehended in a given idea taken in and for itself – not by comparing it with something else – we can mention only a few of the distinctions that belong to that subject. The first is the one according to which we can divide ideas of properties into those that represent a *genuine*, *intrinsic* or *absolute* property, also called a *quality* and those that represent a merely *extrinsic* or *relative* property, also called a *relation*. In order to be able to explicate this distinction, I must first define the two interrelated concepts of a *property* and of *having* somewhat more precisely than I have so far, often as I have used them.

(1) I note, then, that what I mean by the word *property* is entirely the same as what is meant by it in ordinary usage when we take it not in a narrower sense but in the broader sense in which even a transient state, even a fleeting change, no matter how rapid, yields at least a *temporary property* of the object concerned. From *this* standpoint everything that *pertains* to an object, whether for a long time, or no matter how short a period, or even at just a single moment, is at precisely this point in time one of its properties. Common usage also takes the word, *to have*, in both a narrower and a wider sense. In the first sense, we can say of every property to be found in a certain object that it *has* it, as in the proposition, the human soul has immortality. In the second sense, however, what we understand by it is mere *possession* (i.e. the capacity to make some use) of a certain object, as when we say, man has hands. On the first interpretation, what is had is always a property; on the second, it can also refer to an object, which is not a property at all, e.g. hands, gold, and the like. I remind you now that I use the verb, *to have*, in the first or broader sense every time, and consequently that I apply it only to

properties throughout. Accordingly I can say that *whatever is had* (*quodcumque habetur*) *must be a property*. If a question is raised about what components I think these two so intimately interrelated concepts are composed of, I confess that I am still somewhat uncertain about their true components and the relation that obtains between them in this respect. One of them, it seems to me, is thoroughly simple and the other composed of it and a few other parts only. But whether the concept of having is the completely simple one, and the concept of a property derived from it, so that a property is always only that which something has; or whether it is the other way around and property is the simpler concept and having compounded from it, that I do not feel confident to decide, although to me the former is more probable.

(2) The concept of a property can be defined somewhat more sharply still, and at the same time distinguished from another broader concept, of *determinations generally*, with reference to the components every proposition divides into. These are to be demonstrated later on (§ 127). Namely, it will be shown in what follows that every proposition is based on the form, *A has b*, where *A* and *b* signify a pair of ideas, of which the former is called the *subject-idea* and the other one the *predicate-idea*. The idea that appears in *b*'s place (the predicate-idea) must be genuinely the *idea of a property*, whenever the proposition is supposed to be true. And conversely, every genuine idea of a property must be capable of serving as predicate-idea in a true proposition. – We can now say of any idea of a property that insofar as it enters into a proposition as *b* (as predicate-idea), it constitutes a *determination*, namely of the object the subject-idea represents. But the converse does not hold. Not every determination of an object has to occur by way of the predicate-idea in a proposition in which this object is the subject. On the contrary, there are also ideas that serve to *determine* an object without being *properties* of it. This is true of ideas that have the distinctive feature of never being able to take the place of the predicate-idea (*b*), but can only make an appearance as parts of the subject-idea (A). In particular, the *spatial and temporal determinations* of existing things belong to this class. The time in which some real thing exists, while some property can truly be ascribed to it, is no property of this thing. For just this reason, the idea of that time does not appear in the predicate, but in the subject-idea of the proposition. The like is also true of the place-determinations of things.

(3) After these preliminary remarks about the concept of a property, I will attempt to explicate that of a *relation*, from which the division of properties into *intrinsic* and *extrinsic* will then follow of itself. It is easy to suppose that every particular object has its own particular properties. Now since a whole consisting of several objects, *A*, *B*, *C*, *D*, ... as its parts, is as such also an object in its own right, essentially different from its individual parts, it is easy to understand how every whole can have some properties that do not also belong to its parts. Now if I am not mistaken, the properties we call *relations between these parts* are properties of this kind. This is particularly the case when we think of the two, both the objects, *A*, *B*, *C*, *D*, ... on the one hand, and the property *x*, belonging to the whole, on the other as *variable*. That is, when we conceive of other objects, *A′*, *B′*, *C′*, *D′*, ..., which are only of the same kind as *A*, *B*, *C*, *D*, ..., having a property which is also not the same as *x*, but only of the same kind. There is, then, for example, a property which belongs to neither of the lines *A* and *B* themselves, but belongs to the whole constituted by them, that one of them, *A*, is twice as long as the other, *B*. Furthermore, if we put other lines in the place of these, the new whole will not always have the same property, but only one that is similar to it, for example, that the one line is three times as long as the other. Therefore we call the one line's being twice as long a relation obtaining between these lines. In just the same way, we call the circumstance in virtue of which it can be said that Alexander the Great was King Philip's son a relation existing between the former and the latter, because once again this situation is a property neither of the one nor the other alone, but only of both of them. It would be altered if we substituted any other persons whatsoever for Alexander and Philip.

(4) Now on this definition a relation, *x*, that obtains among objects, *A*, *B*, *C*, *D*, ..., is basically a property of the whole constituted by *A*, *B*, *C*, *D*, ... as such, and it belongs only to it. Nevertheless we can say with all truth of each of the individual parts, e.g. *A*, at least this much: it is a property of *A* that it unites with things *B*, *C*, *D*, ... to form a whole that has the property *x*. It is that property of *A* we give the name *extrinsic* to. So what we mean by an extrinsic property of an object is a property of it that consists merely in the fact that it has a definite relation to some other object. We call the fact that a line has a length of two inches an

extrinsic property of this line, then, because this is the case only insofar as the relation mentioned obtains between this line and an inch. Properties that are not extrinsic, i.e. do not consist in a relation of the object in question to another, we call *intrinsic* properties or *qualities.* Now according to these definitions, where there is a relation (namely among several objects A, B, C, ...), then from another point of view (namely with respect to the whole $A+B+C...$) there is only an intrinsic property. The converse is also true; wherever an intrinsic property occurs, even of an entirely simple object A, from another point of view there is a relation. If b is an attribute of A, then the fact that the object A and the property b together form a whole that consists of an object and a property belonging to it, or the fact that A is precisely that object which has the property b as its property is a relation obtaining between A and b. But that not all relations are of exactly this kind is shown by the examples already cited. I call an idea that represents an intrinsic property an *idea of a quality* (*Eigenschaftsvorstellung*). Ideas, on the other hand, that represent an extrinsic property, I call *ideas of relations* (*Verhältnissvorstellungen*).

(5) It may be opportune to add the remark here that we distinguish two kinds of relation. The one kind holds where the objects A, B, C, D ... have an equal part in the property of the whole that is constructed out of them. The other kind holds when this is not so. I call the first kind of relations relations of *symmetry* or *reciprocal* relations, those of the second kind relations of *asymmetry* or *asymmetrical, one-sided.* The distance two given points in space have from each other is a relation of symmetry, for both points contribute to this relation in the same way. The direction in which one of these points stands from the other, on the other hand, is a relation of asymmetry, for the two points contribute to the determination of this direction in different ways. If a relation of symmetry holds between the objects A, B, C, D, ..., then it must be possible to grasp them by means of an idea in which the ideas of A, B, C, D ... all occur in one way, i.e. in the same connections, and so on. The opposite is true of relations of asymmetry. Therefore, we speak, for example, if we wish to express ourselves correctly, not of the distance of *a from b*, but of the distance *between* points *a* and *b*. The first expression would sound as if the point *a* had a different role in the relation obtaining between it and *b* than *b* does.

§ 84

Concepts of Sets and Sums

What the connection is among the individual parts of a group we are thinking of is a question that has remained unresolved in the collective ideas that have been considered so far. If something is now supposed to be settled on this point as well, it is easy to assume that still more complex ideas must come into the picture. The simplest case arises when nothing else is stipulated except that "*the kind of connection among the parts is to be regarded as something completely indifferent.*" That we often feel ourselves moved to conceive of groups with this explicit qualification no one will deny. For example, we say of a pile of coins to be received or given away that the order of the individual pieces of money that make it up is indifferent to us. In enumerating the members of many social organizations we find ourselves moved in the same way to note explicitly that the order in which we mention these persons is to regarded as completely indifferent, and that we are not acknowledging any rank order among them at all. Common language calls such collections *sets*; only it usually attaches many other subsidiary ideas to this expression, especially in thinking of a considerable number of parts. But who does not see that this restriction on the concept would be of no use for the purposes of science. I permit myself, then, to call any group you please, in which the nature of the connection among the parts is to be regarded as an indifferent matter, a *set*, even if it should contain only a very few parts, even only two. This makes it obvious what kind of components I conceive the concept of a set to be composed of.

(2) The example of the pile of coins shows that even with a collection in which the nature of the connection among the parts is indifferent, there can be grounds for not regarding the parts of its parts as parts of the whole itself or substituting them for it. For if we would put in the place of one or another of the pieces present in the pile of money the parts into which it could be broken down by mechanical or chemical means, the value of the whole could very well alter. However, there is no lack of groups of which both statements can be made at the same time, namely that the nature of the connection among the parts is to be regarded as thoroughly indifferent and that the parts of a part are to be regarded as parts of the whole. Such is the case, for example, with the

length of a line. For if we take into account merely its length, we are considering the line as made up of smaller lines. The nature of their connection is indifferent and the parts of which they consist (insofar as these are lines) can once again be regarded as parts of the whole line. Groups of this kind, in which not only can the nature of the connection among the parts be regarded as something indifferent but the parts of the parts can be regarded as parts of the whole, I permit myself to call by an expression borrowed from mathematics, *sums*. This is because it is also true of any sum, and presupposed by the very concept of it, that it is not changed if we modify the order of the parts as we please or replace one of these parts with *its* parts.

§ 86

Concepts of Unity, Plurality and Universality

There are some further concepts we may not leave unmentioned here, on account of the frequent application we will have to make of them in what follows.

(1) The first is the concept of *unity* that occurs so often. Every object that has some property *a*, or (what comes to the same thing) falls under the idea of *something which has a* (*or A*) is called a *unit of kind A*, in the *concrete* sense of the word unit, or a *concrete unit* of kind *A*, or still more briefly, *one A*. The difference that makes its appearance when we sometimes accent the *one* in this expression and sometimes leave it unaccented is in my opinion nothing but the heightened clarity we find especially necessary when the word standing in the place of *A* is of such a kind that the simple number is not differentiated from the plural number in it. The attribute of a thing by virtue of which it can be regarded as a concrete unit of kind *A*, or placed under the idea of *A* as its object, we call the *abstract unity of kind A*. By *unity in general in the abstract sense* we understand, consequently, nothing but that property of anything as a consequence of which there is some idea it is capable of falling under as object. That is why we ascribe unity to almost every object only *in a certain respect*, i.e. only in reference to some idea under which it can be comprehended as an object, while we grant that this very object has in other respects (i.e. in reference to other ideas) no unity.

(2) We call a group of which the parts are (concrete) units of kind *A*

a *plurality*, and so a *concrete plurality of kind A*. As an alternative we can also make use of the word, *number*, and even (where there is no concern over misunderstanding) the word, *set*. (§ 84) The attribute of a plurality by virtue of which it is a plurality is called *plurality in the abstract* sense or *abstract plurality*. A plurality of *A* that contains nothing but an *A* and an *A* is called *two* of kind *A*. A plurality that consists of two *A* and still another *A*, and contains nothing else, is called *three* of kind *A* and so on.

(3) Finally, a group in which every object that falls under the idea of *A* is present, but which has no other component, or (as this can be expressed more briefly), a group in which every *A* is present as a part and every part is an *A* is called the *group of all A*, or *the totality, the whole of A* in the *concrete* sense. The quality which makes a concrete totality or whole what it is is called *universality* in the *abstract* sense or *abstract universality*. Since we often represent the totality of *A* just by the expression, *all A* (*omnes A*), and so by the same expression that we also use (according to § 57 (2)) to signify quite a different idea, simply *A*, we are accustomed to distinguish these two senses of one and the same expression by saying that it is understood *distributively* in § 57, but *collectively* in the present section.

(4) Moreover, it is not to be forgotten that the last two concepts defined here, of plurality and universality are also subject to an expansion of meaning like the one mentioned in § 83 (4). And it is not always sufficiently indicated in the expressions themselves whether one is speaking of plurality or universality in the sense defined or in that broader sense. But we are speaking of a *plurality of A* in the broader sense when we conceive of a group which contains several *A* as parts without specifying whether it does not also contain other parts that are not *A* along with such *A*'s. We are speaking of a *totality of A* in this broader sense when we conceive of a group in which every *A* is present as a part without adding that conversely every (simple) part of this collection also has to be an *A*. For example, if we say "what all men can not accomplish together, no single man will be able to do," we are taking the words, "all men together," in the sense just defined. On the other hand, if we say, "the set of all points equidistant from a given point constitute the surface of a sphere," we must certainly include in this collection every point at the given distance, but no others besides

these. In the contrary case, it would surely not be true that the thing in space that group offers would have to be a spherical surface.

§ 87

Concepts of Quantity, Both Finite and Infinite

(1) The concept of *quantity* also occurs so very often in logical investigations that we can not leave it unmentioned here. What I believe is that we say an object is a *quantity* insofar as we conceive of it as belonging to a class of things any two of which exhibit one of the following two relationships to each other: either they are *equal* to each other or one of them figures as a *whole* which includes a *part equal* to the other. Therefore in comparing two quantities of any kind we always presuppose, as something that does not need to be proved, that one of the two must be the case, that these two quantities are either equal to each other or that one of them is the *larger*, i.e. that one contains within it a part equal to the other.

(2) This account makes it clear that *pluralities* or *wholes*, and also the *units* they are based on, can themselves be regarded as quantities, and under what circumstances. Namely, we are regarding a given plurality or totality as a quantity insofar as what we are paying attention to in it is just that property of it that does not change when we substitute another unit of the same kind for any of the units in it and we consider the way in which its parts are combined to be indifferent. On this presupposition, for every two pluralities of the same kind that we compare with each other, we will always encounter one of these two situations: either they equal each other or one of them contains a part equal to the other. The former will be the case if merely by exchanging every single unit that one contains for a unit of the other we are in a position to transform one plurality into the other. The other situation will hold if after all of the units of which one plurality consists have been exchanged for units of the other, we still have units left in the latter. After considering pluralities of a certain kind in this way, anyone can see for himself how unity and totality pertaining to this kind can be understood as quantities of the same kind.

(3) Every plurality of kind A that occurs as a member of the series we arrive at by making the plurality, two A, the first member and then

deriving each successive member from the preceding one by adding on a new *A* to it (or rather to a plurality equal to it), is called a plurality of *finite quantity*, or merely a *finite plurality* of kind *A*. A plurality of kind *A*, on the other hand, constructed so that every *finite* plurality of kind *A* occurs only as a *part* of it, i.e. that for every finite plurality of kind *A* there is a part in it equal to that plurality, I call a plurality of *infinite* quantity, or an *infinitely large* or *infinite plurality* of kind *A*.

(4) Let us construct a series of which the first member is a unit of any kind *A* you please, and every other member is a sum which emerges by our combining a thing equal to the immediately preceding member with a new unit. I call every member of this series a *number* insofar as I think of myself as apprehending it by way of an idea which tells how it has been formed. It is obvious that any finite plurality can be represented in terms of its quantity by a *number*, but that no number can be given for *infinite* plurality. For that reason we also call it *non-denumerable.*

§ 90

Symbolic Ideas

I will mention only one more class of ideas. What is distinctive about them is that the concept of an *idea* is present in their content *itself*, and in the place of the principal part (§ 58). They must, then, be included under the general form, "an idea which has (the property) *b*." I call such ideas, for want of a more appropriate term, *symbolic* ideas or *ideas of ideas* (*Vorstellungsvorstellungen*). All of the concepts of different kinds of ideas introduced in this part of logic belong here, such as the concept of a simple or compound idea, and the like. If the property *b* ascribed to the idea is neither contradictory in itself nor contradictory to the nature of an idea in particular, then the idea "of an idea which has the property *b*," is real (§ 66) and denotative, i.e. there is truly an object that corresponds to this idea. For example, the idea signified by the expression, "a simple idea," is of this class, for I believe I have shown in § 78 that there are such simple ideas. Thus ideas of this kind can be ideas of an idea or ideas of ideas in a proper sense of the word. There can also be symbolic ideas that are imaginary, however; and it is precisely these we call *merely* symbolic ideas, or symbolic ideas in the *narrower* sense (likewise only ideas of apparent ideas, mere signs of them, symbols), because there is no object corresponding to them at all (because there is no idea at all with the properties they require). The idea "of a concept which is at the same time a whole judgment as well," and the like would belong to this class. In a completely different sense we understand by symbolic ideas ideas of mere *signs*, for example the idea of the word, Abracadabra, and the like.

§ 91

There Are No Two Completely Identical Ideas. Similar Ideas

(1) The first question that presents itself to us if we take note of the differences in the relations among ideas is this: whether the relation in which two ideas stand to each other can ever be the relation of full identity, i.e. whether there can be two thoroughly identical ideas? Now this question, in my opinion, is to be answered in the negative, if we understand by ideas not subjective (thought) ideas but ideas in themselves. Of course one can say of subjective ideas that there are several, indeed an infinite number of them that are identical with each other. For one calls such ideas identical if they only have one and the same idea in itself as their material. They may, moreover, be ever so different in many other respects, for example with respect to their clarity, duration and liveliness, also with respect to the thinking being in whose consciousness they occur. But if we are speaking of objective ideas, it seems to me absurd to suppose that two or more of them are identical with each other. In such ideas nothing but the idea itself is considered. Therefore there is no way of saying that they are identical unless all the properties one can take note of in them (their components, the way they are put together, etc.) are the same. But if this is the case, then they can not be differentiated from each other and therefore they can not be said to be several in number.

(2) Even though there are no two *completely identical* ideas, there are nevertheless many ideas which have so many properties in common that it is very easy to *confuse* them with each other, i.e. to take them to be one and the same. Now since we are accustomed to calling things with so many common properties that they are easily confused *similar*, I permit myself to call such ideas *similar ideas* as well. We have an example in the ideas of being wealthy and of being well-to-do, honor and respect, and various others.

Note 1: If some logicians assert the existence of several ideas completely identical with each other, this comes about only because they (a) either do not understand by the word idea an idea in itself, but a subjective idea (held in the mind or thought), or (b) they do not adequately differentiate the idea from its sign or expression in language, or finally (c) they call ideas identical which I call merely equivalent (see § 90). How common the first case is is clear from the explanation most logicians give of the concept of an idea. For, as I have already recalled in § 53, in almost all textbooks on logic ideas are described as only phenomena in the mind of a thinking being. Whereupon it would be quite consistent to say that there are identical ideas, so much so that we could only accuse those logicians of an inconsistency who denied it. For this can be thrown up at them, that they gradually departed from the concept of an idea assumed in their definition and in what followed put in its place the concept of an idea in itself. That the second and third cases have also arisen, however, can be seen from so many examples some logicians have given of concepts supposed to be identical with each other. Thus it is often said that the definiendum and the definiens, e.g. 'a triangle' and 'a space bounded by three straight lines', are identical ideas, yet this is only one and the same idea expressed in different words. Likewise, it is sometimes said that the concepts of an equilateral and an equiangular triangle, and others like them, are identical with each other, yet these concepts are in fact very different in content and are the same only in extension, i.e. they are *equivalent* (§ 96).

Note 2: On this occasion it may not be out of place to warn beginners of the confusion we often allow with the words *identical* and the *same*. *Sameness* (or *identity*) is from my point of view the concept that arises in our mind when we consider the same object several times and take note of the fact that it is the same object. Specific identity (*Gleichheit*) or *exact likeness* (Gleichartigkeit) on the other hand, is the concept that arises when we consider *several* objects and find that they fall under the same objective ideas. Understandably this can always be said only of some, but not of *all* ideas. For if every idea we formed of the one object also fitted the other, we would not be able to *know* that we have two objects. And if it were not at all possible to cite any idea which fitted the one object and not the other, it would not even be true that the one object is distinct from the other, for this very proposition sets the one object under an idea which the other does not fall under. Specific identity is therefore always only *partial*; and if we declare two or more objects to be *specifically identical* or *exactly alike*, we must, if we wish to speak with precision, always cite the respect in we find them to be exactly alike, i.e. we must designate the idea we put them under. *Sameness* is opposed to mere *plurality*, but *exact likeness* is opposed to *unlikeness*, often called *difference*. But we often take the word *difference* in such a broad sense that we understand by it the mere opposite of sameness, the mere plurality of objects, without wishing to presuppose any unlikeness among them. We permit ourselves to do this all the more willingly because basically any two things (at least real ones) are unlike in some respect, and so at least in this respect they can be called different. – If we are speaking of objects with real existence, e.g. of substances, finite ones; we should take further note of the fact that the same object (the same simple substance or the same collection of several) can take on different properties at different times, not only extrinsic properties but even intrinsic ones. For just this reason, it does not continuously fall under one and the same idea, whereupon we are accustomed to saying the object has become *unlike itself*; it would be more correct, however, to say it has *changed*. – Moreover what we apply the concept of identity to, not only in connection with several objects, but also in connection with the same object, is different in different circumstances. With objects composed of parts what is required for identity is often only that most of its parts, or the most important of them remain. So, for example, I say "the watch I see here is the same as the one that was stolen from me some years ago" – if I only wish to say that most of its parts and the most important ones are the same in it, even though some of them, e.g. the watchglass and a few little wheels could be different. Sometimes, as with the ship of Theseus (which the Athenians had repaired again and again), we understand by the identity of an object *A* with an object *B* (which in terms of its substance is really quite different) only that the latter has developed out of the former by a very long series of imperceptible or unimportant changes (of which none was so great for the object in the sense attached to it previously to be called different from what it had just become). With things for which *place* is the most important feature, identity is often referred to this place alone. So we say, for example, "the water we are sitting by now is the same water we were sitting by this morning" – when all we want to announce is that it is water flowing in the same bed. With organic things (plants and animals), by asserting "what we are now perceiving is the same as what we had perceived on a previous occasion" we do not mean to say anything but that the latter has arisen from the former by some series of intervening causes, not by a kind of generation or propagation. In this sense we say that the oak tree in whose shadows we are sitting grew out of an acorn a hundred years ago, when all we mean is that this oak tree, so different in all of its parts from that seed, nevertheless developed from it, and not by way of a new propagation.

§ 92

Relations among Ideas with Respect to Their Content

(1) After the question of whether there are two perfectly identical ideas has been answered in the negative, the second question arises, whether there are not at least ideas that are identical with each other in a *certain respect*? The first thing one might think of in this connection is with respect to the *content*. I ask, then: are there two or more ideas that are identical with each other with respect to their content? If these ideas are supposed to be simple, it is obvious that an identical content can not be ascribed to them. For with simple ideas what would be their content is one and the same with the ideas themselves. To say, then, that they are of identical content is to say that they are just perfectly identical. But if we are concerned with compound ideas and if we understand by their content the mere collection (sum) of all of the parts (whether they are more proximate or more remote) that compose them, without taking account of the way in which they are put together; we have already called attention in § 56 to the fact that two or more ideas can have the same components and yet be distinguished by the way in which these components are combined with each other. Thus, to provide a new example, the two ideas, "being permitted not to speak" (i.e. the permission to remain silent) and "not being permitted to speak" (i.e. the obligation to remain silent), are essentially different, even though they obviously have the same components. I permit myself, then, to call such ideas simply *ideas with the same content*. But if the way the components are combined to make up an idea was mentioned, along with the com-

ponents themselves, then it is obvious that the whole idea would already be defined. From this it follows that there can never be two objective ideas which have between them the very same components and the very same form of combination as well. Ideas that have not even a single component in common, one could call *ideas of entirely different content*.

(2) With some ideas, if not all, several parts of their content can nevertheless be common and united in the same way besides. Thus for example, the ideas, equilateral pentagon and equiangular hexagon, have in common with each other the concept of a polygon, which occurs as a component of the ideas of a pentagon and of a hexagon. Furthermore, they share the concept of equality, which appears as a component of the ideas of equilateral and equiangular. Since it can be important for many purposes to note that given ideas have such common components and connections, the trouble of specifying for them a word of their own will repay us. I will call ideas that have one or several parts in common, particularly if these parts are combined in the same way in them, *related* ideas, then. I will say further that the relationship between two ideas is the more intimate the larger the number of parts they have in common or the more parts in both of the ideas follow upon each other in the same order. With the establishment of this definition there would be an important difference to be made out between *related* and *similar* ideas (§ 91). For ideas of the form, *A* and not *A*, which no one will regard as similar, i.e. easily to be confused, would be related ideas, i.e. in possession of some common components. On the other hand ideas could have an observable similarity without having a single common component, i.e. without being related. Thus the idea of the morally good has so much in common with the idea of serving the general welfare, or of the honorable, and the like, that in fact the latter have frequently been confused with the former, although they have scarcely a single component in common with it.

§ 93

Relations among Ideas with Respect to Their Breadth

Still more noteworthy relations among ideas make their appearance when we consider their *extension* instead of their content. But in considering the extension of an idea we can look first only to the *set* of objects it comprehends, i.e. to its range, and then also to these objects themselves. Both the one and the other present notable relationships.

If we look first only to the breadth of two ideas, what can happen is that we find an *equal* breadth in both of them or it may be that the breadth of one is larger than that of the other, or it may emerge that in breadth they are both infinite, so that we are no more justified in saying that they are equal than that they are unequal. In the first case we can call the ideas *equally broad*. In the second, we can call them of *unequal breadth*, the one the *broader*, *more comprehensive* or *more extensive*, the other the *narrower* or *less extensive*. In the third case, we must admit that the two ideas are *not at all comparable* in their breadth. An example of a pair of equally broad ideas is given by the ideas, 'human soul' and 'human body', for although the objects of the one are not at all the objects of the other, the set of the former and the set of the latter can be represented by one and the same quantity, at least insofar as for every human soul there is one and only one human body. The idea, 'human finger', on the contrary, is without doubt to be called a broader or more extensive idea than that of 'human hand', for there are certainly more fingers than hands. We have an example, finally, of two ideas that can not be compared in their breadth in the concepts, sphere and tetrahedron. It is easy to reach the opinion that we can expand the concept of the *extension* of an idea defined in § 66 by broadening it in a certain way also to cover whole groups of ideas. We do so by understanding that extension to be the sum of all of those objects that are so constituted as to be represented by some one of the ideas included in the group. In this sense, every inhabitant of the earth will belong to the domain of the set of five ideas, European, Asiatic, African, American, Australian. Given this expansion of the concept, we shall compare not merely single ideas but whole sets of them, A, B, C ... on the one side and M, N, O ... on the other. In considering breadth we shall either be able to find that they are of *equal* breadth, or that the one collection has a *larger* breadth, the other a *smaller*, the former the *broader*, the latter the *narrower*, or finally that they are not comparable in breadth at all. If the relation between ideas or whole sets of them just discussed is to be represented by a *symbol*, it is illuminating to symbolize equally broad domains by spaces of equal size and a broader domain, on the other hand, by a space that is also relatively larger.

A final comment is to be made that we at times call one concept *broader* than another when it does not represent *more* objects, but only *larger* ones, i.e. objects such that the ones in the former set include the ones in the latter as parts. So we say that what is understood by hand in the broader sense is the whole arm, in the narrower only a part.

Note: If the set of objects encompassed by two ideas is finite in both cases, there is no difficulty about defining with perfect exactitude the relation these ideas have to each other with regard to their breadths. The very numbers that represent the sets of the objects also give the relation of the ideas' breadths. Thus, for example, the two ideas, 'sons of Isaac' (of whom there were two), and 'sons of Israel' (of whom there were twelve) have the relation of exactly 2:12 or 1:6 with respect to breadth. Even if the set of objects is infinite for the one idea, but finite for the other, the relation in which their two breadths stand to each other can be defined, not exactly, by means of a pair of numbers, but there is no doubt about which of the two ideas is the broader. But if (as is usually the case) both ideas encompass an infinite set of objects, one might believe there is no possible way of making any quite definite statement about the relation that prevails between their breadths. There are, however, if I am not mistaken, some cases even here in which that relation can be defined. In some of them it can be defined only imperfectly, i.e. only by saying which of the ideas is the broader, but in some it can be defined with complete precision. Whether there are more triangles than syllogisms is a question that can only be answered by saying that two infinite sets are not to be compared. But it seems to me we can correctly respond to the question as to which is broader, the concept of the circumference of a circle or of a circular surface, by saying that the breadth of the two is the same. For there is a circumference for every circular surface and *vice versa*, so that if there were only a finite number of circumferences of a circle we would not hesitate to say that there were just as many circular surfaces and that consequently the two concepts were of the the same breadth. But this relation between their breadths does not alter when we expand the set of objects that fall under both by as many as we choose. We will, then, since there is no other circumstance that leads us to infer an inequality in the relation of these breadths, be justified in assuming they can be represented by an equal quantity even when those sets are infinite. But if this has been granted, we may admit for similar reasons that the concepts, center of an ellipse and focus of an ellipse, are related as 1:2 with respect to their breadth, because for every center there are two foci. Of the concepts, circumference of a circle and diameter of a circle, we can say, on the other hand, that in their breadth they relate as $1:\infty$, since for every circumference there are an infinite number of diameters, etc. Anyone who did not want to grant these conclusions only because infinite sets are not measurable, could not admit that a higher concept, e.g. triangle, is broader than the lower one, right-angled triangle, either. For these two concepts also encompass an infinite number of objects. It is nevertheless true, however, that we can not measure infinite sets as such, and that makes it quite clear that in this connection we understand by *breadth* something quite different from the mere set of objects contained under an idea. As I have already mentioned in § 66 (4), breadth really means any quantity that can be derived from the set of objects falling under a certain idea according to such a rule that it is equal to the sum of those quantities derived by the same rule from the parts into which that set is divided. Upon such an explication the above determinations can be adequately justified. The sets of objects falling under the concepts of the circumference of a circle and of a circular surface are of course not determinable in and of themselves. But the relation in which circumferences and circles stand to each other makes it permissible for us to set them equal to each other if they are supposed to be compared, because there is no reason for assuming inequality. Thus if we take the breadth of one of these concepts to be unity, we may also set the breadth of the other as $=1$. Further, if we assume the breadth of the concept, "center of an ellipse," to be unity, then we may set the breadth of the concept, "focus of an ellipse," as $=2$, for the set of objects falling under the latter can be divided into two parts, each of which is equal to the set of objects of the former concept, etc. If this is correct, it contradicts the claim by Kant (L. § 13) and Kiesewetter (W.a.d.L., Part I, p. 124) "that concepts can be compared in terms of their extensions only if they are subordinated to each other; because otherwise we can not know which of them encompasses more objects." I believe I have just demonstrated that we *can* know this on many occasions, and not just in connection with concepts that encompass only a finite set of objects, but even in connection with concepts that have infinitely many objects.

§ 94

Relations among Ideas with Respect to Their Objects

(1) If we direct our attention merely to the *objects* to which certain ideas refer, what emerges is either that they have certain objects in common or that the opposite is the case. Both cases are sufficiently noteworthy to deserve their own designations. So I call ideas that have one or several objects in common with each other *compatible* or *harmonious* ideas or ideas that agree. Those, on the other hand, that do not have even a single object in common I call *incompatible* or *disharmonious*. Thus the ideas, 'something red' and 'something fragrant', are compatible, for the two represent certain objects in common, e.g. the rose. The ideas, 'solid' and 'surface', on the other hand, are incompatible, for no object that falls under the one also falls under the other. At times there is the noteworthy relationship that among a given set of ideas, A, B, C, D ... only a definite number of them, e.g. n, are compatible. Thus each of the four ideas, roots of the equation $(x-a)(x-b)=0$, roots of the equation $(x-a)(x-c)=0$, roots of the equation $(x-b)(x-d)=0$ and roots of the equation $(x-c)(x-d)=0$, is compatible with exactly two others.

(2) If *several* ideas A, B, C, D ... are supposed to stand in the relation of compatibility to each other, then the *lesser* number A, B ..., which is only a part of the former set, must also stand in the relation of compatibility. For if this were not the case, i.e. if there were no object commonly represented by A, B ..., much less could there be an object commonly represented by all of them, A, B, C, D .... Looking at the matter from the other direction, even if the lesser number of ideas, A, B, ..., are compatible with each other, it could nevertheless be true that the larger number, A, B, C, D ..., which includes the former ideas within it, stand in the relationship of incompatibility. For even if the ideas, A, B ..., have an object in common, it does not have to be common to the rest of the ideas, C, D ....

(3) If a pair of ideas of properties a and b are compatible with each other, then the concreta A and B (§ 60) are also compatible with each other. But we can not draw the converse conclusion from the fact that a

pair of concreta A and B are compatible with each other that their abstracta a and b are also compatible. For if a and b are harmonious, there must be some property x that is a as well as b. But then an object that has the property x is an A as well as a B; consequently A and B are also harmonious. But merely from the fact that A and B agree it does not follow that a and b agree. For although the object that is A as well as B must unite both of the properties a and b within itself, it does not follow at all that the same property in it that belongs under a also belongs under b. Thus the abstracta, cleverness and caution, are harmonious, and therefore their concreta, clever and cautious, are too. On the other hand, the concreta, pious and learned, are harmonious without their abstracta, piety and learnedness, being so, for no kind of learnedness is also to be called a kind of piety.

(4) Just as a single idea can stand in a relationship of compatibility or incompatibility with another idea, so can a whole set of ideas A, B, C, D ... with another set M, N, O ... or just with a single idea M. It is the former relationship when there is some object that falls under one of the ideas A, B, C, D ... and under the idea M as well or one of the several ideas M, N, O, ... It is the second relationship when this is not the case.

(5) If an entire set of ideas A, B, C, D ... stands in the relationship of *incompatibility* with an entire set of different ideas M, N, O ..., then every single one of the ideas A, B, C, D ... must stand in this relationship to every single one of the ideas M, N, O .... But if the two sets stand in the relationship of *compatibility*, then it is not necessary for each of the ideas A, B, C, D to stand in this relationship to M, N, O .... It is sufficient for only one of A, B, C, D ... to stand in it with one of the ideas M, N, O ....

(6) If a single idea A or an entire set of several ideas A, B, C ... is compatible with one or several M, N, O ..., and the latter is in turn compatible with one or several R, S ..., it does not follow at all from this that the first, A or A, B, C ... must be compatible with the last, R, S .... For the objects that A, B, C ... and M, N, O ... have in common with each ofther can be different objects from the ones M, N, O ... and R, S ... have in common. In the opposite case, if A, B, C ... is incompatible with M, N, O ... and M, N, O ... with R, S ..., it does not follow that A, B, C ... must also be incompatible with R, S ..., either.

(7) If the relation of compatibility is supposed to be represented sym-

bolically, then the space that symbolizes the domain of one or several ideas *A*, *B*, *C*, *D* ... must have something in common with the space by which we choose to illustrate the domain of the single idea or of several different ideas *M*, *N*, *O* ..., that are compatible with the former.

§ 95

Special Kinds of Compatibility: (a) Inclusion

(1) If the relationship of compatibility is understood as in the preceding paragraphs, then there are several forms of this relationship which we must yet describe, since they are worth taking note of. If a pair of ideas *A* and *B* stand in the relation of compatibility with each other, then it can be the case that not only some, but all of the objects that fall under one of the ideas, say *A*, also fall under the other one. If it is not presupposed in this case that this is true conversely as well, i.e. that all of the objects falling under *B* also fall under *A*, and consequently it is supposed to remain undetermined whether *B* has some objects falling under it besides the objects that fall under *A*, I permit myself to call this relationship between *A* and *B* the relationship of *inclusion*. I say that the *domain* of idea *B*, or simply idea *B* itself, *includes A*. I call *B* the *inclusive*, *A* the *included* idea. Thus I say the idea of man is included by the idea of an inhabitant of the earth, because every object that falls under the idea of man also falls under the idea of an inhabitant of the earth.

(2) Anyone can see for himself how this relationship can be extended to the case where instead of a single idea, there is an entire set of ideas on one or both sides. I shall say that the ideas *A*, *B*, *C*, *D* ... are *included* by the ideas *M*, *N*, *O* ... if every object that falls under one of the ideas *A*, *B*, *C*, *D* ... also falls under one of the ideas *M*, *N*, *O* ....

(3) If an idea *A* is included by an other idea *B*, at least it can be *no broader* than the latter. For if *A* represented more objects than *B*, how should it be possible for all *A* also to be represented by *B*? The like holds of entire sets.

(4) If idea *A* is included by idea *B* and *B* included by *C*, then *A* will also be included by *C*. The like holds of entire sets of ideas.

(5) If the relationship of inclusion is supposed to be represented symbolically, the space by which we choose to illustrate one or more

included ideas must lie within the space that symbolizes the domain of the inclusive ideas.

Note: This relationship of inclusion has already been set forth by some other logicians. In Maass' *Grundriss der Logik*, §80 states: "One concept *a includes* another concept *b* insofar as all *a* are also *b*." – In fact, this relationship is already worthy of having some attention shown to it because it is the one in which the subject idea of any true proposition stands to the concreto corresponding to its predicate idea. Namely, presupposing that all propositions fall under the form, *A* has *b* or *A* is *b*, then their truth obviously requires that the idea *B* include the idea *A*.

§ 96

(b) The Relationship of Mutual Inclusion, or Equivalence

(1) In view of the way we defined the concept of inclusion in the preceding paragraph, this relationship can also hold mutually between a pair of ideas *A* and *B*; *A* can be included by *B* and *B* by *A*. Namely, this is the case when not only does everything that falls under *A* fall under *B*, but also everything that falls under *B* falls under *A*. Still more briefly, when both ideas have exactly the same objects. I call this relationship between ideas *mutual* or *exact inclusion*, also *equivalence*, and the ideas themselves I call *equivalent ideas*. The two concepts of equilateral and equiangular triangle are an example.

Since, as a consequence of the definition given, equivalent ideas are ideas with the same extension, the question arises whether there are also also equivalent ideas that have, along with the same extension, *the same content* as well, i.e. whether ideas can still be different if their content as well as their extension is the same. And this question may, I believe, be answered in the affirmative, presupposing that by sameness of content we understand sameness of ultimate components, but not also sameness in the way they are combined. For the character of an idea would already be completely determined by both of these. But that ideas of which the more remote parts are the same, can have them connected in a different way and nevertheless represent just the same objects is plain enough from the following examples. A virtuous man who is at the same time clever and a clever man who is at the same time virtuous are a pair of ideas which contain the same components only in a different order, and surely they both refer to the same objects. We have a mathematical example in the two concepts, 2^4 and 4^2.

(3) If someone should ask, on the other hand, whether ideas that are *both simple* can also be equivalent ideas, that would have to be denied. For in order to distinguish two things, one must be able to assert something different of them. But all assertions about an idea can only refer, at least so it appears, to one of two things: either to the object they represent or to the ideas themselves. But in the latter case the only questions are whether this idea is simple or compound and then again out of what parts it is composed, how these parts are connected, and so on. Now equivalent ideas can not be distinguished by different assertions about their objects, for they refer to the same objects. They can only be distinguished, therefore, (it seems) either by saying that the one is simple and the other compound, or that the one is constituted by these parts and the other by those, or if they are constituted by the same parts, the one formed in this way out of these parts and the other in that way. But if both ideas are simple then there will be no apparent distinction to be found here either; so we shall scarely be able to look on them as different ideas. Of course it is not necessary that *both* of them have to be compound. Thus, for example, every pure intuition is a simple idea, and if by adding some of the properties its object has we form a redundant concept of this object, then we have an idea equivalent to the first. If one wants an example of a pure concept which, although simple, has equivalent ideas nevertheless, I cite the concept of 'something', which is equivalent to the double negative concept, 'not not something', and to every similar concept containing an even number of negations. Thus, for every simple concept *a* that has an object, there are infinitely many compound concepts from the class of redundant concepts, not only of the form, 'not not *a*' but also of the form, '*a* which is *a*' and so on. All of them can be regarded as equivalent concepts to *a* itself.

(4) But the only equivalent ideas that arise in this way are such that one of them is redundant. Therefore, let us now consider some of the simplest cases in which equivalent ideas without any redundancy can come into the picture. (a) In the first place, it may be that all objects falling under a certain idea *A* also fall under the two ideas *B* and *C* and that these two ideas have no other object in common. In this case, then,

the idea A, which can be simple, moreover, has an equivalent in the ideas: 'a B which is C', or a C which is B', or 'something which is both B and C'. We have an example when A signifies the moral (or what ought to be done), B what is possible as such, and C what agrees with the general welfare. (b) Secondly, it can be the case that an idea A has certain objects in common with B and has just the same objects in common with C, a different idea from B. In this case, the two ideas, 'an A which is B', and 'an A which is C', will once again be equivalent ideas. We obtain an example when we let A mean a heavenly body, B something 50 times smaller than our earth, and C something that illuminates our earth at night; for then the sole object to which both compound ideas refer is the moon. (c) In the third place, it can be that the ideas A and B have certain ideas in common with each other which similarly the ideas C and D, completely distinct from them, also have in common with each other. In this case, we will obtain equivalent ideas in the ideas, 'an A which is B' and 'a C which is D'. For example, under A think of living in wisdom, under B something in accord with the laws of morality, under C a means to happiness, and under D something that never betrays, and so on.

(5) On the other hand, from the fact that A and A' are a pair of equivalent ideas it could not be inferred that a pair of ideas formed in the same way, one from A and the other from A', i.e. that arise by combining them with the same other ideas, would likewise have to be equivalent ideas. Thus 2^4 and 4^2 are a pair of equivalent ideas and the two ideas, the root of 2^4 and the root of 4^2, are constructed in the same way from the two of them; nevertheless, they are in no way equivalent ideas themselves, since the one has 2, the other 4 as its object.

(6) If idea A is an equivalent of B, and idea B is an equivalent of C, then A and C are also equivalents each other.

(7) If, for the sake of brevity, we symbolize the idea of "something which has (the properties) a, b, c, d..." by [something] $(a+b+c+d...)$, and express the idea of something which has (the property) x, following §60, by X, then all of the ideas expressed by the following symbolic formulae are equivalent ideas, insofar as they are denotative ideas at all: [something] $(a+b)$, $[A]\,b$, and $[B]\,a$; and still more generally: [something] $(a+b+c+\cdots)$, $[A]\,(b+c+\cdots)$, $[B]\,(a+c+\cdots)$, $[C]\,(a+b+\cdots)$.

(8) If we extend the concept of equivalence to entire sets of ideas, we shall have to say that the set of ideas A, B, C, D... is *equivalent* with the

set of ideas *M, N, O...* if every object that falls under some one of the ideas *A, B, C, D...* also falls under one of the ideas *M, N, O...* and *vice versa.* So the two ideas, right triangle and oblique triangle, taken together, are equivalent with the three ideas, equilateral triangle, isosceles triangle and triangle of unequal sides. Likewise, the single idea, flower, is equivalent with the several ideas following: flowers that grow wild where we live, flowers we can cultivate in gardens, and flowers that only thrive in foreign parts.

(9) In order for a relationship of equivalence to hold between ideas *A, B, C, D...* on the one hand and ideas *M, N, O...* on the other, it is not at all necessary either that any one of the ideas *A, B, C, D...* should by itself be equivalent with one of the ideas *M, N, O,...*, or that the sum of the extensions of ideas *A, B, C, D...* should equal the sum of the extensions of ideas *M, N, O....* We can see the first point from the examples chosen in (8). The following example will demonstrate the second. The two ideas, a member of the series 1, 2, 3...10 and a member of the series 2, 3, 4...11, taken together are surely equivalent to the following: a member of the series 1, 2, 3...5 and a member of the series 6, 7,...11. The extent of the first series, however, is without doubt =10 and that of the second is just the same. The sum of the two is therefore =20. The extent of the third idea, however, is =5 and that of the last =6, so that the sum of the two is only 11.

(10) If the relation of equivalence between two single ideas *A* and *B* is supposed to be represented symbolically, we shall have to specify that the same space by which the domain of the one is represented also stands for the other. And if we want to indicate that an entire set of ideas *A, B, C, D...* is equivalent to another *M, N, O...*, then the space through which the domains of the particular ideas *A, B, C, D...* extend must taken altogether be the same as the space through which the domains of the particular ideas *M, N, O...* extend. That may in no way hinder us from taking the spaces by which we indicate the domains of ideas *A, B, C, D...* as other circumstances may require, to be such that the sum of the spaces assigned to *A, B, C, D...* is sometimes larger, sometimes smaller than the sum of the spaces allocated to *M, N, O....* If a diagram is not only supposed to illustrate the fact that a pair of ideas are equivalent, but also make sensible how this relationship comes to hold between them, it is obvious even in the simplest cases that can occur here (5) that this could

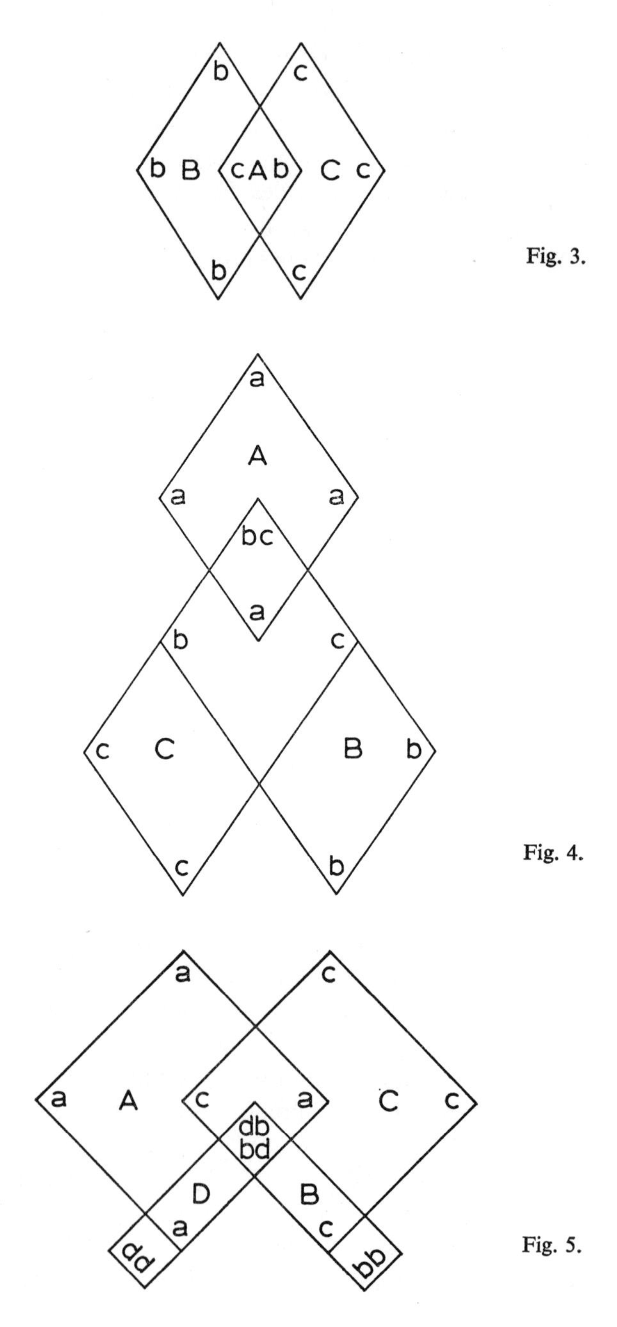

Fig. 3.

Fig. 4.

Fig. 5.

not be carried out if we wanted always to represent any given idea's domain by a circle or even a spherical space. For example, the pictorial representation of even the first of our above cases, in which the ideas of an *A* and of a *B* which is *C* are equivalent ideas, requires some space for the domain of *A* that has the spaces assigned for ideas *B* and *C* common to it. Now since the space that two circles or spheres have in common with each other is, as is well known, never a circle or a sphere, it is clear from this example how impossible it would be to give the domain of every idea a space of such limited shape. On the other hand, if we lay it down only that the space which indicates the domain of an idea should be a plane surface, then we can select any figure we please for the domain of *A* and then draw figures for *B* and *C* that have space *A* in common. The design in Figure 3 accomplishes this, for example. The second and third cases, however, or that the ideas, *A* which is *B* and *A* which is *C*, are equivalent ideas; as are *A* which is *B* and *C* which is *D*, are expressed by Figures 4 and 5.

§ 97

(c) The Relationship of Subordination

(1) The second case that can emerge in connection with the relationship of inclusion (§95) is the one in which this relationship is *not reciprocal.* If one idea *A* is included by another idea *B* without the latter being included by the former, then *B* must have one or more objects besides those that are represented by *A*. This relationship is called subordination and idea *B* is said to the *higher*, idea *A lower*, or *subordinate* to it, *falls under it*. Thus the ideas of man and of living creature stand in a relationship of subordination, and living creature is the higher, man the lower idea, because every object that falls under the idea of man also falls under the idea of living creature, but is it not conversely the case that each of the latter falls under the former.

(2) Several ideas *A, B, C, D*... on the one side stand in the relationship of *subordination* to one or more ideas *M, N, O*... if every object represented by one of the first set is also represented by one of the latter set, but conversely it is not true that each of the latter is represented by one of the former. We shall then call the set *A, B, C, D*... the *lower*, *M, N, O*... the *higher*.

(3) The higher idea must also be a broader idea, but it is not conversely

the case that every broader idea must also be a higher one. The like holds true of entire sets.

(4) If *A* is lower than *B*, *B* lower than *C*, then *A* is also lower than *C*. The like with entire sets.

(5) In order to illustrate the relationship of subordination by a diagram, we shall have to select for the area symbolizing the lower idea's domain a part of the area that represents the domain of the higher idea. Yet the terms *higher* and *lower* betray the fact that we ordinarily illustrate the relation we are concerned with here in quite a different way from the one just considered, according to which we would have to indicate their domains by areas of which one was a part of the other. According to the picture implied in the above terms, with which the locution, that the lower idea *falls under* the higher, is in agreement, we conceive of the higher idea as something in space that is *higher* and of the lower idea as a thing in space that is *lower*, and so lies under the former. It would be a mistake to pay absolutely no attention to this way of illustrating the relationship between certain ideas, a way that in many cases is even more convenient than the first one. But the spatial things we make use of in this new kind of illustration as signs of the ideas will most appropriately be the written terms themselves, or some other written signs we have assigned them at some time. So, in order to indicate, for example, that of the ideas, animal, bird, bird of prey, each successive one is lower than its predecessor, we shall write their names one below the other in the following way:

Animal
Bird
Bird of Prey.

(6) The final point is still to be mentioned, that at times we take the words, *higher* and *lower*, just as we do the words, broader and narrower, (§93 (5)) in quite a different and very vague sense. Frequently a *higher* concept means nothing but a concept that has a higher, more respectable, or even more important object, and the like.

§ 98

(d) The Relationship of Intersection or Concatenation

(1) The form of compatibility considered in §95, the relationship of

inclusion, of which the two relations discussed in §§96 and 97 can be regarded as mere subtypes, came into the picture when we supposed that of two compatible ideas *A* and *B* at least one of them was of such a character that all of the objects falling under it also fall under the other. Now if this is not the case, if it can be said of neither of the two compatible ideas that its objects all fall under the other idea, then a relationship arises which I call *intersection* or *concatenation* or, a word others have already used, disparation. Thus I say that the ideas, learned and virtuous, intersect each other, because, along with certain objects they have in common, each of them also has some that do not fall under the other.

(2) The mere fact that a pair of ideas *A* and *B* stand in the relationship of intersection with a third idea *C* implies nothing at all for the relationship in which they stand to each other. They can be mutually exclusive, or they can intersect each other also, or they can be subordinated to each other or they can even be equivalent ideas. We obtain examples of all these cases if we let *M* mean a man, *A* and *B* respectively morally good and morally evil, wise and sick, temperate and virtuous and deserving of happiness.

(3) The first case, in which certain ideas *A*, *B*, *C*, *D*... that intersect with some idea *M* are mutually exclusive, is, as one might well suppose, worth paying some attention to just because *M* imposes a certain *connection* among ideas *A*, *B*, *C*, *D*... that would not be there without it. Yet knowledge of it may at times be of importance. For example, it is certainly important to know whether, given two learned societies that have no member in common, there is not a third that has some members in common with each of them, for that puts them all into a sort of cohesive relationship with each other. We could say that in such a case the ideas *A*, *B*, *C*, *D*... are *linked indirectly* by *M*. If each successive one of the several ideas *A*, *B*, *C*, *D*... intersects with its immediate predecessor, yet stands in the relationship of incompatibility with all those earlier [in the series], we can confer upon this series of ideas the title, *chain*. The following ideas provide an example: "men who lived in the first, the second, the third milennium," and so on. For surely there are some men who lived both during the first and the second milennium and some who lived both in the second and third, but no one who lived in the first milennium was still living in the third, etc. If all remains as before except that the last idea once again intersects with the first, I might call the

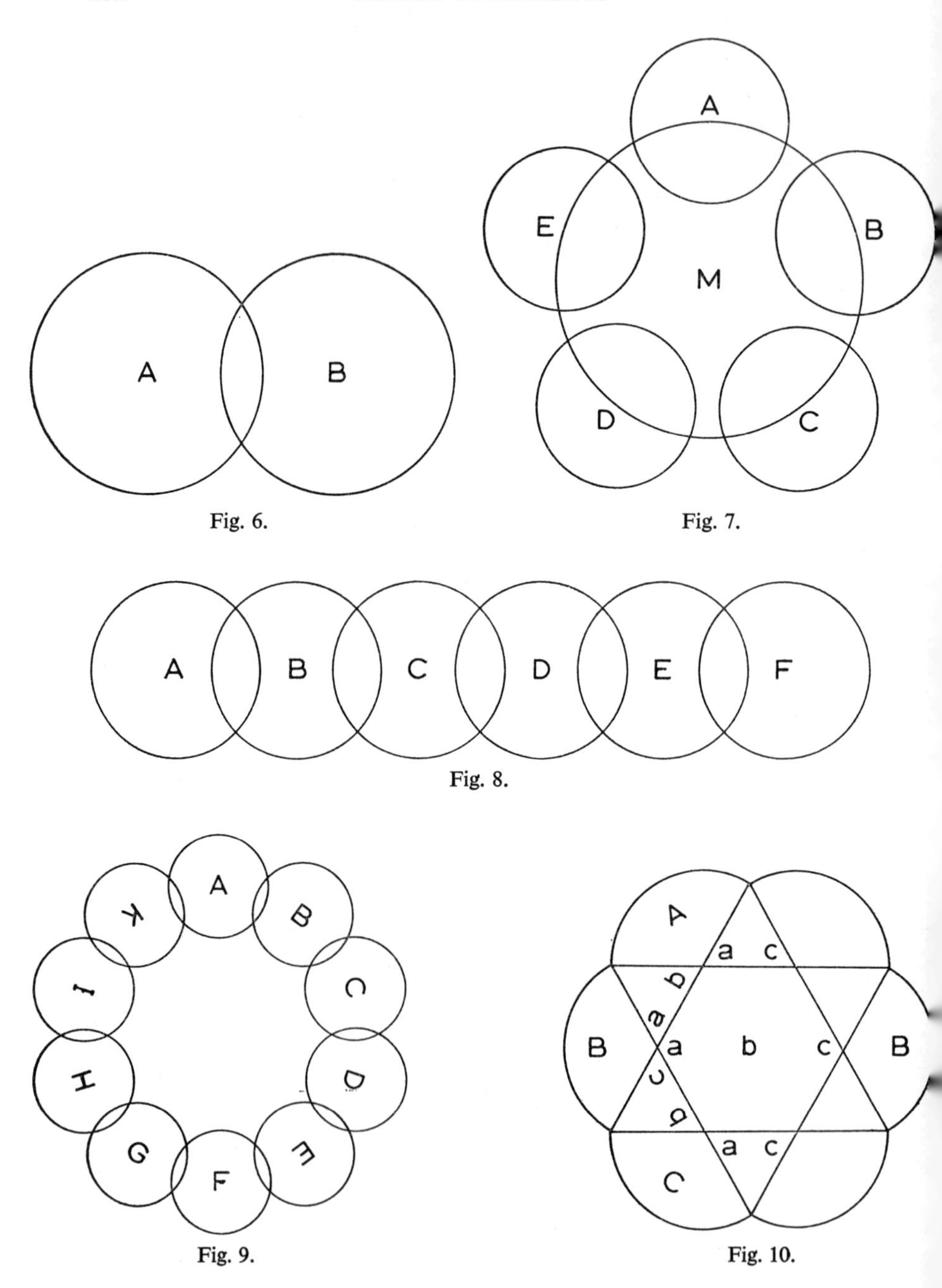

Fig. 6.

Fig. 7.

Fig. 8.

Fig. 9.

Fig. 10.

entire collection of these ideas a *recurrent* or *closed* chain. One example is offered by the seven ideas, sounds that can be counted as *c*, *d*, *e*, *f*, *g*, *a*, *h*; another by the ideas, red, orange, yellow, green, light blue, dark blue violet, and may others.

(4) A still more noteworthy relationship, one moreover that comes up much more often than the one just considered, holds among ideas *A*, *B*, *C*, *D*... when every pair of them whatever stands in the relationship of intersection. No pair has completely the same objects, but each pair has some objects in common. In this particular case the idea of something which is *A* as well as *B* as well as *C*, etc., or the idea, [something] $(a+b+c+\cdots)$, is a denotative idea in which none of the parts *a*, *b*, *c*, *d*... are *redundant* (§69), since the ideas that emerge as soon as we take away one of these parts, e.g. [something] $(b+c+\cdots)$, [something] $(a+c+\cdots)$, are all broader than the idea, [something] $(a+b+c+d\cdots)$. We could give this relationship the title, *omnilateral intersection*. We have an example in the ideas, polygon, equiangular and equilateral, which for just this reason can be united into the non-redundant concept of an equiangular and equilateral polygon.

(5) The thought comes quickly that the relationship of intersection described here, which we have just considered only as holding between *single* ideas, can be extended to entire sets of them as well. Thus we shall say that the set of ideas *A*, *B*, *C*... *intersects* with the set of ideas *M*, *N*, *O*... if there are objects that fall under one of the ideas *A*, *B*, *C*... as well as one of the ideas *M*, *N*, *O*..., but there are also others that each of these collections represents all by itself, etc.

(6) In order to represent the relationship of intersection between ideas *A* and *B* by a diagram, we must symbolize the domains of each of them by areas that have something in common and something distinct, approximately as in Figure 6. If the ideas *A*, *B*, *C*, *D*... intersect with each other only *indirectly*, namely by way of idea *M*, this will give a diagram roughly like Figure 7. Extended or recurrent chains are pictured in Figure 8 and 9, and the relationship of omnilateral intersection in Figure 10.

§ 102

No Finite Set of Standards Is Sufficient to Measure the Breadths of All Ideas

(1) What has been said so far puts us in a position to make somewhat

more precise judgements about the differences among ideas with respect to the breadth of their domain than we have already done in §68. As we saw there, there are ideas with a domain that encompasses only a *finite* number of objects, indeed only a single object. In order to measure the breadth of such a domain, one must choose as a measure an idea that similarly embraces only a finite number of objects. The most natural thing is to take unity as the measure of such as have only a single object. Nevertheless, we have seen in the very same place that there are also ideas that encompass infinitely many objects. Their breadth, then, can be measured only by taking as a unit the breadth of an idea of the sort that similarly has infinitely many objects. Now it could be believed that with this single unit we could succeed in measuring all domains of this kind. But this is not so; on the contrary, it is evident that there are an infinite number of ideas so constituted that one of them will be surpassed in breadth by another an infinite number of times. From this fact it follows that the measure which serves for measuring one of them could not be applied in measuring the other, and consequently that no finite set of measures is sufficient to measure the breadths of all ideas. The truth of this claim is demonstrated, it seems to me, by the following example, to which many others could easily be added. If we designate any whole number whatever by the letter n as an abbreviation, then the numbers n, n^2, n^4, n^8, n^{16}, n^{32}, express concepts, each of which undoubtedly encompasses infinitely many objects (namely infinitely many numbers). It is equally obvious, furthermore, that every object which falls under one of the concepts following n, e.g. n^{16}, also falls under the immediately preceding concept n^8, but that going the other way there are very many objects which fall under the preceding concept n^8 but are not included in its successor n^{16}. Of the concepts, n, n^2, n^4, n^{16}, n^{32}, each successive one is subordinate to its predecessor, consequently. But furthermore it is equally undeniable that the breadth of every one of these concepts surpasses the breadth of its predecessor *an infinite number of times*. For let us say the largest number we want to extend our calculations to $=N$. Then the largest number the concept n^{16} can represent is N and consequently the number of objects it encompasses $=$ or $<N^{1/16}$. And likewise the number of objects the concept n^8 represents, $=$ or $<N^{1/8}$. The relationship of the domain of the concept n^8 to that of the concept n^{16} would be therefore $=N^{1/8}:N^{1/16}=N^{1/16}:1$. But since $N^{1/16}$ can become

any given quantity, if we may take N to be large enough, and since we may take N to be as large as we please – indeed we will approximate the true relationship that obtains between the breadths of the concepts n^8 and n^{16} more and more closely, the larger we take N to be – it follows that the breadth of the concept n^8 surpasses that of the concept n^{16} an infinite number of times. Now since the series n, n^2, n^4, n^{16}, n^{32} can be extended as far as we please, we have in *it* an example of an infinite series of concepts, each of which is an infinite number of times broader than its successor.

(2) It is immediately evident from this that it would be entirely in vain to choose to express the various relationships that can obtain among ideas, even with respect to their breadth only, with complete precision by relations of *spatial things*. In order to be able to succeed in that, there would also have to be an infinite series of quantities in space, each of which exceeded another an infinite number of times. But it is well known that there are no such things to be found. So we could never picture the relationships there are among the breadths (and heights) of ideas exactly, even if we chose to call all three dimensions of space into play. Here we must be satisfied, then, with nothing but an approximate kind of presentation in any case; therefore it is probably best to stay with the diagram in a plane that has been recommended up to this point. For actual representations by physical space or even mere diagrams supposed to represent it are much too inconvenient and weary the powers of imagination. Representations by means of mere lines (as Lambert has tried), however, would not have the intuitive power that is the case with surfaces. The inclusion of one line (as a part) within another, as is necessary for representing the relationship between the domains of a higher and a lower idea, can not be made as visible as the inclusion of one area in another. To be sure we must be satisfied with representing ideas of which the one many exceed the other in the breadth of its domain by an infinite number of times by means of areas of which one is only somewhat larger than the other. It is still less possible to present intuitively by our diagram the case in which two ideas come so close together in their domain that there is ostensibly no intermediate idea between them; for it is always possible to draw between two areas, one larger than the other, an area of intermediate size. But happily such relationships come up only infrequently and their pictorial representation is even less frequently necessary. One

can see for himself, finally, that the two kinds of pictorial representation I mentioned in §95, which are only capable of expressing the relations of *heights* of ideas anyway, would offer no greater precision. Of course we can diagram the distance between the higher and the lower idea larger or smaller, so as to indicate that the one idea exceeds the other in extension by more or by less. But that the extension of the one is an infinite number of times greater than that of the other is something we can never present intuitively in that way, because what we would need for that would be to diagram one distance an infinite number of times larger than another one.

§ 103

Particular Kinds of Incompatibility among Ideas

(1) The relationship between ideas I called *incompatibility* in §94 also admits of a much more precise definition. If it is said of several ideas *A, B, C, D*... that they stand in a relation of incompatibility, all this means according to the explication given in §94 (1) is that there is not a single object that falls under *all* of these ideas. It can still be common to a part of them, e.g. *B, C*... without *A*, or *A, B*,... without *C*. In that case, then, the ideas *B, C*... are compatible among themselves, and likewise *A, B*...; it is only *A, B, C, D*... altogether that are not. But if the opposite is true, i.e. if the various ideas *A, B, C, D*... are incompatible in such a way that not even two of them are compatible with each other, so that no single object that falls under one of these ideas also falls under another, then I will call this relationship *omnilateral incompatibility* or, in order to have a shorter expression, *exclusion*. If we are comparing only two ideas with each other, then, it is all the same whether we say they are *incompatible* or that they *exclude each other*.

(2) This relationship of exclusion can also be extended to entire *sets* of ideas. The sets *A, B, C*..., *M, N, O*..., *R, S, T*..., etc. are said to be mutually exclusive if not a single object that falls under one of them falls under a second as well. But if we compare only two sets *A, B, C*... on the one hand, *M, N, O*... on the other, then it is indifferent whether we call them incompatible or exclusive.

(3) For two ideas to be mutually exclusive all that is required is that the domain of each of them should be composed solely of objects that are not present in the domain of the other. It does not demand, however,

that all objects that are absent in the one appear in the domain of the other. Still, this case is also possible. For if idea *A* does not already include within itself all objects whatsoever (i.e. if it is not equivalent to the idea, object in general), then the idea, "something (anything you please) which is only not *A*," will have a domain that embraces everything which does not fall under *A* itself. Certainly the relationship of such ideas is particularly worthy of note and therefore we call them by a name of their own: *opposite* or *contradictory* ideas. For example, the two ideas, blue and anything at all which is only not blue, are of this kind. All mutually exclusive ideas which are not at the same time contradictory, so that the one does not take into its domain any object you please that the other excludes, we call *contrary* to one antother, or *conflicting*. So, for example, the ideas of blue and yellow, and also blue and witty, and others like them, are not contradictory, but merely contrary. For they do exclude each other; what is blue can not be yellow, nor witty. But not every object lacking in the one is found in the domain of the other; not everything which is not blue must therefore be yellow or witty.

(4) Understandably for every given idea, if it has an extension at all and not the broadest possible extension, a contradictory can be found. This can come about in the way indicated in (3) as long as the two conditions just mentioned hold, because then the two ideas, *A* and not *A*, both have an extension, are mutually exclusive, and in such a way that every object that does not fall under the one will be found under the other. If, on the other hand, the given idea *A* had no extension itself, it could not be called an exclusive idea; but if it had the broadest possible extension of the idea of something in general, the second idea could have no extension. But if an idea has one contradictory, then, according to §96, it also has an infinite number of them. Thus, if the ideas A', A''... are equivalents of *A*, not only would the idea of something which is not *A* be a contradictory of idea *A*, but so would the ideas, something which is not A', something which is not A'', and so on. For example, then, not only the idea of something which is not God is contradictory to the idea of God; so is, something which is not omnipotent, also something which is not omniscient, and so on. Yet all ideas that are contradictory to one and the same given idea must always be equivalents of each other. For because they must include all of the objects the first idea does not encompass, they must all encompass the same objects and consequently all have the same extension. This

is not the case with ideas that are merely *contrary*. For every idea that has an extension, only not the broadest extension, there will be an infinite number of ideas contrary to it. These differ in many ways not only in their content but also on their extension, since their domains can take in this or that part of the objects absent from the first idea. Thus the idea, round, has an infinite number of ideas contrary to it, differing very much in their extensions, e.g. triangular, virtuous, etc. But what all of these ideas contrary to a given idea have in common is that their domains are only parts of the domain of the contradictory idea, or, what comes to the same thing, they are all subordinated to the latter as the idea above them. Thus the ideas, triangular, virtuous, etc. which are all contrary to the idea, round, are all subordinate to the idea of something which is not round.

(5) The concepts of *contradiction* and *contrariety* can also very easily be expanded so that the relations signified by them can enter in not just between single ideas but between entire *sets* of them. Namely, if a pair of sets of ideas, *A*, *B*, *C*, *D*... on the one hand, *M*, *N*, *O*... on the other, are mutually exclusive and at the same time so constituted that every object there is whatsoever (anything you please) either falls under one of the ideas *A*, *B*, *C*, *D*... or under one of the ideas *M*, *N*, *O*..., then we shall be able to say that the relation of *contradiction* obtains between these two sets. If the characteristic last mentioned is absent, then we shall say that they are merely *contrary* to each other. Thus, for example, the set of ideas, "an unconditionally real substance" and "a substance with only conditional reality," stands in the relationship of total contradiction to the set composed of the following two: "a real thing that exists as a mere property of another being" and "things which have no reality at all." If ideas *A*, *B*, *C*, *D*... on the one hand and *M*, *N*, *O*... on the other are supposed to stand in the relationship of contradiction, then the idea of "something which is neither *A* nor *B* nor *C* nor *D*..." must stand in the relationship of equivalence to the set of ideas *M*, *N*, *O*... and likewise the idea of "something which is neither *M* nor *N* nor *O*..." must be equivalent to the set of ideas *A*, *B*, *C*, *D*....

(6) Finally as to the question how two contradictory ideas should be represented by a diagram: it is plain from the definition that the areas representing their two domains must together cover the entire area assigned to the broadest idea, namely that of something in general. So if one assumes the domain of this latter idea to be the entire infinite plane,

and if one of the two contradictory ideas, e.g. *A*, can be represented by the bounded area *A*, then what belongs to the other or the idea of "anything at all that is not *A*" is all the rest of the infinite plane that lies outside of *A*. But whether it would be appropriate, and in what cases, to assign to one of the two ideas, and which one, an area that is only finite, and to which one an infinitely large area; or whether it would not be more appropriate to take the areas for both of them as infinitely large and perhaps in a finite relationship to each other: all of that must be determined by the natures of the given ideas themselves. In order to diagram two ideas that are merely contrary to each other, all that is required is for the areas representing their domains to lie outside each other, and that taken together they still do not fill up the entire space allocated to the idea of an object generally.

§ 108

How the Relationships Discussed in §§ 93ff Can Be Extended to Objectless Ideas as Well

According to the explications given, the relationships among ideas that we were considering from §93 on only extend to ideas that have a *denotative character*. All the same it is certain that we apply several of the relations enumerated so far to ideas for which there are no objects corresponding, to which ideas perhaps *can* not correspond at all, because they ascribe to them certain conflicting attributes. We do so even in the language of ordinary life. For example, we do not hesitate to say that the two ideas, "a mountain which is golden" and "gold which forms a mountain," are *equivalent*, even if we doubt whether there are any objects corresponding to these ideas. But according to the explication of the relation of equivalence set forth in §96, this could only be said insofar as such a mountain exists. Likewise, mathematicians do not demur at calling the concepts of a body with *five* equal plane surfaces and of a body with *seven* a pair of *mutually exclusive* concepts, even though they know there is no body either of the first or of the second kind. Likewise we say that the concept of a being with not a single good property and the concept of a man with not a single good property stand in the relationship of *subordination*, and that the former is *wider* than the latter. And consequently anyone who proved to us that there was no being of the first kind would already have proved that there is no man of the kind described. Here, then, we ascribe a relationship of subordination to ideas on the very occasion of saying of them that they are objectless ideas. This could not happen if we did not take the relationship just mentioned in a certain *extended* sense. And now what is it? We have already discovered in §69, in connection with the concept of *redundancy*, which was originally so defined that it applied only to denotative ideas, a device for extending it in a way that suited our purposes. What we did was to imagine that certain components *i, j*... present in the given idea were *variable*. This device helps in our present case as well. We can immediately extend all of the relations considered in §§93–107 to objectless ideas as soon as we are permitted to regard certain components present in them as variable. Then it is only necessary to turn our attention to the infinite number of new ideas derived from the given ideas when whatever other ideas replace the variable parts *i, j*.... Now what relation we observe these new ideas to have with each other, *whenever they are denotative*, we also ascribe to the given objectless ideas. It is obvious that we do so only *conditionally*, insofar as it is precisely the parts *i, j*... which are permitted to vary. Thus we shall say that a pair of objectless ideas *A*, *B* are *equivalent ideas*, just *with respect to* the variable parts *i, j*..., if the ideas produced by putting in the place of *i, j*... any other ideas we please are *equivalent* to each other in the narrower sense explained in § 96, whenever they become denotative ideas. We shall say that *A* is *higher*, *B lower* if the new ideas derived from *A* and *B*, when they are denotative, are subordinated to each other in the sense explained in § 97, and so on. First of all, for example, the two ideas, a mountain which is golden and gold which is a mountain, are to be considered equivalent, with respect to the variable parts, mountain and gold, because all of the ideas attained by exchanging the two parts just named for any others we please stand in the relation of equivalence defined in § 96. That is, they have the same objects, whenever they have an objects at all. Likewise, we say that the idea, "a being which has not a single good property," is higher than the idea, "a man who has not a single good property" – in that we conceive of the component present in both of these ideas, "having not a single good property,' as variable, whereupon it then becomes obvious that the idea derived from the first is always higher, in the sense defined in § 97, than the idea derived from the second, whenever we put in the place of that variable part something such that what we attain thereby are denotative ideas.

Note: The mathematician makes the most important use of the extension of these relationships, particularly in the theory of equations. According to their original concept these are nothing but expressions of the equivalence of two ideas. Saying that $4+5=11-2$ means only to say that the idea of $4+5$ has the same objects as the idea of $11-2$. But if we adhere to this concept, we could never set up equations such as $2-2=0, 1/\sqrt{-1}=$ $=-\sqrt{-1}$ and others like them. Their meaning is easily explained in accordance with what has been said.

§ 120

On the Rule that Content and Extension Stand in an Inverse Relationship

The result of what has gone before could be that even in that structuring of the theory of ideas with the greatest appearance of regularity to it, introduced by Kant, there are still very many important deficiencies. To be sure we should now also take into consideration the most outstanding other structures, e.g. by Platner, Maass, Reinhold, Twesten, Beneke and others. But since space forbids this, I shall permit myself only to do one more thing, namely to direct my readers' attention to that lack of precision in stating a certain proposition concerning the theory of ideas of which almost all treatises on logic have been guilty since the appearance of the *Ars cogitandi.* It is the rule that with every idea or at least with every *concept content and extension stand in an inverse relation.*

It is of course self-evident that the expression, *inverse relation*, could not be taken in the strict mathematical sense here. Is anyone supposed to believe that the extension of an idea is reduced to exactly a half or a third when the number of components it is composed of is doubled or tripled? But what the beginner must think to himself when we present that proposition to him without further explanation is at least both of the following: "Every idea that has a larger content than another (so that it is composed of the latter's parts and some others besides) has a smaller extension than it does (so that its extension is a part of the other's extension). And conversely every idea that has a smaller extension than another (so that its extension is a part of the other's extension) has a larger content than it does (so that its content is composed of the latter's content and some other parts besides)" – I venture to say that neither of these two propositions is true. (a) The first is not true because an idea's content can be increased without its extension diminishing. All that requires is for us to add components from which no new properties (not already implied by the prior ones) of the objects represented follow, as this comes about with so-called redundant concepts. (§ 69) Thus the content of the concept of a round sphere is larger than that of a sphere simply, although the extension of both concepts is exactly the same. But I will even say that there are additions to an idea by virtue of which its domain is increased at the same time as its content. So, by the addition of 'living', there arises from the idea, "a man who understands all European languages," the idea, "a man who understands all living European languages." The latter certainly has more content and also a larger extension than the former. Likewise, the idea "of a color that can be prepared from blue plant juices" has without a doubt a broader extension than the idea, 'blue', which is only a single component of it. For not only blue, but other colors (e.g. red, green) can be prepared from blue plant juices. And so on. (b) These examples prove at the same time that the second proposition is also incorrect, i.e. that in order to reduce the extension of a given idea or to construct an idea subordinate to it it is not always necessary first to make some addition to its content. This would only be necessary if every idea subordinated to another had to be composed of its parts and something else besides. But we have already seen in § 64 that the objects included under a certain idea *A* can have properties of which the ideas by no means appear as components of idea *A*. Now suppose that *b* is such a property belonging to all of the objects that fall under A. Then all of the objects that belong under *A* also belong under idea *B* or "of something which has the property *b*." Consequently *A* is either equivalent to *B* or (in case *b* belongs to still more objects than those falling under *A*) is subordinate to it. Nevertheless *A* does not need to be composed of *b* at all and still less so of *B*. Examples have already been given many times. Moreover, is it so very inexplicable how this mistaken rule arose and spread? The first one to introduce the concepts of content and extension into logic, the author of the *Ars cogitandi*, had already provided the occasion for the development of the basically erroneous view here by presenting the concept of content in such a way that he counted every property that necessarily belongs to the object of an idea (every *attribu-*

tum) within the idea's content. If all of the necessary properties (or rather their ideas) of an object already had to make their appearance as components of the idea of it, then to be sure the extension of an idea could only be diminished by increasing its content. For if the idea *B* is supposed to be narrower and lower than *A*, the objects comprehended under *B* must necessarily have some properties of their own besides all of the properties that are also common to the objects comprehended under *A*. On that presupposition, then, idea *B* would have to have some other components (namely the ideas of the properties peculiar to the objects comprehended under *B*) besides the components of *A* (namely the ideas of the properties of the objects comprehended under *A*). *B* would thus have to be more complex than *A*. And with that the one part of the rule (namely the second proposition) would be proved. But once the belief was held that the extension of an idea only becomes narrower when the content increases, there was already an inclination to believe that the former becomes narrower *whenever* the latter is increased. And since this is actually true, except for certain peculiar cases like those cited in (a), it is understandable that these exceptions were either not even thought of or they were not looked upon as a refutation of the principle. For example, if it should be noted that the extension of an idea does not diminish when its content is increased by a mere redundancy, that rule was not viewed as mistaken because of this, because the difference between redundant and non-redundant ideas was not looked for in the ideas in themselves, but in our mere thoughts or expressions of them. If I am so fortunate as to have avoided a mistake here which remained unnoticed by others, I will openly acknowledge what I have to thank for it, namely it is only the distinction Kant made between analytic and synthetic judgments, which could not be if all of the properties of an object had to be components of its idea.

VOLUME TWO

CHAPTER TWO

ON PROPOSITIONS IN THEMSELVES

§ 122

No Proposition in Itself Is an Existent

That propositions in themselves by no means belong to that class of things we call *existing* or *real* has already been brought up several times. Therefore I mention it here only for the sake of order, because I hold that this is the first property which must be attributed to all objective propositions. I believe, however, that this truth will be granted me as soon as one has understood the concept of a proposition in itself, without any demand for proof, as I would not know how to give any such proof. Existence belongs only to propositions *thought*, and also to those *held to be true*, i.e. to judgments, but not to the propositions in themselves which are the *material* a thinking being grasps in his thoughts and judgments.

§ 123

Every Proposition Necessarily Contains Several Ideas. Its Content

A second property that belongs to all propositions generally is in my view that they are *compound*, and by virtue of their complexity can be analyzed into certain components, which – however different they may be, still have some generally applicable properties and relations in common with one another. For the present I maintain only that every proposition is a composite and includes *ideas* as its parts. Scarcely anyone will dispute that. For even in the simplest proposition, even if its linguistic expression consists of a single word, we shall nevertheless on closer inspection become aware of many parts, which are nothing other than individual ideas. Who could fail to see, for example, that contained in the simple proposition, Come!, besides the concept of coming, expressed most distinctly in this case, there is also the concept of some obligation

[*Sollen*] and the concept of some person who is supposed to [soll] come. It is equally unmistakable that the proposition expressed by the Latin *sum* also includes the idea, I, along with the concept of being, just as the English expression actually requires the two words, I am. So it is with the proposition represented by the Latin word, *tonat*. It is not only that we see ourselves compelled in English to use the words: It is thundering. Is it not apparent to anyone that in this proposition we throw in the idea of a certain time (namely the immediate present) along with the mere concept of thundering?, and so on. Now if every proposition contains several ideas and we may consequently consider it something composite, I permit myself to call the sum of all its more proximate or its more remote parts its *content*. Thus, for example, the proposition, God has omniscience, is composed of the ideas, God, has, and omniscience, combined. So I call the sum of these three ideas its content. But since the idea of omniscience, as the idea of a knowledge that extends to all truths, is still complex itself, and includes, for example, the ideas, knowledge, truth, and many others, I count these ideas also within the content of the above proposition. The content of a proposition relative to a proposition is thus the same as the content of an idea relative to it, in § 56. In that connection it is only to be noted that because not every idea is composed of parts, as every proposition is, the concept of content must be somewhat expanded in connection with *ideas*, if we are to be able to say of every idea, even of every simple idea, that it is not empty of content.

§ 124

Every Proposition Is Capable of Being Considered as a Component of Another Proposition, or Even of a Mere Idea

The comment was made about ideas in § 62 that there was no limit to the possibility of combining them, i.e. that every one was capable of being a component of another. The same thing is true of propositions, that every one can be considered as a component of another proposition, indeed of nothing but a mere idea. We can join any proposition *A* we please with any other *B* we please in a complex, and in thinking of this complex we have an *idea*. And if we assert something of this complex, e.g. that it is a complex, we have a *proposition*, in which *A* is present as a component.

§ 125

Every Proposition Is either True or False and True or False in All Times and at All Places

This attribute of propositions has also been mentioned earlier and it belongs so uniquely to propositions and so evidently that it is used if not to define, still to make the concept of a proposition intelligible. If nevertheless we sometimes hear it said of a proposition that it is both at the same time, true as well as false, or the contrary, that it is neither, neither true nor false, or only half true, what is being spoken of is never a proposition in itself, but only a mere *linguistic expression* of a proposition. What is meant is that this linguistic expression admits one interpretation on which it has a true sense and another on which it has a false sense, or that it is so indefinite that we do not find ourselves justified in either the one or the other of these interpretations, or that we could gather from the words before us something true, but also something false.

Finally, how it comes to be that many propositions are said to be true only for a certain time or a certain place was already explained in § 25. And so I hope no one will take it seriously that the truth or falsity of propositions is a property of them that varies with time and place.

§ 126

Three Components that Are Undeniably Found in a Large Number of Propositions

(1) It will be conceded to me that in connection with very many propositions, if possibly not all of them, and especially in connection with all true propositions there are certain *objects* they are concerned with, i.e. in reference to which something is asserted in these propositions. Thus, for example, in connection with the proposition, God is omnipotent, God is obviously the object with which it is concerned. On the other hand, the proposition, all equilateral triangles are also equiangular, has an infinite set of objects with which it is concerned. But if there are certain objects with which a proposition is concerned, there must be within it an idea that refers to these objects, encompasses them. For it is only by virtue of this that it comes to be concerned with those objects and no

others. Thus we could not say that the proposition, God is omnipotent, is concerned with God, or has God as its object, its *unique* object, if it did not contain an idea referring, and referring uniquely to God. And we say of the proposition, all equilateral triangles are also equiangular, that it concerns all equilateral triangles only because it includes an idea that refers to all equilateral triangles as one of its parts. Now let me call the idea in a proposition of the objects it asserts something about the proposition's *object-idea* or *subject-idea* or its *basis*.

(2) The further concession will be made to me that, if not all, surely very many propositions assert a certain *property* of the objects they concern. Of such propositions it is obvious that along with the idea just discussed, which represents their objects, they must also contain an idea of the property they ascribe to them. Moreover, they must also contain a distinct idea by which those first two are *connected* with each other, an idea, then, which indicates that the objects considered in the proposition *have* the property cited in it. So, in the proposition, God is omnipotent, or rather in this one, God has omnipotence, besides the concept of God as that object with which the proposition is concerned, there appears the concept of a certain property, namely omnipotence, a property which is ascribed to that object, and finally there is the concept of having itself. I will be permitted to call the idea of the property asserted of the proposition's objects the *property-idea* or *predicate-idea* occurring in it, or the *assertive part*. The idea that ties these two together, however, I call the proposition's *connective* or *copula*. Three components that undeniably occur in a large number of propositions are, then, a subject-component [*Unterlage*], a predicate-component [*Aussagetheil*] and a copula [*Bindeglied*].

§ 127

Which Components Does the Author Assume for All Propositions?

As many propositions as there are in which the three components just mentioned are so *visibly* present that even their expression in language includes *separate symbols* for them, it is nevertheless not to be denied that an even larger number of propositions appear under entirely different linguistic forms. For all that, if we attempt to raise the thoughts we really want to express by such varied forms to the clearest consciousness possible, it seems to me that the longer we do so the more convinced we shall be that the following holds true of all propositions generally. The concept of *having*, or still more specifically, the concept signified by the word *has*, is present in all propositions. Besides this one component two others are present in all propositions, which are connected with each other by that *has* in a way indicated by the expression: *A has b*. One of these components, namely the one indicated by *A*, stands as if it was supposed to respresent the object with which the proposition is concerned. The other, *b*, stands as if it was supposed to represent the property the proposition ascribes to that object. Therefore I permit myself to call the one of these parts, *A*, wherever it may be, the *basis* or *subject-idea*. The other one, *b*, I call the *assertive part* or *predicate-idea*. In order to persuade my readers that these claims are correct, I know of no more productive means than to ask them to try for themselves, with any proposition that occurs to them, whether it can be analyzed into the parts identified. With a considerable number of propositions, or rather whole classes of them, namely all those I shall come to speak of in what follows, I intend to point out the parts of which they consist according to my idea of them myself, and then the question will be whether the readers can agree with my analysis. Beforehand I can only limit myself to the remarks below.

(1) Grammarians all acknowledge that a specific *verb* must be present in every complete sentence in their sense, i.e. in any *linguistic expression* of a proposition. From this alone it can be inferred that the concept of having appears in every proposition. For every verb, if it is not itself the verb *to have*, contains the concept of having as one of its components. To prove this I will appeal to a presupposition that will be granted me more readily. Any particular verb distinct from the word *is* can, without any essential change of meaning, be replaced by a participle derived from the given verb, tied by the word *is*. *A does* is entirely equivalent with *A – is – doing*. But if *is* occurs in a sentence, then it has either the form, *A is*, or *A is B*, depending on whether we designate the rest of the sentence's parts by *A* or by *A* and *B*. In both cases, it seems to me, it is not difficult to convince oneself that the proposition contains the concept of *having*. It is obvious that the meaning of a proposition of the form, *A is* (an existential proposition), such as *God is*, is not different from: *A – has – existence*. And so the concept *has* is present in it, of course, just as there is also an object, signified by the idea *A*, with which the proposition is

concerned, and the property ascribed to this object is in the present case existence. (Of an objection that the scholar, not common sense, could raise here, later in § 142.) It is still more evident that propositions of the form, *A is B*, never have any other meaning but what the expression, *A has b*, indicates as well, insofar as *b* represents the abstractum that belongs to the concretum *B*. Namely, everyone will grant that the *is* in such propositions has by no means the same meaning it has in the existential propositions mentioned just now. There is no intention of asserting existence here, as is already evident from the fact that such a proposition can be true even if the object *A* does not belong to the things that have and can have existence at all. Thus, for example, concepts in themselves do not have existence, yet we do not hesitate to express the judgment: "The concept of a triangle *is* complex," because it is not at all our intent in using this *is* to make it known that the concept of a triangle is something existent. On the contrary, it seems evident to me that the sole meaning of the proposition cited is: "The concept of a triangle – has – complexity." But from this latter expression it follows that the connecting link in our proposition is none other than the concept indicated by the word *has*. Propositions of the form, *A is B*, would always be more clearly and correctly expressed as follows: *A* – has – (the property of a *B* or) *b*." According to this expression, their common connecting link would be just the concept of having. Why we use the form with the verb *is* so much more often than the one with the word *have*, why we almost always even run into some difficulty when we choose to apply the latter is sufficiently explained by the fact that in the form, *A has b*, the part of the expression, *b*, must always be a mere abstractum (the idea of a property), while in the form, *A is B*, what appears in *B*'s place is a concretum. But every language is incomparably richer in symbols for concreta than it is in designations for the abstracta that belong to them. Besides, when the latter do exist, they consist for the most part of long inconvenient words, even compounds of a number of words. Now if the further question is raised, why it is that in all languages concreta more frequently are given symbols of their own than their related abstracta, and for the most part simpler ones, my reply is that this originates only in the fact that to the greatest extent concrete ideas have objects that fall on the senses and occupy our attention more powerfully than abstracta. This has the twofold consequence that on the one hand symbolizing them is that much more pressing a need and on the other hand that we can much more easily come to understand the meaning of these symbols. Thus, for example, we find that a body by its brightness, its beautiful yellow color, its considerable weight, its extraordinary ductility, etc. draws our attention to itself so powerfully that we forthwith judge it important enough to invent a name for it, gold. On the other hand, it seems to us redundant that we should distinctly designate *in abstracto* all of the properties that make this body gold. Se we stop with the symbol for the concretum (gold); but for the abstractum (goldness) either we form no word of its own at all or we symbolize this concept when necessary by a combination of several words, such as "the nature of gold." And just in order to make it easier to get along without this symbol for the abstractum, we devised the form, *A is a B*, instead of the form, *A has b*. In it we can use the same symbol that serves to express subject-ideas in a proposition for symbolizing predicates as well.

(2) But of course there is still reason enough to doubt whether the form just indicated holds for *all* propositions. First of all, there are propositions in which besides the idea of the object they say something about there is only, as it appears, a single entirely simple idea: e.g., *A should*, *A effectuates*, *A wills*, *A senses*, etc. If the ideas of the words, *should*, *effectuates*, *wills*, *senses*, etc. were in fact simple, then it would be proved that not all propositions contain the concept of having and two other ideas besides. But I say that the ideas cited are all compound and that in them the concept of *having* does appear, bound up with another, respectively, of *obligation*, of *efficacy*, of *will*, of *sensation*, etc. What I believe is that the above propositions, if they are supposed to be expressed in such a way that their components stand out most distinctly, would have to be rendered in approximately these ways: *A* – has – an obligation, has efficacy, a will, a sensation, etc. In general I can adopt only one of two opinions: either *should*, effectuates, etc. are, as I have just said, compounded from the concepts of obligation, efficacy, will, sensation, etc. or conversely the latter concepts are compounded from the former. Now the first is much more probable to me if only because the derivation of the former concepts from the latter can easily be accomplished by way of the concept of having, while the converse generates a great many difficulties. For even if the belief could be held that the concept of an effect is perfectly well explicated as the concept of something which

is effected, it would obviously be incorrect to follow this analogy and pass the concept of will off as the concept of something which is willed. What someone wills is the object of his will, not the will itself. Indeed it seems to me that the attempted explication of effect also fails to render the abstract concept of efficacy (effectiveness), but only the concrete concept of what is effected. The most obvious objection to my point of view is that precisely the same relation that holds between the two concepts of *having* and of a *property* also obtains between the two concepts of *willing* and a *will*, of *effectuating* and *efficacy*, etc. Now if the concepts of willing, effectuating, etc. were not simple, but compounded out of the concept of will, efficacy, etc. and the concept of having (as I have just assumed), then the similar relationship makes it seem that the concept of having should not be simple either, but compounded of the one property and another. If willing means the same thing as having a will, effectuating the same thing as having effectiveness, etc., then having would have to mean the same as having a property. But the latter is absurd, because the concept of having can not possibly be identical with that of having a property, but at most equivalent with it. One might well conclude, therefore, that the concepts, will and has a will, effectuates and has efficacy, are not identical either, only equivalent. – I reply that the presupposition we started from here, that the two concepts, property and having, must be related to each other exactly as the concepts, will and willing, efficacy and effectuating, etc., is incorrect. For while it can be maintained of the concept of a property that it is, if not identical, nevertheless equivalent with the concept of something had, it can certainly not be said of the concepts of will, efficacy, etc. that they are identical or even equivalent with the concepts of something willed, effectuated, etc.

(3) But there are propositions for which it is still less obvious how they are supposed to fall under the form, A has b. Among these are so-called hypothetical propositions of the form, if A is, then B is, and also disjunctive propositions of the form, either A or B or C, etc. I shall consider all of these propositional forms in greater detail in what follows, and it is to be hoped that it will then become clear to the reader that there are no exceptions to my rule in these cases.

(4) A new reservation may be aroused, however, by noting that even in those sentences in which it is present explicitly, the word *having*

does not always occur in the same form, but is variously modified according to the person and number (in many languages according to gender as well) of the first term (i.e. the subject idea). Someone could want to conclude from this that the concept that word signifies in its various forms is not always the same either. This reservation vanishes, nevertheless, when one takes it into consideration that those variations in form belong to the merely arbitrary features of language and are introduced only for the sake of greater clarity of expression and perhaps also to achieve more variety. Therefore there are also languages that do not have any such modifications. Language loves a kind of pleonasm, in that it gives *repeated* expression to many concepts, so that they will be the more definitely understood, in one and the same proposition. It is a pleonasm of this sort when we say, I *have b*, instead of I *has b*, and so indicate a second time, by an intentional modification of the connective part, that the subject of the proposition is the person speaking, which is already fundamentally obvious from the symbol, I.

(5) But language uses the form it gives the word *having* in a proposition, or the *verbs* that include this concept, not only to express the person and number of the subject, but also designations of time. It does this so generally that for this very reason it was given the title (in German) *time-word* (*Zeitwort*). Should one not infer from this that it is not the pure concept of having which consititutes the true connecting link, but rather this concept of having tied up with the specification of the time something is had? I reply that we do express a specification of time (namely the present time) by the word, having, even if we are speaking of objects that are not in time at all, for example, in saying: Every truth – has – an object with which it is concerned." From this alone it can be seen that we are by no means justified in inferring from that precise union into which language brings time-specification with the concept of having that there is an essential connection between them. We have already remarded in §§ 45 and 79, and in what follows this will become clearer and clearer to us, that the specifications referred to belong essentially to the proposition's subject-idea. A proposition of the form: "Object *A* – has at time *t* – the property *b*," must always be expressed, if its parts are supposed to come out as clearly as possible, as follows: "Object *A* at time *t* – has – (the property) *b*." For it is not at time *t* that property *b* is ascribed to object *E*, but we ascribe property

b to object *A*, insofar as we think of it as something existing at time *t* (hence invested with this specification).

(6) Once we have been convinced on this point, we shall hardly hesitate to affirm of other specifications linguistic expression ties up with the verb that all the same they do not belong to the connecting link. I mean the specifications: often, seldom, always, and others like them, and also the degree of *probability* we want to assign to the proposition. We say: *A – probably has – b*; *A – certainly has – b*. Yet it is clear that these specifications have nothing to do with the way the predicate *b* belongs to the subject *A*, but only have to do with the relation in which the entire proposition, *A* has *b*, stands to our cognitive faculties or to other propositions. *A* probably has *b* obviously means nothing but: The proposition that *A* has *b* has – probability. The situation is quite similar with specifications of *necessity* or *contingency*, which we likewise often tie to the copula, saying: *A* – necessarily has *b*, etc. We shall find out the true meaning of such propositions for the first time later on (§ 182), where it will become evident that they too have no copula other than the one I have generally assumed. In its place (§ 136) it will be shown that even the concept of *negation*, however intimately language ties it to the verb, is nevertheless not a component of the copula, but of the predicate-idea, since the proposition, *A – non habet – b*, really only means, *A – habet – defectum* τοῦ *b*.

(7) Now if what is present in every sentence as its connecting link is only the concept signified by the word *has*, then there is no doubt that there must also be at least *two* other components present besides that one. For neither *has* nor *A has* nor *has b* expresses a complete proposition all by itself. It follows from the definition already encountered in § 48, however, that there are only two alternatives. Parts *A* and *b* can either be identified as mere ideas or as entire propositions themselves. If the proposition in question is supposed to be a *true* one besides, it might very well turn out that the two parts, *A* and *b*, have to be *ideas*, and *denotative* ideas at that, and that *b* in particular has to represent a property. But if it is not prescribed that the proposition, *A* has *b*, should be a truth (as we are speaking presently only of the characteristics that must belong to all propositions indifferently, even false ones), then I just do not see why *A* and *b* would have to be nothing but ideas, and on top of that denotative ideas, and *b* in particular the idea of a

property. Should it not be permissible for any combination of the form, *A* has *b*, to be called a proposition, without regard to whether the symbols *A* and *b* signify mere ideas, and what kind of ideas, or whether they signify whole propositions as well? But even (because the matter is in fact apparently indifferent) if we restrict the concept of a proposition to the case in which *A* and *b* are mere ideas, in no case may we require (I say) that these ideas both be denotative, and the latter a genuine idea of a property besides. For why, for example, may not the combination the following words express: "A body bounded by five identical surfaces is not bounded by triangles," be called a proposition? We would say of anyone who actually combined these ideas that he was making a judgment, even though there is no body bounded by five identical surfaces. According to § 66, it depends on the most accidental of circumstances whether or not an object corresponds to a given idea. With the idea of a golden tower it depends on whether someone or other has in fact constructed a tower out of gold. Now if it was supposed to belong to the essence of a proposition that its basis is a really denotative idea, then the answer to the question whether a certain combination of ideas deserves to be called a proposition would have to depend on such an accidental circumstance as whether the basic idea present has a real object corresponding to it. The words, a golden tower is costly, would express a proposition provided that a golden tower had actually been erected somewhere; if this had not happened, they would not only represent no true proposition, but they would represent no proposition at all.

§ 130

The Extension of a Proposition Is Always the Same as the Extension of Its Base

According to what has been said, there are certain *objects*, perhaps not in connection with every proposition but with most of them and with all true ones, they assert something about. That property of a proposition by virtue of which it concerns just these objects and not any others I call its *domain* or *extension* or *sphere*. Thus in order to give a complete specification of a proposition's extension we must not only specify *how many* but also *which* objects it is concerned with. If we only look at the first, i.e. inquire merely whether a given proposition has only a single object or several of them and how many, then we are inquiring about the *size* of its domain, i.e. the proposition's *breadth.*

But in my opinion the part of a proposition that makes known to us its extension is the *base,* and no other. It is this part that has to represent all of the objects it is supposed to be concerned with, and in fact the proposition does concern all of the objects it represents, i.e. it says something about them. The extension of a proposition, then, is always the same as the extension of its base. If the latter is an objectless idea, then the proposition itself has no object with which it is concerned either, it too is *objectless.* Therefore everything I have said in § 66 and elsewhere about the way to determine an idea's extension and to symbolize what has been discovered can be applied to the determination and symbolizing of a proposition's extension or breadth.

§ 133

Conceptual Propositions and Empirical Propositions

No matter how anyone conceives of the parts of which any proposition must be composed, he will scarcely dispute the fact that there are propositions, even true propositions, that consist merely of pure concepts, without containing any intuition. It is quite obvious, for example, that the following propositions are of this sort: God is omnipresent, Gratitude is an obligation, the square root of the number 2 is irrational, etc. We shall first see in what follows how propositions of this kind, particularly if they are true, differ in very essential points from others which contain intuitions as well. I find therefore that for intellectual proposes it is indispensable to designate them with a name of their own. I shall call them *propositions from pure concepts, conceptual propositions* or *conceptual judgments,* and also if they are true, *conceptual truths.* Other propositions, which thus contain some one or more intuitions, might all be called because of this very circumstance *intuitive propositions.* They are also called *empirical propositions, perceptual propositions,* and the like. Thus, for example, I shall call the proposition, This is a flower, Socrates was an Athenian by birth, empirical propositions because each of them contains an intuitive idea, indeed probably several of them.

Note: The principal reason I find the division into conceptual and empirical truths, as here interpreted, so important is because the truths set forth in an intellectual discourse must be handled quite differently if they consist of mere concepts than if they also contain intuitions. This is especially true if what is called for is not only making them certain but also citing their objective grounds. We can only seek for the ground of a pure conceptual truth in certain other conceptual truths. The ground of an empirical truth, however, can also lie at least in part in the objects to which the intuitions it contains refer. But the more important the distinction between conceptual and empirical truths is, the stranger it would have to be for it to have escaped earlier logicians. It has not; but if we might complain that this distinction is not set forth in the usual logic textbooks with all the clarity we could wish for, still we must admit that even the most ancient philosophers knew of it and discussed it in a number of ways. Plato revealed a very important distinction between pure concepts (νοήσις) and Ideas transcending all experience on the one hand and merely empirical ideas or intuitions (φαυτασίαι) on the other. He required of a rational science (ἐπιστήμη), especially of pure science (χαθαρά), which concerns the immutable, that its theorems be derived not from experience, but from pure concepts. He looked on pure thought as being occupied with mere concepts, in which they are analyzed, combined, and the like without any attention being paid to sense perception (intuition). What is absent, then, except that he did not designate the concept of propositions composed of mere νοήσες and the concept of other propositions in which this is not the case with names of their own, the better to fix them as concepts? And from him, who did not go into any precise specification of the components of propositions and who generally only spoke of concepts or ideas instead of propositions, that was not something to have been expected anyway. Of Aristotle, on the other hand, we have adequate knowledge that he did distinguish between universal propositions (προτάσεις χαθόλυ, which are surely nothing but propositions consisting of pure concepts) and others (as, for example, in Anal. post. I, Chap. 7). And on that basis he insisted that we might not believe propositions such as have merely a perishable thing (φθαρτόν, empirical thing, intuition) as their object to be capable of a proper proof. Locke (*Essay*, B. 4, Chap. 3, § 31. Chap. 4. § 6. 16 and elsewhere) not only conceived of the distinction between conceptual and empirical propositions with full clarity, but he also distinguished between the sciences to which they are respectively indigenous and maintained that the former gave complete certainty, the latter never anything but probability. Now although I can not agree unconditionally with this claim, it seems to me nevertheless that it discloses a very correct view of the distinction between conceptual and empirical propositions. All perceptual judgments (and these constitute the largest and most important class of empirical judgments) are by virtue of their derivation mere judgments of probability, because they result from a major premise which has probability only. Conceptual judgments, on the other hand, can be merely probable only in an accidental way, insofar as we are not completely sure that we have not made a mistake in their derivation or to the extent that we have obtained them from mere perception. – Crusius (W. z. G. §§ 222. 231.) not only spoke of the distinction between this twofold kind of judgments, but also proposed their own names for them and recommended that they receive more attention from philosophers. He calls "propositions with a subject which is an individual or a collection of individuals, e.g., the earth is round, or the Greeks captured Troy, *individual propositions*, the others *universal propositions*." Of the latter he says "that they are either of infinite scope, if they speak of general concepts from which all individuality has been removed, e.g. all bodies are compound" – (these are the propositions I call conceptual propositions), "or of finite scope, if they are conjoined individual propositions, e.g. all of the planets in our skies are smaller than the sun. It is important to take note of this distinction," he says further, "because universal propositions of infinite scope" (conceptual propositions) "must be proved in a different way. Their truth must either be proved from the concepts themselves" (which is really the only way of objectively establishing them) "or from an external but necessary basis, namely from one such that its necessity can be proved from the concepts themselves" (so its truth is once again proved from concepts) "or from the attributes of God." (These will likewise only be known from concepts.) It is unmistakably evident from these statements that Crusius had his eye on the distinction between conceptual and empirical propositions. In my opinion the only mistake he made was that, lacking a clear concept of intuition, he confused them with singular ideas and also believed that intuitions could never find a place except in the subject of the proposition, that they could not sometimes enter into the predicate. In modern times, when the understanding of the distinction between concepts and intuitions has been sharper than before, surely much more light would have been shed on the distinction between conceptual and empirical propositions if the following circumstance had not caused its correct explication to miscarry. (At least that is my belief.) The division of our knowledge into knowledge we can become convinced of (as we usually say) only by experience and knowledge that requires no experience was already familiar to the ancient philosophers, but they did not pay sufficient attention to it. By Leibniz and Kant especially it was brought into prominence as one of the most important distinctions. Now it happens, however, that this division of our knowledge almost coincides with the division of propositions into conceptual and empirical propositions, since the truth of most conceptual propositions can be decided by mere thought without any experience, while propositions that include an intuition can in general be judged only on the basis of experience. Therefore what happened is that it came to be believed that the essential difference between these propositions was to be found not so much in the nature of their components as in the way we can become convinced of their truth or falsity. On this view the former were defined as those which can be known without any experience, the latter, however, as those which require experience, and accordingly they were given the names: *a priori* and *a posteriori* judgments. (See for example the Introduction to Kant's *Critique of Pure Reason.*) I too find the distinction made here important enough to be retained forever, but I believe we should not let attention to it suppress another one which does not depend on the mere relationships of propositions to our cognitive faculties, but on their intrinsic character. I refer to their distinction into those which are composed merely of pure concepts and others for which this is not the case. Indeed, I permit myself to maintain that basically it has really been the latter division they had in mind when they were concerning themselves with the former, without being clearly aware of it. For if what was conceived of under the title, judgments *a priori*, was really rendered correctly by defining them as cognitions that are independent of all experience, then it would scarcely have been necessary to add on right away a pair of attributes by which *a priori* judgments were supposed to be recognizable, namely *necessity* and *universality*. Now whether a proposition is strictly universal or not and whether it could be said that the predicate it ascribes to its subject belongs to it with necessity or not, these are all matters that depend on the intrinsic character of the proposition itself and concern not at all its accidental relationship to our cognitive faculty. There is also no doubt that since Kant explicitly says that all mathematical propositions belong to [the class of] *a priori* judgments, he would also have assigned to this class of judgments mathematical propositions which given our present limitations we are not capable of knowing, for example a formula for deriving all prime numbers, and that on the other hand he would have assigned certain other propositions, for example the answer to the question what are the inhabitants of Uranus doing right now, to the class of empirical propositions, even though there are no experiences that lead us to make a decision on this question. Beck (*Logik*, § 67) explicitly brings to our attention that a judgment can be *objectively a priori* although it is *subjectively* present as merely *a posteriori*; I do not believe he will be contradicted on this point very much. But from this it becomes apparent that whether a judgment is *a priori* or not is considered an objective characteristic in the judgment itself. Consequently it should be defined in an objective way, not in a way drawn from the judgment's mere relationship to our cognitive faculty. That the two criteria of universality and necessity are not apt for this purpose, however, is already clear from the fact that they can be applied only to *true* propositions. Besides that, all logicians declare the proposition, some numbers are prime numbers, to be a particular proposition, and most of them declare the proposition, all finite beings are fallible, to be problematic. Yet both are pure *a priori* propositions. Indeed, in § 182 I hope to show that if the concept of *necessity* is to be defined, it presupposes the distinction between *a priori* truths and others.

§ 137

Various Propositions about Ideas:
(a) Assertions of the Denotative Character of an Idea

There is an extremely noteworthy class of propositions we make use of not only in every scientific enterprise but even in the business of everyday life. It comprises the propositions that deal with *ideas* and secondly those that deal with *entire propositions*. From the innumerably many kinds of propositions that belong to these two classes, I am going to pick out, however, only those which not only occur most frequently but also have the distinctive feature that the way they are usually expressed in language does not make it easy to know what components they are composed of.

If the object with which a proposition is concerned is a mere *idea*, understandably its subject-idea or base must be the idea *of* an idea. Now since we call such ideas (§ 90) *symbolic* ideas, it is a general feature of propositions concerned with mere ideas that their base is a *symbolic* idea. But the first thing that can occur to us in considering an idea is probably the question whether it also has an object corresponding to it or (to use the abstractum) denotative character, denotation. – If we answer this question in the affirmative, we are asserting a proposition that generally will be included under the form: "*The idea A – has – denotation.*" Let me be permitted to call such propositions affirmations or assertions of denotation.

Now propositons that language actually expresses in the form just given, e.g. "the concept of an angel has denotation," are not the only ones that belong to this class. If I am not mistaken, so do all of those propositions with a linguistic expression of the form: "There is an *A*," such as "There is a God," "There is a highest moral law," "There is a body bounded by four identical surfaces," and the like. The fact that what we intend to assert by the words, *there is*, in propositions of this latter form, is not always the real existence of the object to which idea *A* refers is already plain from the fact that we use them even with objects which can have no being in reality, for example with the highest moral law, which as a mere truth in itself is not and can not be anything existent. The true meaning of such propositions, then, is just that idea *A* has an object corresponding to it. It is only when it is implicit in idea *A* itself that the corresponding object must be an existent, as with the concept of God, that the proposition, there is an *A*, is not yet the same as, to be sure, but is equivalent to the proposition, *A* has existence. But I shall go still further and say that even propositions of the form, *A certain A has b*, or *Some A are B*, or *Several A are also B*, are basically nothing but assertions of a denotative character. At least this is so if they are supposed to be understood as they are generally understood in scientific essays, especially in treatises on logic. What I refer to is that these expressions are not supposed to specify anything at all as to *how many A* have the property *b*, whether just a single one, or in fact several, or even all of them. In such a case I would not know how else to render what the propositions cited specify except that the idea of an *A* which is at the same time *B* (or has the property *b*) was a denotative idea. So if an expression is called for which brings out clearly enough what the logical parts of these propositions are, the only one that will serve (I believe) is this: "The idea of an *A* which has the property *b* – has – denotation."* If all the same someone should find it too improbable that the interpretation of those linguistic expressions attempted here really rendered the very thought we attach to them, since they are composed of entirely different words than should be the case according to this interpretation, I will not argue the point any further. All I ask is that he at least concede to me that the meaning attached to those forms of speech and the meaning given by my interpretation are *equivalent* to each other in the sense that whenever the one proposition is true, the other is also. This will hardly be disputed. (Cf. § 173.)

§ 138

(b) Denials of the Denotative Character of an Idea

If we answer the question raised in the preceding paragraph, as to whether the idea we are considering has an object, in the *negative*, then we are making a proposition of the form: "Idea *A* – has – no denotation." I permit myself to call such propostions *denials of denotation.* I count among them, however, not only those propositions of which the linguistic expression has precisely the form just indicated, e.g. "The concept of a round square has no object," but also all propositions of the form, *There is (are) no A.* For I believe we do not intend to indicate anything by this expression but that idea *A* is objectless. With propositions expressed in language under the form, "No *A* is *B*," it can be doubtful whether they are to be counted as denials of the denotative character of an idea and so interpreted as if they read, "The idea of an *A* which is also a *B* (or has the property *b*) has no denotation," or whether they have to be regarded as merely negative propositions of the form, "All *A* are non-*B*," *i.e.* "Every *A* has the property not *b*." But if, as happens not infrequently, we take the expression, "No *A* is a *B*," in such a sense that we do not

* Locke recognized this sense of particular judgments (*Essay*, Book 4, Chap. 9, § 1). J. G. Fichte (Posthumous Works, Vol. I, p. 375) also noted that the judgment, Some *A* are *B*, really has the meaning: *A* can also be a *B*.

intend to presuppose at all whether there even is an *A*, then we may never declare this expression to be equivalent to "Every *A* is a non-*B*." For if idea *A* had no objects at all, then what the words, "Every *A* is a non-*B*," signify, even if it might be a proposition, could never be called true, because there would be no objects with which it is concerned. In such a case, the real meaning we intend to express by the words, "No *A* are *B*," would only be the first one mentioned: "The idea of an *A* which was also a *B* has no object." For example, if a geometer, to prove that there are no bodies bounded by five identical surfaces started out from the propositions, "A body with five identical surfaces is not bounded by triangles, it is also not bounded by quadrangles, and so on," he does not intend these expressions to be understood as if there were a body bounded by five identical surfaces, because precisely what he wants to prove is that such a body is impossible. What he intends to say is only that the idea of a body bounded by five identical surfaces has no object once you think of the surfaces as triangles, likewise once you think of them as quadrangles, and so on, and from that it finally follows that the idea of such a body has no object at all.

§ 139

(c) Further Propositions that Define the Extension of an Idea More Closely

(1) Once we have recognized that a given idea *A* is denotative, then the further question arises, how many objects does it have. If we discover that it has several, then this idea, according to the definition in § 68, is to be called a *general idea*, and a proposition that expresses this should be called the *assertion* or *affirmation of a general idea*. For example, "The idea of a triangle has several objects," would be of this sort. Now the question is, what are the components of such a proposition? Or in case there should be several equivalent propositions by which such an assertion can be made, I prefer to ask what are the parts of the simplest of them? There is no disputing that the commonest and shortest *expression* language uses for symbolizing this thought is: "There are several *A*." But it has become clear from so many previous examples that the shortest linguistic expression does not always disclose the components of the proposition in itself most distinctly, nor even signify among all those

that are equivalent *the* proposition which is the very simplest. Nevertheless I will make a beginning with the analysis of this sentence in order to arrive at the expression I am seeking. "There are several *A*," surely does not intend anything other than what we can express somewhat more clearly so: "There is a plurality of *A*." And this latter, according to the explication of plurality we obtained in § 86, would have once again only the following meaning: "There is a set of things of which each part is an *A*." Now if I was right in § 136 in explicating propositions of the form, There is an *A*, as mere assertions of the denotative character of an idea and reducing them to the expression, "The idea of *A* has denotation," then we will have to express the proposition, "There is a plurality of *A*, or there is a set of which each part is an *A*," in the following way: "The idea of a plurality of *A* (or the idea of a set of which each part is an *A*) has denotation." Lengthy as this expression is, as many parts as the proposition it signifies consequently has, still I can find no other expression that does not contain still more complex ideas when I try to attain clarity about every sign in it. And so I believe we shall do best to stick to this interpretation.

(2) Not quite so worthy of having notice taken of them as the kind of propositions just considered, which assert the existence of a general idea, are those contrasting propositions which *deny* the existence of a general idea, or propositions by which we assert that an idea A is not a general idea. Nevertheless, they are important enough to deserve being cited. I will call them *denials of a general idea*. The proposition, "The idea of the universe does not have several objects," would be an example. Anyone satisfied with the way I have just attempted to analyze *assertions* of a general idea will raise no further question about how the *denial* of a general idea should be analyzed. It is obvious of itself that its analysis could only read: "The idea of a plurality of *A* has no denotation."

(3) When we say of an idea *A* that it is not a general idea, what we are presuming is not quite precisely that it has at least one object. It can also have none at all. Nevertheless, the opinion comes easily to mind that it is a case particularly worthy of attention when an idea does have denotation, yet does not have several objects, but only a single one, and so (according to the nomenclature in § 68) is a *singular* idea. Propositions in which this property is asserted of an idea, I will call *affirmations of a singular idea*. For example: There is only one God,

there is only a single highest moral law, etc. Anyone can see for himself that I shall recognize no interpretation for this sort of propositions besides this one: "Idea *A* is a singular idea," i.e. "Idea *A* has denotation and the property that the idea of a plurality of *A* does not have denotation." Unbelievable as it may seem that such a complex thought composed of so many parts should be contained in the few syllables, "There is only one *A*," I confess that I am unacquainted with any briefer interpretation of this expression. But just as there are *affirmations* of a singular idea, so are there *denials* of them. Because they have only slight importance, I can pass them by, however.

(4) With an idea that really represents several objects it is sometimes possible, as we saw in § 63, to specify the *number* of them. Propositions that do this can be called *specifications of the breadth of an idea* or *assertions of number*. Their general form will be: "The totality of objects that fall under the idea *A* (or perhaps more briefly, the totality of *A*) has the number *n*" etc.

§ 146

Objectless and Denotative, Singular and General Propositions

If I am correct in supposing that there are also propositions that have a base representing no object at all, it will be permissible to call such propositions *objectless*. Other propositions, for which an object they concern can be indicated, we can call *denotative* propositions or propositions with an *extension*. If they have only a single object, they can be called *singular propositions*, and if they have several of them, in the ultimate case an infinite number, *general propositions*.

§ 147

The Concept of the Validity of a Proposition

It is such a well-known fact that propositions can be divided into *true* and *not true* and it has already been assumed so many times in this book, that we can be satisfied with just mentioning it here. But it is undeniable that any given proposition can be only one of the two, and permanently so, either true and then true for ever, or false and, again, false forever. (§ 125) Unless we should *alter* something about it, and so no longer be considering the proposition itself but another proposition in its place.

We often do that, without being distinctly aware of it, and precisely this is one of the causes creating the illusion that the same proposition can be sometimes true, sometimes false, depending on how it is related to various times, places and objects. So we say that the proposition, wine costs 10 thalers the pitcher, is true at this place and this time, but false at another place or another time. Also that the proposition, this flower has a pleasant smell, is true or false, depending on whether we are using the *this* in reference to a rose or a carrion-flower, and so on. But does anyone fail to see that it is not actually one and the same proposition that reveals this diversified relationship to the truth? That we are considering several propositions, which only have the distinguishing feature of arising out of the same given sentence, in that we regard certain parts of it as variable (just as we did with ideas in §§ 69 and 108), and substitute for them sometimes this idea and sometimes that one? In the first example, it is the implicit condition, added in the mind, as to time and place that produces sometimes a true, sometimes a false proposition, depending on whether it is this time and place or that time and place. In the second example, we change the idea that is meant by the word, *this*, referring the proposition at one time to a rose, at another time to a carrion-flower, and consequently do not have before us the same proposition, but two essentially different ones. But even if as these examples show, we do suppose certain ideas in a given proposition to be variable, often without being clearly aware of it, and then consider the relation to the truth that follows for this proposition upon filling those variable places with whatever different ideas, it is always worth the trouble to do this consciously and with the definite intention of becoming the more precisely acquainted with the nature of the given propositions by observing this relation of theirs to the truth. Namely, if we consider in a given proposition not merely whether it is itself true or false, but also what relation to the truth follows for all the propositions that develop out of it when we assume certain of the ideas present in it to be variable and permit ourselves to exchange them for whatever other ideas, we shall be led to discover many extremely remarkable properties of propositions. For example, if we look on the idea of Caius in the proposition, "The man Caius is mortal," as one that is arbitrarily variable, and so replace it with whatever other ideas, e.g. Sempronius, Titus, rose, triangle, etc., the special characteristic manifests itself that all of the new propositions brought out in this way are true, whenever they have any denotation at all, i.e. whenever the idea that forms their base is really the idea of an object. For if we replace the idea of Caius by ideas that signify real men, e.g. Sempronius, Titus, then a true proposition arises. But if we replace it with another idea, e.g. rose, triangle, then the proposition that emerges not only has no truth, it does not even have denotation. It is easy to see that this result would not come about if the given proposition were a different one. Thus, for example, if the proposition read: "The man Caius is omniscient," the exact opposite would be the result. Every one of the propositions that emerged from any substitution whatsoever of another idea for the idea of Caius would lack truth. On the other hand, if the proposition originally read, "The being Caius is mortal," then among the propositions that could be produced in the manner described, there would be some true ones, some false ones that are nevertheless denotative. This would depend on our replacing the idea of Caius sometimes with the idea of a being to which the predicate of mortality belongs, and sometimes another. This makes it clear that the relationship to the truth of all the propositions that can be produced from a given proposition by assuming one or more of its parts to be variable, may be looked upon as a property by which the nature of this proposition itself can become better known.

By far the most propositions, however, are so constituted that the propositions which can be derived from them are neither all true nor all false, but only some of them true and some of them false. In this case we can raise the question, how *many* true and how *many* false ones there are, or what relation the one set has to the other or to the entire set? If we are supposed to be permitted to replace the ideas regarded as variable with any other ideas whatsoever, though, the set of true propositions and the set of false propositions produced from the given proposition will both be infinite in every case, once there is even one of them. For suppose that i' is an idea that when put in the place of the variable i in proposition A makes it true (or false); then every idea equivalent to i' will have precisely the same effect, if not always, still in most cases surely. But there are an infinite number of such equivalent ideas (according to § 96). If we did not choose to restrict the kind of ideas that are supposed to replace the one regarded as variable in some way, then, the relation of the set of true or false propositions to the entire set that can be produced generally could rarely be determined. On the other hand, if we impose some restriction and stipulate, for example, that we can never substitute for the ideas *i*, *j*... regarded as variable ideas that are equivalent to each other (in the sense of § 96 or the still broader sense of §108), and if we require further that only such ideas can be selected as produce a denotative proposition, then the set of ideas which can still be selected, and consequently the set of propositions which can be created, will be diminished considerably. It will become possible much more often than it was previously to define numerically the relation in which the set of true or false propositions stands to the set of all of them. For example, if in the proposition, "that the ball numbered 8 will be among those drawn in the next lottery," we regard the single idea 8 as variable, but require that it be replaced by no idea equivalent to another that has already been selected and generally only with ideas that form a denotative proposition, then the total number of propositions that can be produced $=90$, if the lottery consists of 90 numbers. For now we will only be able to substitute for 8 in the proposition given one of the numbers 1 to 90. Every other number or idea we put in this place would convert the base of the proposition, or the idea of "the ball numbered *X*" into an objectless idea. This example shows us at the same time that the restriction by which we reduce to a smaller number the set of propositions that can be produced from a given proposition can frequently be contrived in such a way that the propositions remaining are precisely and only those worthy of our attention in some respect, just those we want to see assembled and counted up. Thus, for example, it doesn't matter to us what sorts of propositions can be engendered from the one just cited merely by substituting for the idea of 8 a large variety of ideas equivalent to it, but it is useful to know how many propositions come into the picture if we proceed with the restriction adopted in this case. It is particularly useful to know what relationship the set of true propositions that emerge in this way is to the entire set. Namely, this relationship defines the degree of probability the given proposition takes on in certain circumstances. So let me be permitted to designate the concept of the relationship of the set of true propositions that can be engendered from a given proposition by substituting for certain ideas in it to be regarded as variable certain other ideas, according to some rule, to the set of all propositions produced in this way by a title of its own. I choose to call it the *validity* of the proposition. The extent to which a proposition is valid, or how much validity it has, is consequently supposed to mean exactly the same as how the set of true propositions derived from it when we substitute for certain ideas in it to be regarded as alterable other ideas according to a given rule relates to the set of all of them. The degree of validity will be presented by a fraction, in which the numerator is to the denominator as the former set to the latter. Thus, for example, the preceding proposition's degree of validity is $5/90=1/18$, if five balls are drawn, for then among all of the 90 propositions that come into the picture in this case, there are only 5 which are true.

It is obvious that the validity of one and the same proposition must turn out differently depending on whether we look on this or that, just a single one or several ideas within it as variable. So, for example, the proposition, this triangle has three sides, has the property of being constantly true insofar as the single idea we look upon as variable in it is the idea, this, on the presupposition that we only replace it with ideas that produce a denotative proposition. If, however, we look on the idea of a triangle along with the idea of this as also variable, or instead of either of them the idea of a side, then the degree of its validity will turn out to be another one entirely. In order to be able to judge a proposition's degree of validity correctly, then, we must indicate in every case which ideas in it are supposed to be considered variable.

If the proposition A is of such a character that the propositions that can be derived from it when only denotative propositions may be constructed and ideas $i, j \ldots$ alone are looked on as variable are all true, then its degree of validity with respect to $i, j \ldots$ is the largest there is, $=1$. We can call the proposition a *universally* or *completely valid* proposition. If in the opposite case the propositions derived from A are all false, its degree of validity is the least possible, $=0$. We can say, therefore, that it is a *universally* or *totally invalid* proposition. We can also call completely valid propositions propositions which are *generically* or *formally true* and those that are completely invalid generically or formally *false*, if we were to think of the generic type we were speaking of as the set of all propositions which differ only in ideas $i, j, \ldots$.

If proposition A is generically true or false with respect to ideas $i, j, \ldots$, then the denial of A, or the proposition that A is false, represents a proposition which is generically false or true in the same respect.

(8) In somewhat the same way as it was shown in § 66 that the *extension of ideas* is one of their intrinsic properties, it becomes clear that the validity of a proposition is also one of its intrinsic properties.

§ 148

Analytic and Synthetic Propositions

(1) It is clear from the preceding section that there are propositions which are generically true or false when we take certain parts of them to be variable, but that the same proposition, which has this characteristic when just ideas $i, j, \ldots$ are regarded as variable, would not retain it when different ideas are regarded as variable, or more of them. In particular, it is easy to understand that no proposition can be so constituted that the property we are speaking of would remain even if we chose to consider *all* the ideas of which it consists to be variable. For if we could change all the ideas in a proposition as we pleased, then we could transform it into any other proposition whatsoever, and consequently we could surely sometimes make a true proposition out of it, sometimes a false one. But suppose there is just a *single* idea in it which can be arbitrarily varied without disturbing its truth or falsity, i.e. if all the propositions produced by substituting for this idea any other idea we pleased are either true altogether or false altogether, presupposing only that they have denotation. This property of the proposition is already sufficiently worthy of attention to differentiate it from all those propositions for which this is not the case. I permit myself, then, to call propositions of this kind, borrowing an expression from Kant, *analytic.* All the rest, however, i.e. in which there is not a single idea that can be arbitrarily varied without affecting their truth or falsity, I call *synthetic propositions.* So, for example, I shall call the propositions, "A morally evil man deserves no respect" and "a morally evil man nevertheless enjoys eternal happiness," a pair of analytic propositions. In both of them there is a certain idea, namely the idea of a man, for which you may substitute any idea you please, e.g. angel, being, etc., in such a way that the first (if it only has denotation) is true and the second false in every case. On the other hand, I could not point to a single idea in the proposition, "God is omniscient" and "a triangle has two right angles," which could be arbitrarily varied with the result that the

former would remain constantly true and the latter constantly false. Consequently for me, these would be examples of synthetic propositions.

(2) We have some very general examples of analytic propositions which are also true in the following proposition: *A* is *A*, *A* which is *B* is *A*, *A* which is *B* is *B*, Every object is either *B* or not *B*, etc. Propositions of the first type, or those included under the form, *A* is *A* or *A* has (the property) *a*, we are used to identifying by a name of their own as *identical* or *tautological* propositions.

(3) The examples of analytic propositions I have just cited are differentiated from those given in (1) by the fact that nothing is necessary for judging the analytic nature of the former besides logical knowledge, because the concepts that make up the invariant part of these propositions all belong to logic. But judging the truth or falsity of propositions like those in (1) requires quite another kind of knowledge, because concepts alien to logic exert an influence in them. To be sure, this distinction has its ambiguity, because the domain of concepts belonging to logic is not so sharply demarcated that no dispute could ever arise over it. At times it could be useful to pay attention to this distinction, and so we could call propositions such as those in (2) *logically* analytic or analytic in the *narrower* sense and those of (1), on the other hand, analytic in the *broader sense*.

Note 1: Making a judgment as to whether a proposition as it is expressed in language is analytic or synthetic often demands more than a cursory glance at the words. A proposition can be analytic, logically analytic as well, even identical, without any indication of it by its verbal expression. Again, many a proposition can read in words just like an analytic, even an indentical proposition and still be synthetic in meaning. Thus one might not see right away that the proposition, "every effect has its cause," is identical, or in any case analytic, as it really is. For since we always understand by an effect something effected by something else, and by the locution, having a cause, the same thing as being effected by something else, what that proposition really means is: "What is effected by something else is effected by something else." – The like holds of the propositions: If *A* is larger than *B*, then *B* is smaller than *A*; if $P = M \cdot m$, then $M = P/m$; and many others. On the other hand, there are many propositions that have even been elevated to the status of everyday proverbs, which sound quite analytic or even tautological without really being so. For example, it is often said that "what's bad is bad," and literally this is an empty tautology, of course. But what someone is really thinking of with these words, what he wants to have them understood to mean could be something quite different and could even be different things in different circumstances. One person might want to indicate by those words that he could not make up his mind to declare anything he finds bad to be anything but bad. Someone else could have the intention of calling to mind the fact that the effort to put the bad in a good light is always a vain effort, because sooner or later one always knows it for what it is, and so on. The same is true of the proposition Leibniz introduces (*Nouv. ess.* L. IV. Ch. 8) as an example of an identical (or rather analytic) proposition "which would not be without its use:" "Even a learned man is a man." In the sense in which one interprets it when one finds it useful, it is not analytic. The interpretation in that case is this: Even a learned man is fallible. The like holds true of Leibniz' second example: *cuivis potest accidere, quod quicam potest.* Thus, too, Pilate's, ὃγέγραφα, γεγραφα was no tautology, but had the sense: What I have written I will not change. Moreover, others have already taken note of the fact that apparently identical propositions were not always actually identical, in particular Reusch (*Syst. L.* § 435.)

Note 2: A number of logicians, Krug (L. § 62) himself among them, apply the word, *identical*, by which I designate an *intrinsic* property of some propositions, to signify a mere relationship among several propositions, calling judgments I call *equivalent* in § 156 identical. Among the logicians who attach the same concept as I do to this word, the ones who come closest to the explication I have given are those like Wolf (L. § 213), Maass (§ 220) and others who say that identical propositions are propositions in which subject and predicate are the same ideas. The only reason I can not adhere to this definition without modification is that, as I see it, the proposition, *A* is *A*, does not have the idea *A* for its predicate idea, but rather the corresponding abstractum. Therefore I say that a proposition is identical if its base is the concretum and its assertive part the abstractum corresponding to it. But when others, e.g. Lambert (Dian. § 124), Kiesewetter (W. A. d. L. § 109 and § 134), Klein (§ 158), put the essence of identical judgments in the fact that their subject and predicate are *equivalent ideas*, i.e. ideas with the same extension, this is completely mistaken. For on this definition, for example, the judgment, "A triangle is a figure with an angle sum of two right angles," would have to be called identical, which we surely do not intend. Still others say that identical propositions are *identifications*. But obviously they are not, if we mean by identification what is understood in common linguistic usage by expressions of identity or sameness, for the concept of identity or sameness does not need to enter into them at all.

Note 3: It is remarkable that while some philosophers would proclaim all of our thinking to be nothing but a collection of identical (or yet analytical) judgments, others took the opposite position and went so far as never to concede the title of propositions to any identical propostions at all. Maimon (L. P. 55) does this insofar as such propositions can provide no ground for others. He may not have been mistaken, but why deny them the name of propositions on that account? We also read in Keckermann's System L. 1. 2, sect. 1: *Identicae propositiones per se non sunt*. And so on.

Note 4: A distinction more or less similar to the one I am making here between analytic and synthetic judgments was already known to the ancient logicians. For example, Aristotle (*Soph.* c. 3) cites it as an error into which the art of the sophists seeks to draw us down: τὸ πολλάκις ταὐτὸ λέγειν. Locke (*Ess.* B. 4, Ch. 8) sets up the concept of trifling propositions, which he defines as those which do not instruct us, and numbers among them (a) all identical propositions and (b) all propositions that assert a part of a complex idea of its object. It is to be seen that what is meant here are analytic judgments and they are defined almost more clearly than we shall find further on in connection with Kant himself. But an important error follows, when Locke adds that all propositions in which the species is subject and the genus predicate are of this kind. For not every concept of a species is compounded of the concept of the genus. – Schmidt has already noted that Crusius (W. z. G. 260) understood the distinction between analytic and synthetic propositions just as Kant did, on occasion. But even if it is true that this distinction was mentioned before at times, nevertheless it was never properly pinned down and fruitfully applied. The merit of having been the first to have done that indisputably belongs to Kant. All the same, it would seem to me that the explications of this distinction one encounters, whether in Kant's own writings or those of others, still fall somewhat short of logical precision. For example, if we read in Kant's *Logic* (§ 36) that analytic propositions are those the certainty of which rests on the identity of the concept of the predicate with the notion of the subject, this is applicable to identical propositions at most. If it is said, as in the *Critique of Pure Reason* (Intro. § 4) and elsewhere that in analytic judgments the predicate is contained in the subject (in a concealed manner), or does not lie outside of it or already occurs as a component of it; or with Fries (Syst. d. L. P. 184) "that the predicate only repeats ideas of the subject;" or with Ulrich (Log. § 9), Jakob (L. § 659), E. Reinhold (L. § 82) and others, that the predicate is already (*implicite*) conceived in the subject; or with Gerlach (L. § 70) that the subject is only defined with respect to one of its parts in particular; or with Rösling (L. § 68) that the predicate is a partial idea or in a negated judgment a negation of the subject concept, and the like; these are in part merely figurative forms of expression that do not analyze the concept to be defined, in part expressions that admit of too wide an interpretation. For everything that has been said here can also be said of propositions no one would take for analytic, e.g. The father of Alexander, King of Macedon, was King of Macedon; A triangle similar to an isosceles triangle is itself isosceles, and the like, namely that the predicate idea is nothing but a repetition of one of the components of the subject idea, is included in it, lies within it, is conceived in it, etc. This unfortunate state of affairs could be avoided if, along with Eberhard (Phil. Mag. V. I, § 3, No. 4), Maass (ibid. V. II, §1, No. 2, cp. Log. § 210), Krug (L. § 67, Note 1) and others, one made use of the expression that in analytic judgments the predicate is one of the essential parts of the subject or (which comes to the same thing) constitutes one of its essential attributes, understanding these to be constitutive attributes, i.e. such as are present in the concept of the subject. But this definition is applicable to only one kind of analytic judgments, only those of the form: *A* which is *B* is *B*. Should there not be others as well? Should we not count the judgment, *A* which is *B* is *A*, and also the judgment, Every object is either *B* or not *B*, among analytic judgments? In general it seems to me all of these definitions fail to place enough emphasis on what makes this sort of judgment really *important*.

This, I believe, consists in the fact that their truth or falsity does not depend on the particular ideas of which they are constituted, but remains the same no matter what changes are made in some of them, presupposing only that the proposition's denotative character is not itself destroyed. For just this reason I permitted myself to give the above definition, even though I know it makes the concept of these propositions somewhat broader than it is ordinarily conceived to be, for propositions like those cited in (1) are not ordinarily counted as analytic. Besides, I thought it useful to interpret both concepts, of analytic as well as synthetic propositions, so broadly that not only true but false propositions could be included under them. – Still, no matter which definition may be accepted, in no case, I believe, will one be induced to concede that the distinction between analytic and synthetic judgments is a merely *subjective* one, and that the same judgment can sometimes be analytic, sometimes synthetic, as we make this concept or that of the objects to which the subject (or more strictly the subject idea) refers. Yet this is what many logicians, e.g. Platner (Phil. Aph. B. 1, §§ 510, 512), Maass (in Eberhard's Phil. Mag., V. II, Pt. 2, No. 2), Krug (L. § 67, Note 1), Hillebrand (L. § 295, Note 1), Schultz (Prf. d. Kant. Kr., Pt. I, p. 32), E. Reinhold (L. § 82, Note 1), Beneke (L. § 58), Troxler (L., Vol. I, p. 277) and others have maintained. Maass said (loc. cit.): "The triangle can be defined as a figure in which the angle sum is two right angles. In that case, the proposition, the sum of all the angles in a triangle is two right angles, which is regarded as synthetic on the usual definition, becomes analytic." I have a different opinion on the matter. Since what I mean by a proposition is not a mere combination of *words* that asserts something, but the *meaning* of this assertion, I do not admit that the proposition, the sum of the angles in a triangle, etc., remains the same when we attach to the word, triangle, now this concept and now another. About as little as it would be the same judgment if we said the words, "Euclid was a famous mathematician," and at one time understood the name Euclid to represent the man who taught geometry to Alexander under Ptolemy Soter, at another time Euclid of Megara, the student of Socrates. It is true that in the preceding example both ideas referred to one object, and in this case to different objects, but for propositions to be acknowledged distinct it is sufficient that they should only be composed of different ideas, even if they have the same object.

§ 154

Compatible and Incompatible Propositions

(1) The most important relations among propositions first come to light when, as in § 147, we look upon certain ideas contained in them as variable and take note of how the new propositions generated by substituting for those ideas any other ideas whatsoever stand with respect to their truth or falsity.

(2) We already know that almost any proposition can be made true sometimes, false sometimes when we put any other ideas we may choose in the place of certain ideas in it taken to be variable. But if we compare several propositions *A, B, C, D*... and regard certain ideas *i, j*,... common to all of them as the arbitrary ones, then the question arises whether there may well be some ideas substituted for *i, j*,... which are so constituted that those propositions thereupon *all* become true at the same time. If this question is to be answered in the affirmative, I will call the relation obtaining among propositions *A, B, C, D*... a relation of *compatibility* or *concurrence* and the propositions themselves *compatible*, *concurring* or *consistent* propositions. If the question is to be answered in the negative i.e. if there are no ideas substituted for *i, j*,... which make propositions *A, B, C, D*... true all together, I call this relationship of the propositions so identified a relation of *incompatibility* or *inconsistency*, and the propositions themselves I call *incompatible* or *inconsistent*. Thus I call the following three propositions compatible with each other, if I may regard the idea, "this flower," as an arbitrary variable within them: This flower has a red blossom; this flower is fragrant; this flower belongs to the twelfth class in the Linnaean system. For if I substitute the idea of a rose, all three propositions become true. The following three propositions, on the other hand: No finite being has omniscience; man is a finite being; a man has omniscience; I call incompatible, if it is only the ideas, finite being, man and omniscience, that are supposed to be regarded as variable in them. For whatever ideas one attempts to substitute for them, one will never succeed in converting those three propositions into truths at the same time. As often as two of them are made true, the third becomes false.

(3) It is self-evident from the definition given that both the relationship of compatibility and of incompatibility are reciprocal.

(4) The similarity between this relationship among *propositions* and

the relation among *ideas* I designated by a like term in § 94 will also be obvious to anyone, especially after the extension given in § 108 obtains. Namely, what counts with ideas is whether or not they actually represent a certain object and what counts with propositions is whether they have truth or not. And just as I called ideas compatible or incompatible with each other according to whether or not there are certain objects they represent in common, so I now call propositions compatible or incompatible with each other according to whether or not there are certain ideas by which they can be made true all together.

(5) If we regard now these ideas and now those ideas as the variables in the same set of propositions *A*, *B*, *C*, *D*..., they can figure now as compatible, now as incompatible. Thus the two propositions, a lion has two breasts and a lion has two wings, figure as compatible if we look upon the idea, lion, as the variable. For if we substitute for it the idea of a bat, both propositions will be true at the same time. But if it is supposed to be the idea of two that we can vary arbitrarily, then the two propositions present themselves as incompatible, because there is no idea available which when substituted for it makes both propositions true. It can be understood in particular that if we should be permitted to increase at will the number of ideas supposed to be regarded as variable in a given set of propositions, these propositions would always be seen to be compatible. For if we could vary as many ideas as we pleased in a proposition, we could vary every one of them, and so we could transform any proposition into any other proposition, without doubt, therefore, into a truth. Consequently, when we say of a given set of propositions *A*, *B*, *C*, *D*... that they are compatible or that they are incompatible, we must, in order to be definite about it, always add *in what respect*, i.e. in relation to which ideas *i*, *j*,... selected as variables, we are saying this.

(6) All truths are compatible, no matter what ideas are considered variable within them. For the ideas already present in them originally have the property of making them all true.

(7) In every set of propositions that are not compatible with each other there must be at least one false proposition, therefore. There can, however, be several of them all false at the same time.

(8) But there can also be false propositions among propositions that are compatible with each other; indeed they can all be false at the same time.

For the fact that certain propositions are false in terms of the ideas of which they originally consist does not prevent them from all being able to become true at the same time in terms of certain other ideas. Only it is understandable that in that case at least one of the ideas supposed to be regarded as variable should appear in each of the given propositions, because otherwise it could not be altered at all and consequently could not be changed to the state of truth.

(9) If the *n* propositions *A*, *B*, *C*, *D*... are not compatible with each other with respect to ideas *i*, *j*,..., there can still be a relationship of compatibility among any smaller number of these propositions, e.g. among any $(n-1)$, $(n-2)$... of them, with respect to the same ideas *i*, *j*,.... For even if no ideas are available to substitute for *i*, *j*,... which make the *n* propositions *A*, *B*, *C*, *D*... all true, a portion of these propositions, e.g. $(n-1)$, $(n-2)$, of them, can nevertheless be made true at one time. Thus the three propositions, All *A* are *B*, All *B* are *C*, No *A* is *C*, are not compatible with each other with respect to the ideas *A*, *B*, *C*, but any two of them are entirely compatible with respect to those ideas.

(10) However, in the opposite case, if a part of the propositions *A*, *B*, *C*, *D*..., e.g. *A*, *B*... are not compatible with each other with respect to certain ideas *i*, *j*..., then neither is the set of *all* of them a set of propositions compatible with each other with respect to the same ideas. For if there were ideas which when substituted for *i*, *j*... make *A*, *B*, *C*, *D*... all true together, then *A*, *B*... would also be compatible.

(11) If certain propositions are compatible with each other with respect to fewer ideas *i*, *j*..., then they are also compatible with respect to the larger number of ideas *i*, *j*, *k*, *l*,..., in which the former ideas are repeated. And if they are incompatible with respect to a larger number of ideas *i*, *j*, *k*, *l*..., then they are also incompatible with respect to the smaller number *i*, *j*,.... The contrary does not follow. That certain propositions are incompatible with respect to fewer ideas *i*, *j*... does not imply that they are also incompatible with respect to the larger number *i*, *j*, *k*, *l*,.... And that they are compatible with respect to more ideas *i*, *j*, *k*, *l*... does not imply that they are compatible with respect to fewer, *i*, *j*....

(12) It in no way follows from the fact that propositions *A*, *B*, *C*, *D*... as well as propositions *G*, *H*, *I*, *K*... are compatible with propositions *M*, *N*, *O*... with respect to ideas *i*, *j*,... that propositions *A*, *B*, *C*, *D*,... and *G*, *H*, *I*, *K*,... are also compatible *with each other* with respect to the same

ideas. For there could well be certain ideas which made propositions *A*, *B*, *C*, *D*... true at the same time as propositions *M*, *N*, *O*,... when they were substituted for *i*, *j*,... and certain other ideas which made propositions *G*, *H*, *I*, *K*,... true at the same time as propositions *M*, *N*, *O*, and no ideas at all which make propositions *A*, *B*, *C*, *D*... true at the same time as propositions *G*, *H*, *I*, *K*.... Thus each of the propositions, All *A* are *B* and No *A* is *B*, is compatible with the proposition, All *A* are *C*, with respect to the three ideas *A*, *B*, *C*. All the same, the first two propositions are not at all compatible with each other with respect to the same ideas.

(13) It is equally faulty to infer that because neither propositions *A*, *B*, *C*, *D*... *nor* propositions *G*, *H*, *I*, *K*... are compatible with propositions *M*, *N*, *O*,... with respect to certain ideas *i*, *j*,... , propositions *A*, *B*, *C*, *D*... and *G*, *H*, *I*, *K*,... should not be compatible *with each other* with respect to the same ideas. For even if there are no ideas that make propositions *A*, *B*, *C*, *D*,... true at the same time as propositions *M*, *N*, *O*... when they are substituted for *i*, *j*,..., and none which make *G*, *H*, *I*, *K*... true at the same time as propositions *M*, *N*, *O*... , there can nevertheless still be ideas which make propositions *A*, *B*, *C*, *D*... true at the same time as propositions *G*, *H*, *I*, *K*... . Thus the propositions, "The earth revolves on its axis" and "The earth goes around the sun," are both equally incompatible with the proposition, "The earth stands still," when the only idea we are supposed to consider variable is the idea of the earth. All the same, the first two are compatible with each other with respect to the same idea.

(14) It in no way follows from the fact that certain propositions *A*, *B*, *C*, *D*... are compatible with each other with respect to certain ideas *i*, *j*,... that their *negations*, or the propositions, Neg. *A*, Neg. *B*, Neg. *C*, Neg. *D*,... (§ 141), are also compatible with each other with respect to the same ideas. For it could well be that one of the propositions *A*, *B*, *C*, *D*... , e.g. *A*, would not only be made true by some of the ideas that also make all the rest *B*, *C*, *D*,... true, but besides that would be made true by all of the ideas by which one of these propositions, e.g. *B*, is made false and consequently the proposition Neg. *B* made true. In this case there would be no ideas that make both propositions, Neg. *A* and Neg. *B*, true at the same time. Thus the two propositions, "Some *A* are *B*" and "It is false that all *A* are *B*," are surely compatible with

each other with respect to ideas *A* and *B*, but the two propositions generated by negating them, "It is false that some *A* are *B*" and "All *A* are *B*," obviously stand in the relationship of incompatibility with each other with respect to the same ideas, as previously.

(15) From the fact that the *negation* of each of the individual propositions, *A*, *B*, *C*, *D*... is compatible with the rest of them with respect to ideas *i*, *j*,..., it by no means follows that the negation of two or more of of these propositions is also compatible with the rest of them with respect to the same idea. For the fact that there are certain ideas which make propositions Neg. *A*, *B*, *C*, *D*... true at the same time when they are substituted for *i*,*j*... and certain ideas that likewise make propositions *A*, Neg. *B*, *C*, *D*... all true at the same time, and so on, does not imply at all that there are also ideas which make the propositions Neg. *A*, Neg. *B*, *C*, *D*... true at the same time. For this latter result quite different ideas are necessary than for the first, for here propositions *A* and *B* are both supposed to be false, while there it will always be only a single one. Consider the following three propositions: "Caius is a man," "Caius was born either on the ocean or in one the three continents of Europe, Asia or Africa," "Caius was born in either Europe, Africa, America or Australia." They form a system of propositions in which the negation of any one is compatible with the other two, on the presupposition that the single idea, Caius, is considered variable. But the negations of *two* of these propositions are incompatible with the third. Suppose we substitute for the idea of Caius one of the following ideas: Socrates, Tamerlane, Washington, a sea lion. Then sometimes all of the propositions will become true at the same time, sometimes only two of them. Any two of these propositions are consequently compatible with the negation of the third. But let us substitute for Caius an idea that makes the last two propositions false, for example the moon. Then the first will also be false every time, because a being born neither on the ocean nor in one of the continents named certainly can not be a man. The negations of those two at the same time are consequently incompatible with the first proposition.

(16) All propositions in which the *Predicate* is supposed to be considered variable are compatible with each other no matter what subject they have as long as it is a denotative idea. For if each proposition has its own predicate idea distinct from the predicate ideas of the rest of

them, and it is supposed to be considered variable, then it will be easy to give each of them a predicate which makes it true. For each of them we use the idea of a property that is common to the objects its subject refers to. But if some or all of the propositions have one and the same predicate idea, then it is necessary to choose for them the idea of a property common to all of the objects represented by their various subjects. There is always such a property, however, because even the most diverse objects have some common properties.

(17) All propositions with different subjects, supposed to be the variable ideas within them, are compatible with each other no matter what predicates they have, if they are only real *ideas of properties*. For under these conditions ideas can always be found that make all of these propositions true when they are substituted for the given subjects. All we do is use for each proposition the idea of one of the objects which has the property indicated by the predicate.

(18) Propositions that have *the same subject*, which is just what is supposed to be the variable idea in them, are compatible if the concreta corresponding to their predicates are compatible with each other; and they are incompatible if those are. Namely if the concreta of the predicate ideas are compatible, then in every case there are some objects in which all of the properties predicated in these propositions are united. Therefore we shall make all of these propositions true if we make an idea that refers exclusively to such an object the common subject idea. The propositions are compatible, then. In the opposite case, if those concreta are not compatible with each other, then there would be no object which will have all the properties predicated in the given propositions, consequently no idea of an object which could make them all true when substituted for the common subject.

(19) For any proposition you please, if first certain ideas in it are assigned as variables, there can be found an infinite number of propositions incompatible with it; and, if only it is not generically false, an infinite number that are compatible with it. An infinite set of propositions can be formed according to the rule that will be noted in the following: "Proposition A is false," "That proposition A is false is true," and so on. Obviously all of them are incompatible with the given proposition A. For it is plain that there can be no idea which makes proposition A true at the same time as the propositions just constructed. Furthermore, if

A is not generically false itself, there is also an infinite set of propositions compatible with it, for every truth that does not contain the variables $i, j, \ldots$ at all is, because it remains unchanged and surely compatible with A as these vary.

(20) A proposition which is false and which contains none of the ideas $i, j, \ldots$ we regard as variable in a certain set of propositions, stands in the relation of incompatibility with these propositions no matter how they may be constituted. For since it contains none of the ideas $i, j, \ldots$, it remains unchanged no matter what ideas we substitute for $i, j, \ldots$, and so it never becomes true and is accordingly not compatible with those other propositions.

(21) If, then, a proposition F stands in the relationship of compatibility with certain others $A, B, C, D, \ldots$, although it does not contain a single one of the variable ideas $i, j, \ldots$ it must be a true proposition.

(22) Whatever relationship propositions $A, B, C, D, \ldots$ have to each other with respect to ideas $i, j, \ldots$, the propositions $A', B', C', D', \ldots$ derived from them by substituting $i', j', \ldots$ for $i, j, \ldots$ have the same relationship with respect to $i', j' \ldots$. For because propositions $A', B', C', D' \ldots$ are derived from propositions $A, B, C, D \ldots$ merely by exchanging $i, j, \ldots$ for $i', j', \ldots$, and because ideas $i', j', \ldots$ are supposed to be considered variable in them, by a new substitution of $i, j, \ldots$ for $i', j' \ldots$ one can produce propositions $A, B, C, D, \ldots$ once again out of propositions $A', B', C', D' \ldots$. And indirectly by way of them we can produce by continued substitution for i, j whatever propositions would originally have been offered by $A, B, C, D \ldots$. So whatever the relationship of the latter is, the relationship of the former must be just the same.

§ 155

Special Types of Compatibility: (a) The Relation of Derivability

(1) The similarity already mentioned between the relations of compatibility and incompatibility among *propositions* that were just considered and the relations with the same names among *ideas* extends so far that the same subdivisions I adopted for these relationships among ideas can also be made in connection with propositions. Let us consider first of all the relationship of *compatibility*.

(2) When we say that certain propositions $A, B, C, D, \ldots M, N, O, \ldots$

stand in the relation of compatibility, and just with respect to ideas *i, j,...*, then as a consequence of the explication that has been given we are claiming nothing more than that there are certain ideas which on being substituted for *i, j,...* transform all of those propositions into true ones. So far it has been left completely undecided whether, besides these ideas that make propositions *A, B, C, D... M, N, O...* all true, there were others that made only one or another part of them true, but not all, and if so, which of the given propositions can be made true more often than the rest. It could well be understood, however, that these questions are important. Let us then first consider the case that there is a relation among the compatible propositions *A, B, C, D,... M, N, O...* such that all of the ideas that make a certain section of these propositions true, namely *A, B, C, D...*, when substituted for *i, j,...* also have the property of making some other section of them, namely *M, N, O...* true. The special relationship between propositions *A, B, C, D...* on the one side and propositions *M, N, O...* on the other which we conceive of in this way will already be very much worthy of attention because it puts us in the position, insofar as we once know it to be present, to be able to obtain immediately from the known truth of *A, B, C, D...* the truth of *M, N, O...* as well. Consequently I give the relationship which subsists between propositions *A, B, C, D...* on the one hand and *M, N, O...* on the other the title, a relationship of *derivability*. And I say that propositions *M, N, O...* would be *derivable* from propositions *A, B, C, D...*, with respect to the variables *i, j,...*, if every set of ideas which makes *A, B, C, D...* all true when substituted for *i, j,...* also makes *M, N, O...* all true. For the sake of variety, I shall also sometimes say that propositions *M, N, O... follow from* or can be *inferred* or *concluded* from the set of propositions *A, B, C, D....* I shall call propositions *A, B, C, D...* the *antecedents* or *premises*, *M, N, O...* which are obtained from them *consequences* or *conclusions*. Finally, insofar as the relation described here between propositions *A, B, C, D...* and *M, N, O...* is very similar to the relation of more and less comprehensive ideas, I will permit myself to call propositions *A, B, C, D...* the propositions *comprised* and *M, N, O...* the propositions *comprising* them.

(3) The assumption that all of the ideas which make propositions *A, B, C, D,...* true when substituted for *i, j,...* also make propositions *M, N, O...* true does not at all presuppose that this would also have to

be true conversely, i.e. that all of the ideas which make propositions *M*, *N*, *O*... true also make propositions *A*, *B*, *C*, *D*... true. Consequently the relation of derivability does not necessarily have to be reciprocal. Thus every pair of ideas substituted for *A* and *B* that make the proposition, All *A* are *B*, true surely also make the proposition, Some *A* are *B*, true. The latter is therefore derivable from the former. But it is not conversely the case that every pair of ideas substituted for *A* and *B* that make the proposition, Some *A* are *B*, true also make the proposition, All *A* are *B*, true. Thus the latter is not derivable from the former.

(4) If one of the propositions *A*, *B*, *C*, *D*... from which propositions *M*, *N*, *O*... are supposed to be derivable with respect to ideas *i*, *j*,..., e.g. proposition *A*, does not include a single one of the variables, we can dispense with it and say of the remaining propositions *B*, *C*, *D*... that propositions *M*, *N*, *O*... are also derivable from them alone, with respect to ideas *i*, *j*,.... For under these circumstances proposition *A* must be true and remains true all the time, no matter what ideas are substituted for *i*, *j*,.... So whenever propositions *B*, *C*, *D*... are true, *A*, *B*, *C*, *D*... will also be true and consequently *M*, *N*, *O*... as well.

(5) If certain propositions *M*, *N*, *O*,... are supposed to be derivable from certain others *A*, *B*, *C*, *D*... and there is a false proposition among them, then there must be a false proposition among the latter as well. For if *A*, *B*, *C*, *D*... were all true, then *M*, *N*, *O*... would have to be also; otherwise it would not be so that every set of ideas substituted for *i*, *j*,... (namely ideas *i*, *j*,... themselves) which makes *A*, *B*, *C*, *D*... true also makes *M*, *N*, *O*... true.

(6) If *all* the propositions derivable from propositions *A*, *B*, *C*, *D*... with respect to certain ideas *i*, *j*,... are true, then propositions *A*, *B*, *C*, *D*,... must be true themselves. For certainly the propositions, *A* is true, *B* is true, *C* is true, etc., belong among the propositions that can be derived from *A*, *B*, *C*, *D*..., no matter what ideas *i*, *j*,... may be. Then if all the propositions that can be derived from *A*, *B*, *C*, *D*... are true, these propositions must also be true. But if they are true, then propositions *A*, *B*, *C*, *D*... must also be true themselves.

(7) From no proposition *A* is its *negation*, Neg. *A*, i.e. the proposition, *A* is false, derivable, no matter what ideas *i*, *j*... that occur only in *A* may be regarded as variable. For no set of ideas which make proposition *A* true can also make the proposition, *A* is false, true.

(8) All conclusions that flow from certain propositions A, B, C, D... with respect to ideas i, j,... are compatible with all of those propositions compatible with premisses A, B, C, D... with respect to the same ideas, likewise with respect to the same ideas. For if propositions A, B, C, D,... are compatible with propositions A', B', C', D'..., with respect to ideas i, j,..., relative to which propositions M, N, O... are derivable from them, then there are certain ideas which make propositions A, B, C... and propositions A', B', C',... all true together. But whenever propositions A, B, C... become true, so do M, N, O...; thus there are certain ideas which make propositions A', B', C'... and M, N, O... true at the same time.

(9) Propositions which are not compatible are not conclusions from premises which are compatible, always understood with respect to the same variable ideas. For if they were conclusions, they would have to be compatible with each other according to (8).

(10) If the conclusions are not compatible, then the premises can not be compatible either, always understood with respect to the same variable ideas. For if the premises were compatible, then according to (9) their consequences would also have to be compatible.

(11) Conclusions can very well be compatible with each other, however, even if their premises are not compatible, always understood with respect to the same variables. For all that is required for propositions M, N, O,... to be consequences of propositions A, B, C... with respect to ideas i, j,... and for propositions M', N', O',... to be consequences of propositions A', B', C',... with respect to the same ideas is that every set of ideas that makes propositions A, B, C,... true when substituted for i, j,... also makes propositions M, N, O,... true and that every set that makes propositions A', B', C',... true also makes propositions M', N', O',... true. The converse is not required, that whenever propositions M, N, O,... on the one hand and M', N', O',... on the other become true, propositions A, B, C,... on the one side and propositions A', B', C',... on the other must become true. Now if propositions M, N, O,... become true more often than A, B, C,... and propositions M', N', O',... more often than A', B', C',..., then it is possible that there should be some ideas which make propositions M, N, O,... and M', N', O',... true at the same time when they are substituted for i, j,..., while no such ideas are to be met with for A, B, C,... and A', B', C',.... Thus the two

propositions, Caius is stingy and Caius is a spendthrift, are not compatible with each other, if the only idea supposed to be considered variable in them is that of Caius, Nevertheless, with respect to the same idea, the conclusion can be drawn from the first proposition that Caius is not generous and from the second the conclusion can be drawn that sooner or later Caius will no longer be able to be generous. These are a pair of propositions that are certainly compatible with each other.

(12) A proposition which is not *generically true* (§ 147) with respect to ideas *i, j,...* can never be derivable both from a particular proposition *A* and its negation, Neg. *A*. For if a proposition is not generically true, then there are some ideas which make it false when substituted for *i, j,....* But every idea that makes it false must, if it is supposed to be derivable from *A*, according to (5) also make *A* false, and if it is supposed to be derivable from Neg. *A*, must also make Neg. *A* false. Thus the same idea would have to make *A* and *N*eg. *A* false, which is absurd.

(13) But if there are *several* premises from which a certain proposition *M* is supposed to be derivable, e.g. *A, B, C, D...*, it is always possible for the same proposition to be derivable from the negation of some of them as well, perhaps even of all of them. For now from this proposition's falsity, i.e. from Neg. *M*, the negation of each particular one of the propositions *A, B, C, D,...* can not be inferred immediately, but only the fact that they are not all true. The following are an example of a pair of propositions so constituted that the same conclusion follows both from them and from both their negations: From the two propositions, Every *A* is a *B* and It is false that every *A* is a *C*, if the ideas indicated by *A, B, C* are the only ones supposed to be considered variable, the conclusion can be derived with complete certainty that ideas *B* and *C* are not equivalent ideas. Negating these two propositions yields, "It is false that every *A* is a *B*" and "Every *A* is a *C*," from which the same conclusion as before obviously follows.

(14) If a proposition *M* is *compatible* with propositions *A, B, C, D,...* with respect to ideas *i, j,...*, then its negation, Neg. *M*, is surely not derivable from these propositions with respect to the same ideas. For if Neg. *M* were derivable from *A, B, C, D,...* with respect to *i, j,...* then every set of ideas substituted for *i, j,...* that makes *A, B, C, D,...* all true also makes Neg. *M* true, and therewith *M* false. Consequently *M* could not be compatible with *A, B, C, D,...* with respect to these same ideas.

(15) If propositions *A*, *B*, *C*, *D*,... are compatible with each other with respect to ideas *i*, *j*,..., but stand in the relationship of incompatibility with proposition *M*, then the proposition Neg. *M* is derivable from them with respect to the same ideas. For if propositions *A*, *B*, *C*, *D*,... are compatible with each other with respect to ideas *i*, *j*,..., then there must be ideas that make these propositions all true when they are substituted for *i*, *j*,.... But since *M* is supposed to be incompatible with these propositions, the very ideas that make *A*, *B*, *C*, *D*,... all true must make proposition *M* false and hence proposition Neg. *M* true. Consequently Neg.*M* is derivable from *A*, *B*, *C*, *D*,....

(16) If we remove from the premises *A*, *B*, *C*, *D*,... of a conclusion *M* any one of them, e.g. *A*, and replace it with the negation of *M*, Neg. *M*, if it can be combined with the others, then the negation of the missing proposition, i.e. Neg. *A*, can be derived from the set of propositions *B*, *C*, *D*,... and Neg. *M*. For if Neg. *A* were not derivable from the propositions mentioned, Neg. *A* does not have to become true every time they become true. i.e. There must be cases in which propositions *B*, *C*, *D*,... Neg. *M* and proposition *A* are true at the same time, which is absurd. For whenever *B*, *C*, *D*,... and *A* are true at the same time, *M* must also be true, and so Neg. *M* can not be true at the same time.

(17) If the same propositions *M*, *N*, *O*... are derivable with respect to the same ideas *i*, *j*,... both from the set of propositions *A*, *B*, *C*, *D*,... and *X* and the set of propositions *A*, *B*, *C*, *D*,... and Neg. *X*, then they are also derivable from propositions *A*, *B*, *C*, *D*,... alone, with respect to the same ideas. For every set of ideas that makes propositions *A*, *B*, *C*, *D*,... true also makes propositions *M*, *N*, *O*,... true, no matter whether proposition *X* is made true or false.

(18) It would be a different matter if there were in the one set more than one proposition which is the negation of one in the other set. From the fact that proposition *M* can be derived both from the set of propositions *A*, *B*, *C*, *D*,... *X*, *Y* and from the set of propositions *A*, *B*, *C*, *D*,... Neg. *X*, Neg. *Y*, it by no means follows that *M* can be derived from propositions *A*, *B*, *C*, *D*,... alone. For there could be ideas that make propositions *A*, *B*, *C*, *D*,... and only one of the pair *X* and *Y* true. In these cases, *M* would not need to become true for us to be able to say that it is derivable both from the set of propositions *A*, *B*, *C*, *D*... *X*, *Y* and the set of propositions *A*, *B*, *C*, *D*... Neg *X*, Neg. *Y*.

(19) If propositions *M*, *N*, *O*,... are derivable from *A*, *B*, *C*, *D*... with respect to the *larger* number of ideas *i*, *j*, *k*, ..., then they are also derivable from the same propositions with respect to the *smaller* number of ideas *j*, *k*,... (which are a part of the former), if propositions *A*, *B*, *C*, *D*,... stand in the relation of compatibility with respect to this smaller number of ideas. For if the latter is the case, there are some ideas which make propositions *A*, *B*, *C*, *D*,... all true when they are substituted for *j*, *k*,.... But whenever these become true, so do propositions *M*, *N*, *O*,.... Thus propositions *M*, *N*, *O*,... are also derivable from *A*, *B*, *C*, *D*,... with respect to the smaller number of ideas *j*, *k*,....

(20) If in the opposite case propositions *M*, *N*, *O*,... are derivable from propositions *A*, *B*, *C*, *D*,... with respect to the smaller number of ideas *i*, *j*,... they do *not also* have to be derivable with respect to the larger numbers of ideas *i*, *j*, *k*,... (in which the former are repeated), although on this presupposition the propositions surely remain compatible with each other. (§ 154. (11)) For if we assume another variable idea *k* besides *i*, *j*, ..., the set of true propositions that can be formed from those given, *A*, *B*, *C*, *D*,... can increase in number a great deal, and so it is possible for *M*, *N*, *O*,... not to become true every time *A*, *B*, *C*, *D*,... become true. Thus the proposition, Caius is mortal, is derivable from the proposition, Caius is a man, if the idea of Caius is the only one we consider to be variable in them both. But if we chose to consider the idea of man as variable also, the latter proposition would not have the relationship of derivability to the former, as the example demonstrates right away when we put in the place of Caius, God, and in the place of a man, a simple being.

(21) Not every proposition *M*, much less any set of several propositions *M*, *N*, *O*,... can be put into a relation of derivability to any single proposition *A* or even with any set of several propositions *A*, *B*, *C*,... merely by virtue of the fact that we can select as we please which and. how many ideas *i*, *j*,... are supposed to be considered variable in these propositions. For example, suppose the two propositions, *A* has *b* and *C* has *d* have no component in common besides the idea, *has*. Then it is obvious that whichever ideas we declare to be variable in these propositions, a relationship of derivability will never arise between them, since the ideas subsituted in one of them will be completely independent of the ideas substituted in the other.

(22) If propositions *M*, *N*, *O*,... are derivable from propositions *A*, *B*, *C*, *D*,... with respect to ideas *i*, *j*,..., and propositions *P*, *Q*, *R*,... are derivable from propositions *F*, *G*, *H*,... with respect to the same ideas, and *A*, *B*, *C*, *D*,... are compatible with *F*, *G*, *H*,... with respect to the same ideas, then the set of propositions *M*, *N*, *O*,... *P*, *Q*, *R*,... is derivable from the set of propositions *A*, *B*, *C*, *D*,... *F*, *G*, *H*,... with respect to the same ideas. For if propositions *A*, *B*, *C*, *D*,...and *F*, *G*, *H*,... are compatible with respect to ideas *i*, *j*,..., there are ideas which make propositions *A*, *B*, *C*, *D*,... and *F*, *G*, *H*,... all true, when they are substituted for *i*, *j*,.... But precisely these ideas also make propositions *M*, *N*, *O*,... and *P*, *Q*, *R*,... true. So the latter are derivable from the former.

(23) It is also true even in the case where propositions *M*, *N*, *O*,... are derivable from propositions *A*, *B*, *C*, *D*,... with respect to certain ideas *i*, *j*,..., but propositions *P*, *Q*, *R*,... are derivable from propositions *F*, *G*, *H*,... with respect to certain ideas *k*, *l*,..., which may be different from *i*, *j*,... wholly or in part, that then the set of propositions *M*, *N*, *O*,... *P*, *Q*, *R*,... is derivable from the set of propositions *A*, *B*, *C*, *D*,... *F*, *G*, *H*,... with respect to the set of ideas *i*, *j*, *k*, *l*,..., provided only that none of the ideas *k*, *l*,... which are different from *i*, *j*,... appears in propositions *A*, *B*, *C*, *D*,... *M*, *N*, *O*,... and none of the ideas *i*, *j*,... which are different from *k*, *l*,... appears in propositions *F*, *G*, *H*,... *P*, *Q*, *R*,... and provided moreover that the set of propositions *A*, *B*, *C*, *D*,... *F*, *G*, *H*,... represents a compatible set with respect to ideas *i*, *j*, *k*, *l*,.... For if the latter is the case, then there are certain ideas which make *A*, *B*, *C*, *D*,... *F*, *G*, *H*,... all true when they are substituted for *i*, *j*, *k*, *l*,.... But because none of the ideas *k*, *l*,... different from *i*, *j*,... occurs in propositions *A*, *B*, *C*, *D*,... and none of the ideas *i*, *j*,... different from *k*, *l*,... appears in propositions *F*, *G*, *H*,..., no other true propositions will be formed from propositions *A*, *B*, *C*, *D*,... by assuming the variability of *i*, *j*, *k*, *l*,... all together besides those which arise merely by assuming the variability of *i*, *j*,..., i.e. none besides the ones with which *M*, *N*, *O*,... also become true. Likewise, from propositions *F*, *G*, *H*,... no other true propositions besides the ones which are also produced by assuming *k*, *l*,... to be variable, i.e. none besides the ones with which propositions, *P*, *Q*, *R*,... also become true. Thus, whenever *A*, *B*, *C*, *D*,... *F*, *G*, *H*,... all become true, *M*, *N*, *O*,... *P*, *Q*, *R*,... all become true as well. It follows that they are derivable from the former with respect to ideas *i*, *j*, *k*, *l*,....

(24) If propositions *M*, *N*, *O*,... are derivable from propositions *A*, *B*, *C*, *D*,... with respect to certain ideas *i*, *j*,...; and propositions *X*, *Y*, *Z*,... are derivable from propositions *M*, *N*, *O*,... and *R*, *S*, *T*,... with respect to the same ideas, then propositions *X*, *Y*, *Z*,... are also derivable from propositions *A*, *B*, *C*, *D*,... *R*, *S*, *T*,..., with respect to the same ideas. For if propositions *M*, *N*, *O*,... are derivable from propositions *A*, *B*, *C*, *D*,... with respect to ideas *i*, *j*,... then every set of ideas substituted for *i*, *j*,... which makes *A*, *B*, *C*, *D*,... all true also makes *M*, *N*, *O*,... all true. Thus every set which makes *A*, *B*, *C*, *D*,... *R*, *S*, *T*,... all true also makes *M*, *N*, *O*,... *R*, *S*, *T*,... all true. Consequently (because of the derivability of *X*, *Y*, *Z*,... from *M*, *N*, *O*,... *R*, *S*, *T*,...) it also makes *X*, *Y*, *Z*,... true.

(25) Even if propositions *M*, *N*, *O*,... are derivable from *A*, *B*, *C*, *D*,... with respect to ideas *i*, *j*,..., but propositions *X*, *Y*, *Z*,... are derivable from *M*, *N*, *O*,... *R*, *S*, *T*,... with respect to ideas *k*, *l*,... which may be different from *i*, *j*,... wholly or partially; propositions *X*, *Y*, *Z*,... are derivable from propositions *A*, *B*, *C*, *D*,... *R*, *S*, *T*,... with respect to the entire set of ideas *i*, *j*,... *k*, *l*,..., provided only that that none of the ideas *k*, *l*,... different from *i*, *j*,... occurs in propositions *A*, *B*, *C*, *D*,... and none of the ideas *i*, *j*,... different from *k*, *l*,... occurs in *M*, *N*, *O*,... *R*, *S*, *T*,.... For if none of the ideas *k*, *l*,... different from *i*, *j*,... occurs in *A*, *B*, *C*, *D*,..., then the assumption that *i*, *j*,... *k*, *l*,... are all supposed to be variable does not generate any different truths from *A*, *B*, *C*, *D*,... than the assumption that only *i*, *j*,... are supposed to be variable. I.e. it generates no truths besides those with which propositions *M*, *N*, *O*,... also become true. And if none of the ideas *i*, *j*,... different from *k*, *l*,... occur in *M*, *N*, *O*,... *R*, *S*, *T*,... then the assumption that *i*, *j*,... *k*, *l*,... are all supposed to be variable does not generate any other truths from *M*, *N* *O*,... *R*, *S*, *T*,... besides those produced on the assumption that *k*, *l*,... are the only variables, i.e. none besides those with which propositions *X*, *Y*, *Z*,... also become true. Thus *X*, *Y*, *Z*,... are derivable from *A*, *B*, *C*, *D*,... *R*, *S*, *T*,... with respect to *i*, *j*... *k*, *l*,....

(26) If premises *A*, *B*, *C*, *D*,... from which a certain proposition *M* is derivable with respect to ideas *i*, *j*,... are of such a nature that it is not possible to leave out any one of the propositions *A*, *B*, *C*, *D*,..., or even any component of one of them and have *M* still remain derivable from the remainder with respect to ideas *i*, *j*,..., I call the relationship of

derivability of M from A, B, C, D,... *exact*, *exactly proportionate* or *adequate*. In the opposite case I call it *redundant*. Thus the relation of derivability between the two premises, All α are β and All β are γ, and the conclusion, All α are γ, when ideas α, β, γ, ρ are supposed to be considered variable, is exact. We are not able to take away a single component of those two propositions, much less an entire proposition, if the proposition, All α are γ, is still supposed to be derivable from the remainder insofar as ideas α, β, γ continue to be considered the variables. On the other hand, the relation of derivability between the same premises and the following conclusion, Some β are α, I call redundant, because it holds even if we retain only the first of the two premises. The inference is equally redundant if the conclusion, All α are δ, is derived from the two premises, All α are β, All β and γ are δ, for it will be there even if the simpler premise, All β are δ, is chosen instead of All β and γ are δ.

(27) Neither the conclusion nor one of the premises of an exact relation of derivability can be a proposition which is generically true. Not the conclusion, for a proposition which is generically true does not require as a condition of its truth the truth of its premises at all. Nor any of the premises, for we could leave out a premise which is generically true without the derivability of the conclusion from the propositions that are left coming to an end. (4)

(28) If the relation of derivability between premises A, B, C, D,... and conclusion M is supposed to be exact, then the negation of this conclusion Neg. M, must be compatible with any part of the premises, with respect to the same ideas i, j,... which are considered variable in that relation. For if Neg. M were incompatible with any part of propositions A, B, C, D,..., for example, with B, C,..., then according to (15) the proposition Neg. Neg. M, and thus surely M, itself, would already be derivable from propositions B, C,... alone. Consequently the relation of derivability between A, B, C, D,... and M would not be exact. (26)

(29) In the case of an exact relation of derivability, no premise may be derivable from the rest of them with respect to the same ideas which are supposed to be considered variable in the relation itself. For if premise A were derivable from the rest, B, C, D,..., then the entire set of propositions A, B, C, D,... would be derivable from propositions B, C, D,... with respect to those ideas. As a consequence, proposition M, which is derivable from propositions A, B, C, D,..., would also be

derivable from the smaller number *B, C, D,...* (24). So the relation of derivability of *M* from *A, B, C, D,...* is certainly not exact.

(30) In the case of an exact relation of derivability the negation of every premise taken individually must be compatible not only with all the rest of them but also with the negation of the conclusion subjoined, of course with respect to the same ideas considered variable in the relationship. For if the proposition Neg. *A* were not compatible with propositions *B, C, D,..., A* would be derivable from them, and consequently according to (29) propositions *A, B, C, D,...* would not be suitable as premises for any exact relation of derivability. Going beyond that, if the propositions Neg. *A, B, C, D,...* were not also compatible with the negation of the conclusion, i.e. Neg. *M*, then according to (15) Neg. Neg. *M*, i.e. *M* itself, would be derivable from Neg *A, B, C, D,...*. But since *M* is also supposed to be derivable from *A, B, C, D,...*, the truth or falsity of *A* would be a matter of complete indifference for *M*. Consequently it is certain that *M* would also be derivable from *B, C, D,...* alone. It follows that the relation of derivability between *A, B, C, D,...* and *M* is not exact.

(31) Even with an exact relation of derivability the negations of *two* or more premises can stand in the relation of incompatibility with the rest with respect to the same ideas relative to which the relation is supposed to obtain. For the sake of abbreviation let us designate the three propositions, All α are β, All β are γ, and All α are γ, by *A, B,* and *C*. The relationship of derivability in which the three propositions, Neg. *A*, Neg. *B*, Neg. *C*, as premises stand to the conclusion, "The set of three propositions, Neg *A*, Neg. *B*, Neg. *C* is a set of nothing but true propositions," if ideas α, β, γ are supposed to be regarded as variables, is surely exact. For we can not let any of those three premises, not even a single component of them, fall if that conclusion is supposed to remain. However, the negation of the first two premises, i.e. asserting *A* and *B* is incompatible with the third, i.e. with Neg. *C*.

(32) Even if the relation of derivability between premises *A, B, C, D,...* and conclusion *M* and also the relation between premises *M, R, S, T,...* and conclusion *X* are both exact, with respect to the same ideas *i, j,...* it still does not follow that the relation of derivability which according to (24) also holds between premises *A, B, C, D,... R, S, T* and conclusion *X* must be an exact one. Thus the relation of derivability be-

tween the premises, All α are β, All β are γ, and the conclusion, all α are γ, is exact. Again, the relation of derivability between the premises, All α are γ, All γ are β, and the conclusion, All α are β, is equally so. But the relation of derivability between the three premises, All α are β, All β are γ, All γ are β, and the conclusion, All α are β, is not exact.

(33) There is no doubt, however, that *sometimes* when the conclusion M is exactly derivable from premises A, B, C, D,... and the conclusion X from premises M, R, S, T,..., the relationship of derivability of conclusion X from premises A, B, C, D,... R, S, T,... can also be exact. In the preceding example, if we replace the last premise, All γ are β, with the premise, All γ are δ, and the conclusion, All α are β, with the conclusion, All α are δ, all three of the inferences considered there will be exact. Now whenever along with the relations of derivability in which the premises A, B, C, D,... stand to the conclusion M and the premises M, R, S, T,... stand to the conclusion X, the third relation of derivability arising between premises A, B, C, D,... R, S, T,... is also exact, then I say the last relationship is *compounded* from the first two. A relation of derivability which is not compounded in this way, I call a *simple* one.

(34) There must be simple relations of derivability. Thus, for example, the relation of the two premises, All α are β, All β are γ, to the conclusion, All α are γ, could be a simple one. For someone will hardly be in a position to cite a proposition which can be derived from one or both of these premises and would at the same time be so constituted that we could exactly derive the above conclusion once again either from it alone or in conjunction with a second one.

(35) If a pair of propositions, A has x, B has x, have the same predicate, which is the only idea in them to be considered variable, then the second is derivable from the first if the subject idea of the first, A, stands in the relation of inclusion (§ 95) to the subject idea of the second, B. If this is not the case, the relation of derivability does not apply. For if idea A includes idea B, if every B is an A, then every idea that makes the proposition A has x true when it is substituted for x also makes the proposition B has x true. In the opposite case, if A does not include B, if there is some B which is not an A, then there will be some property which belongs exclusively to this B. Let us call it b'. Then the property, not b', is a property that belongs to all A, but not to all B. The idea, "property not b'," then, will make the proposition, A has x, true,

but make the proposition, *B* has *x*, false when it is substituted for *x*.

(36) When a pair of propositions, *X* has *a*, *X* has *b*, have the same subject, which is the only idea supposed to be considered variable in them, the second is derivable from the first if idea *B* (the concretum corresponding to *b*) includes idea *A*; and if this is not the case, there is no relation of derivability. For if idea *B* includes idea *A*, if every *A* is at the same time *B*, then every idea which makes the proposition, *X* has *a*, true when it is substituted for *X* will also make the proposition, *X* has *b*, true. But if, on the contrary, *B* does not include *A*, if there is some *A* which is not at the same time *B*, then there will be some idea referring exclusively to this *A*. If it is *A′*, then the idea *A′* will make the proposition, *X* has *a*, true when it is substituted for *X*, but it will not make the proposition, *X* has *b*, true.

§ 156

(b) The Relation of Equivalence

(1) If the relation of derivability between propositions *A*, *B*, *C*, *D*,... and *M*, *N*, *O*,... holds both ways, with respect to the same ideas *i*, *j*,..., i.e. if every set of ideas substituted for *i*, *j*,... that makes *A*, *B*, *C*, *D*,... all true also makes *M*, *N*, *O*,... all true, and if conversely every set of ideas substituted for *i*, *j*,... which makes *M*, *N*, *O*,... true also makes *A*, *B*, *C*, *D*,... true as well, then I say that propositions *A*, *B*, *C*, *D*,... and *M*, *N*, *O*... have a relation of equivalence to each other and for that reason I call the propositions themselves equivalent, with respect to the same ideas *i*, *j*,.... So I say that the proposition, "Every *A* has *b*," is equivalent with the two propositions, "Idea *A* has denotation" and "The idea of an *A* which does not have *b* has no denotation," if ideas *A* and *b* are regarded as variable. For on this presupposition the last two are derivable from the first and it is derivable from the two of them.

(2) On this definition, for us to be able to look on propositions *A*, *B*, *C*, *D*,... taken *together* as equivalent with propositions *M*, *N*, *O*,... *together* by no means requires that for every single one of the propositions *A*, *B*, *C*, *D*,... there should be some single proposition among *M*, *N*, *O*,... equivalent to it. It is not even true that the number of propositions *A*, *B*, *C*, *D*,... must be the same as the number of propositions *M*, *N*, *O*,..., as the example given has already shown.

(3) If propositions *A*, *B*, *C*, *D*,... are all true, then propositions

M, N, O,... equivalent to them are also all true. In the opposite case, if the former are not all true, the latter are not all true.

(4) If propositions A, B, C,... are together equivalent with propositions A', B', C',... taken together, with respect to ideas i, j,..., and propositions A, B, C,... stand in some one of the relationships previously considered with certain others M, N, O,..., always with respect to the same ideas; then the propositions equivalent to them, A', B', C',... stand in exactly the same relationship with M, N, O,... with respect to the same ideas. This is so whether that relationship is a relation of compatibility or incompatibility; of derivability, either such that M, N, O,... are derivable from A, B, C,... or A, B, C,... from M, N, O,...; or of equivalence also. For since propositions A, B, C,... are equivalent with propositions A', B', C',... with respect to the same ideas that are also supposed to be considered variable in the relationship between A, B, C,... and M, N, O,..., every set of ideas substituted for the variables in A, B, C,... and M, N, O,... which makes A, B, C,... all true also makes A', B', C',... all true, and every set which does not produce this result with A, B, C,... will not do so with A', B', C',... either. Now since this is all that the name of the relation which holds between A, B, C,... on the one hand and M, N, O,... on the other depends on, it is obvious that precisely this relation also obtains between A', B', C',... and M, N, O,....

(5) If propositions A, B, C,... are equivalent to propositions A', B', C',... and propositions D, E, F,... equivalent to propositions D', E', F',..., with respect to the same ideas i, j,..., and if besides that propositions A, B, C,... are compatible with propositions D, E, F,... with respect to the same ideas; then propositions A, B, C,... D, E, F,... together are equivalent to propositions A', B', C',... D', E', F',... together, with respect to the same ideas. For if propositions A, B, C,... are compatible with propositions D, E, F,... with respect to ideas i, j,..., there are ideas which make propositions A, B, C,... D, E, F,... all true when they are substituted for i, j,.... But these very ideas also make A', B', C',... D', E', F',... all true, because all ideas that make A, B, C,... true also make A', B', C',... true and all that make D, E, F,... true also make D', E', F',... true. It is similarly demonstrable that all ideas that make A', B', C',... D', E', F',... all true when substituted for i, j,... also make A, B, C,... D, E, F,... all true. Consequently the two sets have the relationship of equivalence.

(6) If propositions *A*, *B*, *C*,... are equivalent to propositions *A'*, *B'*, *C'*,... with respect to ideas *i*, *j*,..., and propositions *M*, *N*, *O*,... are derivable from propositions *A*, *B*, *C*, with respect to the same ideas, then propositions *A*, *B*, *C*,... *M*, *N*, *O*,... together also stand in the relationship of equivalence to propositions *A'*, *B'*, *C'*.... For propositions *A*, *B*, *C*,... *M*, *N*, *O*,... are all made true by exactly the same ideas, and by no more, as those which make propositions *A'*, *B'*, *C'*,... true.

(7) Merely from the fact that propositions *A*, *B*, *C*,... are equivalent to propositions *A'*, *B'*, *C'*,... with respect to ideas *i*, *j*,... it by no means follows that propositions *M*, *N*, *O*,... and *M'*, *N'*, *O'*,..., which can be derived from those propositions, even if with respect to the same ideas, would also be equivalent to each other with respect to those ideas. For if *M*, *N*, *O*,... are merely derivable from *A*, *B*, *C*,... and not equivalent to them, *M*, *N*, *O*,... could be true more often than *A*, *B*, *C*,... *M'*, *N'*, *O'*,..., however, could be equivalent to *A'*, *B'*, *C'*,... or still more frequently, they could become true in connection with different ideas from *M*, *N*, *O*,.... In that case *M*, *N*, *O*,... and *M'*, *N'*, *O'*,... would not be equivalent by any means. Thus, the two propositions, "This figure is an equilateral triangle" and "This figure is an equiangular triangle," are equivalent to each other, with respect to the variable idea, this. The two propositions that can be derived from them in terms of the order of classification, however, "This figure is equilateral" and "This figure is equiangular," are not at all equivalent to each other with respect to the idea, this.

(8) Thus we may not immediately infer from the fact that the sum of propositions *A*, *B*, *C*, *D*,... is equivalent to the sum of propositions *A'*, *B'*, *C'*, *D'*,... with respect to ideas *i*, *j*,..., and further that a part of the former propositions, e.g., *A*, *B*,... is equivalent to a part of the latter, e.g. *A'*, *B'*,... with respect to the same ideas, that the remaining parts, *C*, *D*,... and *C'*, *D'*,... are equivalent to each other with respect to the same ideas. For propositions *C*, *D*,... could merely be derivable from *A*, *B*,... in some way and propositions *C'*, *D'*,... from *A'*, *B'*,....

(9) If two propositions *A* and *A'* are equivalent to each other with respect to ideas *i*, *j*,..., and neither of them is generically true, then propositions Neg. *A* and Neg.*A'* are also equivalent to each other with respect to the same ideas. For since the one of the propositions, e.g. *A*, is not generically true, there are certain ideas that make it false when

they are substituted for $i, j, \ldots$ and so make the proposition Neg. A true. But these very ideas must also make proposition A' false and consequently Neg. A' true. Otherwise A would not be derivable from A'. It follows in the same way, however, that every set of ideas which makes the proposition Neg. A' true also makes the proposition Neg. A true.

(10) It is not the same when the relation of equivalence holds between whole sets of propositions. It is not always the case that if the set of propositions A, B, $C, \ldots$ is equivalent to the set of propositions A', B', $C', \ldots$ then the set of propositions Neg. A, Neg. B, Neg. C,... has to be equivalent to the set of propositions Neg. A', Neg. B', Neg. $C', \ldots$ with respect to the same ideas. So the two propositions, "All A are B" and "All B are A," are obviously equivalent to the following single proposition: "Every object of one of the ideas A, B is an object of both," if we regard ideas A, B as the variables. But the two propositions that arise from negating the first two, namely "It is false that all A are B" and "It is false that all B are A," are by no means equivalent to the proposition, "It is false that every object of one of the ideas A, B is an object of both." For the latter will be true without the first two being true together, for example when idea A is higher than B, in which case the first proposition will be true but the second false.

(11) If a pair of propositions, A has x, B has x, have the same predicate, which is supposed to be regarded as the only variable idea in them, they are equivalent to each other if their subject ideas A and B are equivalent (§ 96). In the opposite case they are not. For if ideas A and B are equivalent to each other, every object that falls under one of them falls under both, and consequently the same idea of a property that makes one proposition true when it is substituted for x must also make the other true. But if A and B are not equivalent to each other, then one of them, e.g. A, must represent an object the other does not represent. There must, then, also be a property that belongs exclusively to this object. If this is a', the property not a' will belong to all B. And so the idea, "property not a'," will make the proposition, B has x, true when it is substituted for x but it will not make the proposition, A has x, true.

(12) If a pair of propositions, X has a, X has b, have the same subject, which is supposed to be regarded as the only variable idea in them, they are equivalent to each other if ideas A and B are equivalent* and in

* It is not required that the abstracta, a and b, be equivalent themselves.

the opposite case they are not. For if ideas *A* and *B* are equivalent, if every object of the one also falls under the other, then every idea that makes one proposition true when substituted for *X* must also make the other true. But if *A* and *B* are not equivalent, there is some object that falls under one of them, e.g. *A*, that does not fall under the other, and so there is an idea which refers to it alone. If this is *A'*, then when the idea *A'* is substituted for *X*, it will make the proposition, *X* has *a*, true, but not the proposition, *X* has *b*.

(13) Merely from the fact that a pair of propositions are composed in the same way of parts (propositions or single ideas) that are equivalent to each other, it in no way follows that they are equivalent to each other. Thus the two propositions, an equilateral triangle is also equiangular and an equiangular triangle is also equiangular, are not at all equivalent. *Note*: Lambert (Dian. § 146) calls propositions I call *identical* (§ 148) equivalent. For him, then, the word, equivalent, was not the designation of a relationship which holds between several propositions, but the name of a property which can belong to a single proposition taken by itself. Still, even those who only designate a relationship by this word do not always take it in the same sense. Since for the most part a *propositon* has heretofore been understood to be only the linguistic representation of what I call a proposition, it is no wonder that many have defined the concept of equivalent propositions in such a way that merely different *expressions* of one and the same proposition have been given the name of several equivalent propositions. Reimarus (Vernunftl. § 160) does so, for example, and even Metz, although he does distinguish (L. § 112) propositions from their linguistic expression, says nevertheless (§ 119): "Judgments are the same or equivalent which have the same matter and form and therefore differ from each other only in expression." Krug (§ 62), Schulze (§ 53) and others have also declared equivalent propositions to be the same. Several, who come closer to the concept I have brought forward, define equivalent propositions as propositions which can be substituted one for the other. So it is with Crusius (W. z. G. § 248) and Knutzen (L. § 175), who adds the further provision, *salvo sensu*. In my opinion a pair of propositions which although equivalent nevertheless must be different (as a pair) (§ 150) not only can never be exchanged for each other without change of sense (*salvo sensu*), since it is precisely and only that which presents a different *sense* that can be called a dif-

ferent *proposition*, but such an exchange is also impermissible in many other respects. For example, where the proposition, "This is an equilateral triangle," is in the right place, its equivalent, "This is an equiangular triangle," will certainly not always fit. And in a proposition of the following kind: "The content of proposition *A* is such and such," can we really always substitute a proposition equivalent to *A*? – Other logicians, who had a notion that the *expression* of a proposition should not be confused with the proposition itself (i.e. with the sense of the expression), distinguished for this very reason between two kinds of equivalence, *grammatical*, which depends on the expression, and *logical*, which belongs to propositions such that *quae simul sunt verae, vel simul falsae per cogitationum relationem.* Reusch (L. § 493), for example. Maass' definition (§ 227) is similar: two judgments are equivalent if each of them follows from the other, i.e. (according to § 18) if the one must be true as soon as the other is. It is apparent that these definitions are based upon exactly the same concept I adopted too, it is only that no account could be given of its parts. Baumgarten's definition (Acr. § 264) is peculiar to him: *Propositiones ejusdem qualitatis, quibus complete notabiliter idem respondet conceptus, sunt aequipollentes.* § 263 also belongs here: *Conceptus combinando extrema propositionis ortus illi respondere dicitur.* In the first place, the propositions, "*A* is *B*" and "*B* is *A*," would be equivalent to each other, for the concepts, "*A* which is *B*" and "*B* which is *A*," which arise *combinando extrema propositionum*, are *conceptus complete notabiliter iidem* in a sense in which only two concepts generated in such a way can be called so. Mehmel (Denkl., p. 43) also sets up a distinctive concept of (logical) equivalence: "Judgments with logically identical moments of quality, quantity, relation and modality are called equivalent; and it does not at all follow from their equivalence that they can be substituted for each other. Equivalent propositions in which subject and predicate are materially identical besides are called identical." On this definition, the propositions, All men are mortal and All physical objects are heavy, would have to be called equivalent to each other, because they have the same quality, quantity, relation and modality. Would that really be to the purpose? – In Gerlach's *Logik*, § 88, it is said that judgments are equivalent "if they are logically the same even with a difference in external form," and in Hillebrand (L. § 308) "if they agree in content while having a difference in external form," and in Calker (L. § 112) "if they have like content of ideas, but having different form are nevertheless identical in value and meaning," in Rösling (L. § 105) "if so far as the meaning, the sense is concerned they constitute one and the same judgment", and so on. Logical identity, it is apparent, is just another word for logical equivalence, but not a definition of this concept. But it is wholly mistaken that the contents of a pair of equivalent propositions would be identical with each other, if what is understood by content is the sum of the ideas they are composed of, as a number of the examples already given demonstrate. – The expression, value and meaning, is still more vague. For if identical *value* of a proposition is supposed to be a definition of equivalence, we have just substituted one word for another. But if attention is supposed to be directed primarily at the added word, *meaning*, and it is taken in what is otherwise its usual sense, I call attention to the fact that only the linguistic expression of a proposition has meaning, not the proposition *an sich*, since it is the same with itself. Propositions with the same meaning are therefore not several propositions, but just one proposition. – In Kiesewetter's W. A. d. L. (§ 234), *material* equivalence is applied to "locutions with the same meaning," and the example is cited, "Caius is the father of Titus and Titus is the son of Caius." These two expressions appear to me, however, to be in no way just different expressions of one and the same proposition, but expressions of two really distinct propositions, because they have different subjects as well as different predicates. – But it could be still more important than these mistakes in definition that (at least so far as I would know) in no treatise on logic has anyone thought to extend the relation of equivalence to entire *sets of propositions*, although it must be admitted that in a number of intellectual disciplines, mathematics in particular, attention to this relationship is of great importance. There are often distinctive theorems in analysis which assert only that the set of certain equations (equations, however, are only a kind of proposition) is equivalent to the set of certain other equations or it may well be even with a single one.

§ 157

(c) The Relationship of Subordination

(1) Suppose that the relation of derivability between propositions A, B,

C, *D*,... and *M*, *N*, *O*,... is not, as in the preceding section, reciprocal, but only holds on one side, for example if propositions *M*, *N*, *O*,... can be derived from propositions *A*, *B*, *C*, *D*,..., but not the latter from the former, with respect to certain variable ideas *i*, *j*,.... Or (what comes to the same thing) suppose that every set of ideas that makes propositions *A*, *B*, *C*, *D*,... true when substituted for *i*, *j*,... also makes propositions *M*, *N*, *O*,... true, but the converse does not hold, that every set which makes the latter true also makes the former true. Then this relationship between propositions *A*, *B*, *C*, *D*,... on the one hand and propositions *M*, *N*, *O*,... on the other is called a relationship of *subordination*. I permit myself (on account of the similarity of this relation to the one between ideas, § 97) to call propositions *A*, *B*, *C*, *D*,... the *subordinated* or *lower* propositions, or (if that should sound all too objectionable) the propositions of *more restricted* or *less validity* or those which *say* more. Propositions *M*, *N*, *O*, on the other hand, are to be called the *superordinate* or *higher* propositions, or those of *more extended* or *greater validity*, or those which *say less*. It would be still less objectionable to say that propositions *M*, *N*, *O*,... are *unilaterally derivable* from propositions *A*, *B*, *C*, *D*,.... The latter to be called the *unilateral* premises, the former the unilateral conclusions. Thus, for example, the proposition, *A* is *C*, is derivable from the two propositions, *A* is *B*, *B* is *C*, if ideas *A*, *B*, *C* are regarded as variable. But on this same presupposition the two latter propositions, *A* is *B* and *B* is *C*, are not derivable conversely from the former proposition, *A* is *C*. I say, then, that between the propositions, *A* is *B*, *B* is *C*, on the one side and *A* is *C* on the other side, there is a relationship of subordination. I call the first two propositions subordinate, the last superordinate to them and I ascribe to the first two a more limited, to the last a more extended validity.

(2) If propositions *M*, *N*, *O*,... are unilaterally derivable from propositions *A*, *B*, *C*,..., with respect to ideas *i*, *j*,..., then there are always some propositions that are compatible with *M*, *N*, *O*,... without being compatible with *A*, *B*, *C*,... with respect to the same ideas *i*, *j*,.... For since there are ideas which make *M*, *N*, *O*,... all true when substituted for *i*, *j*,... without also making *A*, *B*, *C*,... all true, then there are ideas which make *M*, *N*, *O*,... all true and one of the propositions Neg. *A*, Neg. *B*, Neg. *C*,... true along with them. Let Neg. *A* be such a proposition.

Then *M*, *N*, *O*,... and Neg. *A* are compatible. But that Neg. *A* and *A*, *B*, *C*,... are not compatible is self-evident.

(3) If propositions *M*, *N*, *O*,... are only unilaterally derivable from propositions *A*, *B*, *C*, *D*,... and propositions *R*, *S*, *T*,... again only unilaterally derivable from propositions *M*, *N*, *O*,..., always with respect to the same ideas *i*, *j*,..., then propositions *R*, *S*, *T*,... are also only unilaterally derivable from propositions *A*, *B*, *C*, *D*,..., with respect to the same ideas. That *R*, *S*, *T*,... are derivable from *A*, *B*, *C*, *D*,... is manifest from § 155 (24), but that this derivability is only one-sided is plain from the fact that there are some ideas which make *R*, *S*, *T*,... all true without making *M*, *N*, *O*,... all true. But in terms of these ideas *A*, *B*, *C*, can not all become true, for if they became true, then *M*, *N*, *O*,... would also have to become true.

(4) If a pair of propositions, *A* has *x*, *B* has *x*, have the same predicate, and it supplies the variable idea in them, then the second is unilaterally derivable from the first if the subject idea of the second, *B*, is subordinate to the subject idea of the first, *A*. And if the latter is not the case, neither is the former. For if idea *B* is subordinate to *A*, then according to § 155 (35), the proposition, *B* has *x*, is derivable from the proposition, *A* has *x*, but not *vice versa*. And if the proposition, *B* has *x*, is supposed to be derivable from the proposition, *A* has *x*, but not *vice versa*, then according to the very same section idea *A* must encompass idea *B*, but not conversely idea *B* also encompass idea *A*. And so *B* must be subordinate to *A*.

(5) If a pair of propositions, *X* has *a*, *X* has *b*, have the same subject and it forms the variable part of them, then the second is unilaterally derivable from the first if idea *A* is subordinate to idea *B*. If this is not the case, neither is that. Proved by § 155 (36).

§ 158

(d) The Relationship of Concatenation

(1) There still remains to be considered the case in which there is a relation of compatibility between propositions *A*, *B*, *C*,... and *M*, *N*, *O*,..., but only such a relationship that neither propositions *M*, *N*, *O*,... are derivable from propositions *A*, *B*, *C*,... nor the latter from the former, with respect to the same ideas *i*, *j*,.... In other words, when there are ideas that make *A*, *B*, *C*,... and *M*, *N*, *O*,... all true when they are

substituted for *i, j,*..., but there are also others which make only *A, B, C,*... all true without making *M, N, O,*... all true, and others as well that only make *M, N, O,*... all true without making *A, B, C,*... true. Since this relationship has such a great similarity to the relation I called *concatentation* or *intersection* in connection with ideas (§ 98), I might also call this relationship between propositions by the same name, if one does not prefer to call it a relationship of *independence*.

(2) Every pair of propositions of the form, Every *X* has *y*, Every *Y* has *x*, in which ideas *x* and *y* are supposed to be considered variable, constitutes a pair of intersecting propositions. For surely there are ideas which make these two propositions true at the same time when they are substituted for *x* and *y*. Any two ideas by which *X* and *Y* become reciprocal ideas offer an example of this. But then there are also certain ideas which make only one proposition true by itself when substituted for *x* and *y*. For if we chose for *x* and *y* a pair of properties such that *Y* becomes higher than *X*, then the proposition, Every *X* is *Y*, is true and the proposition, Every *Y* is *X*, false. But if we chose for *x* and *y* a pair of properties which make *X* higher than *Y*, then the proposition, Every *Y* is *X*, is true and the proposition, Every *X* is *Y*, false.

(3) If a pair of propositions, *A* has *x*, *B* has *x*, have the same predicate, and it is supposed to be regarded as the only variable idea in them, they stand in the relationship of intersection if ideas *A* and *B* either intersect or are incompatible with each other. If this is not the case, that is not. For these two propositions both to be capable of being true at the same time all that is required (according to § 154 (16) is that *A* and *B* be denotative ideas. But in order for each of them to be capable of being true by itself, it is necessary that ideas *A* and *B* do not stand in the relationship of subordination, since otherwise (according to § 155 (35) either one proposition or the other would be derivable from the other. And so ideas *A* and *B* must be either intersecting or incompatible.

(4) If a pair of propositions, *X* has *a*, *X* has *b*, have the same subject, which is supposed to be their only variable part, they stand in the relationship of intersection with each other if ideas *A* and *B* stand in the relationship of intersection. If this not the case, neither is that. For if ideas *A* and *B* intersect, then there are objects which fall under both of them; an idea that refers exclusively to such an object consequently makes both propositions true when it is substituted for *X*. But then there

are also objects that fall under only the one idea and not the other. An idea that refers exclusively to such an object will consequently make only one proposition true and not the other. But if ideas *A* and *B* do not intersect, either both of them or at least one is objectless and both propositions or at least one of them is generically false, or ideas *A* and *B* are incompatible and then (according to § 154 (18) so are the propositions, or one is subordinate to the other and then (according to § 155 (36), one of those two propositions is derivable from the other.

(5) There are pairs of intersecting propositions which can not be false at the same time and there are other pairs which can be, always understood with respect to the same variable ideas, with reference to which the relationship holds. The two propositions following are an example of a pair of intersecting propositions which can never both be false: Every *X* has *a* and It is false that every *X* has the properties $a+b$, wherein the only variable is idea *X*, with *a* and *b* a pair of properties that are sometimes connected but not always. That these two propositions are intersecting becomes clear from the following considerations. If we substitute for *X* an idea that exclusively refers to things which have property *a* but not *b*, both become true. Further, the first proposition becomes true by itself if we substitute for *X* an idea which refers only to to objects which have property *b* along with *a*. Finally, the second proposition will be true by itself if we choose for *X* an idea of objects which have neither property *a* nor *b*. We can see, however, that both propositions can never become false at the same time, because the truth of the second necessarily follows from the falsity of the first. If it is false that every *X* is an *A*, it is all the more certainly false that every *X* is an [*A*]*b*.– Finally, the two propositions already considered in (2) are an example of a pair of intersecting propositions which can both become false at the same time: Every *X* is *Y* and Every *Y* is *X*. These two both become false as soon as we substitute for *X* and *Y* a pair of mutually exclusive ideas.

(6) If a pair of propositions *A* and *B* intersect, then their negations, or the propositions Neg. *A* and Neg. *B*, are either independent or incompatible, all with respect to the same ideas. Namely, if *A* and *B* intersect and are such that they can never both become false (5), then propositions Neg. *A* and Neg. *B* are incompatible. But if they can both become false, Neg. *A* and Neg. *B* have the relationship of intersection. For there are ideas by virtue of which they both become true and there are also

ideas which make only the one proposition true by itself. The latter case is provided for by ideas which make only one of the propositions, *A* and *B*, true, the other false.

(7) Propositions equivalent to intersecting propositions are themselves intersecting, all with respect to the same ideas. The proof is as in § 156 (4).

§ 159

Special Types of Incompatibility

(1) Just as the relationship of *compatibility* considered so far offers many noteworthy subtypes, (§§ 155–158), the like is true of the relationship of *incompatibility*. If nothing is said of several propositions *A, B, C, D,...* except that they stand in the relationship of incompatibility with each other, with respect to ideas *i, j,...*, all this says is that there are no ideas that make propositions *A, B, C, D,... all* true when they are substituted for *i, j,....* That assertion of the incompatiblity of *A, B, C, D,...* all together does not say, however, that some of these propositions, e.g. *A, B,...* could not be made true by themselves, without *C, D,...*, or *B, C, D,...* without *A*, in virtue of certain ideas common to them. So somewhat as we investigated the question (§ 155) in considering compatible propositions *A, B, C, D,... M, N, O,...* as to whether there are not some *A, B, C,...* so constituted that every set of ideas substituted for the variables *i, j,...* which makes them all true would also make one or more of the others *M, N, O,...* true, we can raise the question again here as to whether there are not among the incompatible propositions *A, B, C,..., M, N, O,...* some, say *A, B, C,...*, so constituted that every set of ideas substituted for *i, j,...* which makes them all true would make certain others *M, N, O,... false*. If this is what occurs, then the relationship of propositions *M, N, O,...* to propositions *A, B, C,...* is the exact opposite of the relationship we there called *derivability*. I permit myself to call it the relationship of *exclusion*, and I say that one or more propositions *M, N, O,...* are *excluded* by certain others *A, B, C,...*, with respect to the variable ideas *i, j,...* when every set of ideas which makes *A, B, C,...* all true when they are substituted for *i, j,...* makes *M, N, O,...* all false. I call propositions *A, B, C,...* the *excluding* propositions and propositions *M, N, O,...* the *excluded*. For example, I find such a relation of exclusion to hold between the two propositions,

A is *B* and *B* is *C*, on the one hand and the proposition, No *C* is *A*, on the other, when I consider ideas *A*, *B*, *C* to be the only variables. For every set of ideas that makes the first two propositions true makes the third false. So I call the first two the excluding, but the last the one excluded by them.

(2) If propositions *A*, *B*, *C*,... exclude certain other propositions *M*, *N*, *O*,... with respect to certain ideas *i*, *j*,..., then the negations of the latter, or propositions Neg. *M*, Neg. *N*, Neg. *O*,... must be derivable from the set of propositions *A*, *B*, *C*,... with respect to these same ideas. For every set of ideas which makes *A*, *B*, *C*,... all true when substituted for *i*, *j*,... must make *M*, *N*, *O*,... all false, and so make propositions Neg. *M*, Neg. *N*, Neg. *O*,... all true.

(3) Consider the case when the relation of exclusion between propositions *A*, *B*. *C*,... and *M*, *N*, *O*,... is reciprocal, with respect to the same ideas *i*, *j*,..., i.e. when every set of ideas that make *A*, *B*, *C*,... all true makes *M*, *N*, *O*,... all false, and when every set of ideas that makes *M*, *N*, *O*,... all true likewise makes *A*, *B*, *C*,... all false. We could very well call this relationship between propositions *A*, *B*, *C*,... and *M*, *N*, *O*,... a *reciprocal exclusion.* This is the relationship between the two propositions, *A* is as old as *C*, and *B* is three times as old as *C*, and the two following: *A* and *B* combined are seven times as old as *C* and *B* is as old as *A* and *C* combined; if *A*, *B*, and *C* are the only ideas that can be regarded as variable. For if the first two are true, the last two are false, and if the latter are true the former are false.

(4) All that is said by saying propositions *A*, *B*, *C*,... and *M*, *N*, *O*,... reciprocally exclude each other with respect to ideas *i*, *j*,... is that whenever *A*, *B*, *C*,... all become true, *M*, *N*, *O*,... must all become false, and whenever *M*, *N*, *O*,... all become true, *A*, *B*, *C*,... must all become false. In other words, that propositions Neg. *M*, Neg. *N*, Neg. *O*,... are derivable from propositions *A*, *B*, *C*,... and propositions Neg. *A*, Neg. *B*, Neg. *C*,... from propositions *M*, *N*, *O*,.... Nothing has been decided about whether this twofold derivability itself is reciprocal or only unilateral. Now let this be stipulated also. The relation of derivability between propositions *A*, *B*, *C*,... and Neg. *M*, Neg. *N*, Neg. *O*,... and between *M*, *N*, *O*,... and Neg. *A*, Neg. *B*, Neg. *C*,... *is* reciprocal, i.e. every set of ideas substituted for *i*, *j*,... that makes propositions *A*, *B*, *C*,... all true or false, at the same time makes *M*, *N*, *O*,... false or true. In

this case we say that propositions *A*, *B*, *C*,... on the one hand and propositions *M*, *N*, *O*,... on the other stand in the relation of *contradiction* or *contradict* each other. For the sake of brevity we can call propositions *A*, *B*, *C*,... taken together the *contradictory* of *M*, *N*, *O*,... and the latter the contradictory of the former. But if the relation of derivability is only unilateral, we say that propositions *A*, *B*, *C*,... on the one hand and *M*, *N*, *O*,... on the other stand in the relationship of mere *contrariety* or are *contrary* to one another..... The two propositions:

"Every *X* is a *Y*... *(A)*,

and

"The idea of a non-*Y* has denotation"... *(B)*

stand in the relationship of contradiction to the following two:

"It is false that every non-*Y* is a non-*X*"... *(M)*

and

"The idea of an *X* has no denotation"... *(N)*

if ideas *X* and *Y* count as the variables. Here, in the first place, the falsity of *M* and *N* plainly follows from the truth of propositions *A* and *B*. For if it is supposed to be true that every *X* is a *Y*, the idea of *X* must have denotation, and consequently the proposition that denies it this denotation is already false. If the idea of non-*Y* is supposed to have denotation, furthermore, as proposition *B* asserts, then the claim that every non-*Y* is a non-*X* must also be true, and so proposition *M*, which denies this, false. Likewise, from the truth of propositions *M* and *N* the falsity of *A* and *B* follows. For if idea *X* does not have denotation, as *N* says, then the proposition, every *X* is a *Y*, is surely not true. Furthermore if idea *X* has no denotation, then the idea of non-*X* has the widest extension of the idea of something in general, and consequently the proposition, "Every non-*Y* is a non-*X*," can lack truth, as *M* asserts, only because the idea of non-*Y* does not have denotation. And so proposition *B* is false. But it is also evident that the truth of propositions *M* and *N* can be inferred from the falsity of *A* and *B*. That *M* is correct follows immediately from the falsity of *B*, for if the idea non-*Y* has no denotation, it is beyond a doubt that the proposition, Every non-*Y* is a non-*X*, lacks truth. If the idea in question, non-*Y*, does not have denotation, it follows further that idea *Y* has the widest extension of something in general. Consequently proposition *A*, every *X* is a *Y*, can lack truth only because idea *X* has no denotation itself, which is what proposition *N* asserts, and it is therefore true. Finally, the truth of propositions *A* and *B* can also be derived from the falsity of *M* and *N*. For if *M* is false, then it holds true that every non-*Y* is a non-*X*, and consequently the idea of non-*Y* must be a denotative idea, as *B* asserts. If proposition *N* is also false besides, then the idea of *X* must have denotation. Consequently it can be inferred from the proposition just mentioned, that every non-*Y* is a non-*X* thet every *X* is a *Y*, as *A* says. – Perhaps it will be pleasant to many readers to discover a mathematical example of contradiction as well, one which can be duplicated by many others. If the the symbol, $\overset{n}{=}$, symbolizes a mere inequality between two given quantities without specifying which of them is the greater, then the following six propositions:

$$a+b+c=3m; \qquad 2b+c=3m; \qquad 2c+b=3m;$$
$$a+b+d\overset{n}{=}3m; \qquad 2a+d\overset{n}{=}3m; \qquad 2d+a\overset{n}{=}3m;$$

stand in the relation of contradiction, with respect to the variable ideas *a*, *b*, *c*, *d* and *m*, to the following six propositions:

$$a+d=2m; \qquad 2b+d=3m; \qquad 2d+b=3m;$$
$$b+c\overset{n}{=}2m; \qquad 2c+a\overset{n}{=}3m; \qquad 2a+c\overset{n}{=}3m;$$

This is shown by the solution of the 6 equations included in these 12 propositions and in their negations. The example in (3) provides an example of propositions that are merely contrary to each other.

(5) From the definition given in (4) this follows immediately: If propositions *A*, *B*, *C*,... stand in the relationship of contradiction with propositions *M*, *N*, *O*,..., then propositions *A*, *B*, *C*,... must stand in the relationship of equivalence with propositions Neg. *M*, Neg. *N*, Neg. *O*,..., and propositions *M*, *N*, *O*,... with propositions Neg. *A*, Neg. *B*, Neg. *C*,..., in each case with respect to the same variable ideas. Also, if propositions *A*, *B*, *C*,... are all true or false, then propositions *M*, *N*, *O*,... are all false or true.

(6) No proposition that is generically true or false can be present among propositions contradictory to each other, with respect to the same ideas in terms of which they are generically true or false. For by virtue of the definition contradictory propositions each must be capable of being made true as well as false.

(7) Any proposition *A* at all and its negation Neg. *A* stand in the relationship of contradiction with each other with respect to any ideas whatsoever, unless *A* is a generically true or false proposition. For unless this is the case, there are ideas which make *A* true and other ideas which make *A* false. But all ideas which make *A* true (or false) make Neg. *A* false (or true), and *vice versa.*

(8) If propositions *A*, *B*, *C*,... stand in the relationship of contradiction to propositions *M*, *N*, *O*,..., then the propositions, Neg. *A*, Neg. *B*, Neg. *C*,... also stand in the relationship of contradiction to propositions Neg. *M*, Neg. *N*, Neg. *O*,..., with respect to the same ideas.

(9) Propositions equivalent to propositions that have the relationship of contradiction have the relation of contradiction themselves, always with respect to the same ideas. Proved as in § 156 (4).

(10) If propositions *A*, *B*, *C*, *D*,... stand in the relationship of contradiction to propositionss *M*, *N*, *O*,... and propositions *M*, *N*, *O*,... to propositions *R*, *S*, *T*,..., with respect to the same ideas, then propositions *A*, *B*, *C*,... stand in the relationship of equivalence to propositions *R*, *S*, *T*,... with respect to the same ideas. For those particular ideas that make *A*, *B*, *C*,... all true or false make *M*, *N*, *O*,... all false or true, and consequently make *R*, *S*, *T*,... all true or false.

(11) Suppose propositions *A*, *B*, *C*,... have the relationship of contradiction to propositions *M*, *N*, *O*,... and propositions *E*, *F*, *G*,... to propositions *P*, *Q*, *R*,..., with respect to the same ideas. Suppose besides this that propositions *A*, *B*, *C*, *D*,... *E*, *F*, *G*,... are compatible with each other and so are propositions Neg. *A*, Neg. *B*, Neg. *C*,... Neg. *E*, Neg. *F*, Neg. *G*,..., with respect to the same ideas. Then propositions *A*, *B*, *C*,... *E*, *F*, *G*,... taken all together stand in the relationship of contradiction to propositions *M*, *N*, *O*,... *P*, *Q*, *R*,... taken all together, with respect to the same ideas. For since propositions *A*, *B*, *C*,... *E*, *F*, *G*,... are supposed to be compatible with each other, there are ideas which make them all true when substituted for *i*, *j*,.... But since *A*, *B*, *C*,... are contradictory to *M*, *N*, *O*,... and *E*, *F*, *G*,... to *P*, *Q*, *R*,..., the ideas just referred to must also make *M*, *N*, *O*,... *P*, *Q*, *R*,... all false. Furthermore, since Neg *A*, Neg. *B*, Neg. *C*,... Neg. *E*, Neg. *F*, Neg. *G*,... are supposed to be compatible, there are also ideas which make them true all together, but in that case *M*, *N*, *O*,... *P*, *Q*, *R*,... must also become true. In the same way it is shown that the truth or falsity of

M, *N*, *O*,... *P*, *Q*, *R*,... has as a consequence the falsity or truth or *A*, *B*, *C*,... *E*, *F*, *G*,....

(12) The twofold condition laid down here, that both propositions *A*, *B*, *C*,... *E*, *F*, *G*,... and their negations, or propositions Neg. *A*, Neg. *B*, Neg. *C*,... Neg. *E*, Neg. *F*, Neg. *G*,... must be compatible with each other, is by no means redundant. For if propositions *A*, *B*, *C*,... are not compatible with propositions *E*, *F*, *G*,... obviously there can be no talk of the set of these propositions, namely *A*, *B*, *C*,... *E*, *F*, *G*,... standing in the relationship of contradiction with some other set. For the definition already requires those propositions to be compatible with each other. But since it does not follow at all from the compatibility of certain propositions *A*, *B*, *C*,... *E*, *F*, *G*,... that their negations are also compatible with each other (§ 154 (14)) we must lay this down as a condition also, since this too is required by the definition of contradiction.

(13) If propositions *A*, *B*, *C*,... *E*, *F*, *G*,... taken all together stand in the relationship of contradiction with propositions *M*, *N*, *O*,... *P*, *Q*, *R*,... taken all together, and a part of the former set already stand in the relationship of contradiction to a part of the latter *M*, *N*, *O*,... by themselves, with respect to the same ideas, it still does not follow at all that the rest of the propositions *E*, *F*, *G*,... stand in the relationship of contradiction to the rest of the other set *P*, *Q*, *R*,.... For that certain propositions contradict certain others means only that the one set is equivalent to the negations of the other, and the latter with the negations of the former. We may not, however, infer from the equivalence of a whole set of propositions and one part of it the equivalence of what is left over on both sides.

(14) If it is not a single proposition but a set of several, *A*, *B*, *C*, *D*,..., which stands in the relationship of contradiction with some other set *M*, *N*, *O*,..., then conceivably (namely whenever propositions *A*, *B*, *C*, *D*,... are not all equivalent to each other) there are ideas that do not make all of the propositions *A*, *B*, *C*, *D*,... true or false when they are substituted for the ideas regarded as variable *i*, *j*,..., but only one part of them. Now with these ideas, by virtue of the definition, propositions *M*, *N*, *O*,... may neither all become true nor all false either. Now in case there should be only a single one of them, one or the other must always be so. It follows, therefore, that several propositions can only stand in the relationship of contradiction to a single one if those various proposi-

tions A, B, C, D,... always either all become true or all become false upon substitution of any ideas whatsoever for ideas i, j,..., i.e. if they are all equivalent to each other. The examples in (4) show, however, that a whole *set* of propositions can contradict a whole set of other propositions, and in that case it will not be necessary, as those very examples prove, for the propositions in the one set to be equivalent to each other.

(15) If several propositions A, B, C,..., which contradict several propositions M, N, O,..., are not all equivalent to each other, between them and those that contradict them, M, N, O,... there must be some propositions which are compatible with each other, always with respect to the same variable ideas. For if propositions A, B, C,... are not all equivalent to each other, there are ideas which do not make all of them true, but only some of them. With these ideas, however, not all but only some of M, N, O,... may become false as well. Some of these propositions, then, can become true along with some among A, B, C,..., i.e. they are compatible.

(16) If a pair of propositions A and M are equivalent to each other, with respect to certain ideas i, j,... and are not generically true with respect to them, then each of them stands in the relationship of contradiction with the negation of the other, A with Neg. M, M with Neg. A, with respect to the same ideas. For every idea that makes A true or false makes Neg. M false or true, and *vice versa*.

(17) But if a whole set of propositions A, B, C,... stands in the relationship of equivalence to a whole set of other propositions M, N, O,..., it can not be inferred immediately that a relationship of contradiction holds between propositions A, B, C,... on the one hand and Neg. M, Neg. N, Neg. O,... on the other. For all that follows from the equivalence of the former propositions is that every set of ideas that makes A, B, C,... all true makes Neg. A, Neg. B, Neg. C,... all false, and that every set of ideas that makes M, N, O,... all true also makes Neg. A, Neg. B, Neg. C,... all false. But it does not yet follow that every set of ideas that makes A, B, C,... all false also makes Neg. M, Neg. N, Neg. O,... all true, and *vice versa*, as required by the relationship of contradiction. Thus the two propositions, $x+y=a$, $x-y=b$, are undeniably equivalent, with respect to ideas a, b, x, y, to the two propositions, $x=(a+b)/2$, $y=(a-b)/2$. That notwithstanding, we can not say that those first two propositions are contradictory to the negations of the last two, or to

the propositions, $x \overset{n}{=} (a+b)/2$, $y \overset{n}{=} (a-b)/2$. For it in no way follows from the truth of the latter propositions that the first two have to be false.

(18) If the single proposition M is derivable from the single proposition A, with respect to certain ideas, and the single proposition Neg. M is derivable from the single proposition Neg. A, with respect to the same ideas, then the propositions, A and Neg. M, Neg. A and M, are contradictory with respect to the same ideas. For every set of ideas that makes proposition A true or false, makes proposition Neg. M false or true, and every set of ideas that makes proposition Neg. M true or false makes proposition A false or true.

(19) But it is not exactly the same with several propositions. I.e. if M, or several propositions $M, N, \ldots$ are derivable from several propositions $A, B, C, \ldots$, and Neg. M or Neg. M, Neg. $N, \ldots$ derivable from Neg. A, Neg. B, Neg. $C, \ldots$, it in no way follows that there has to be a relation of contradiction between propositions $A, B, C, \ldots$ and Neg. M or Neg. M, Neg. $N, \ldots$. It is true that from the truth or falsity of propositions A, B, C, the falsity or truth of propositions Neg. M or Neg. M, Neg. $N, \ldots$ follows. But all that follows from the truth or falsity of proposition Neg. M or the several propositions Neg. M, Neg. $N, \ldots$ is that $A, B, C, \ldots$ can not all be true or cannot all be false, not that they all must be false or true.

(20) It is a distinctive feature of any two single propositions which contradict each other that one of them must always be true, the other false. For the one, A, must certainly be either true or false. But if A is true, then the other, M, must be false. Even if a whole set of propositions $A, B, C, \ldots$ stands in contradiction with a single proposition, M, only one of two alternatives holds: either $A, B, C, \ldots$ are all true, namely when M is false, or $A, B, C, \ldots$ are all false, namely when M is true. However, if several propositions $A, B, C, \ldots$ stand in contradiction to several, $M, N, O, \ldots$, we may by no means presuppose that the one set consists only of true propositions, the other only of false. The third situation can also occur, that both sets include some true and some false propositions.

(21) Propositions equivalent to contrary propositions are contrary themselves. Proved as in § 156 (4).

(22) If a pair of single propositions A and M or even a pair of entire

sets *A*, *B*, *C*,... and *M*, *N*, *O*,... are merely contrary to each other, all of them can be false.

(23) From every single proposition *A*, as well as from every set of propositions *A*, *B*, *C*,... which excludes a single proposition *M*, propositions can be derived which contradict this latter proposition, presupposing only that it is not generically false, always understood with respect to the same ideas. For if proposition *M* is not generically false, there are ideas that make it true when they are substituted for the variables *i*, *j*,.... Since proposition *A*, or the several propositions *A*, *B*, *C*,..., exclude it, moreover, there are also ideas that make *M* false. And so *M* is neither generically true nor generically false. Consequently, in Neg. *M* we have immediately a proposition which certainly contradicts it. (7) But according to (2), Neg. *M* is derivable from *A* or from *A*, *B*, *C*,....

(24) If certain propositions, *M*, *N*,..., not generically true, are derivable from a single proposition *A*, then the contradictory of *A* must be derivable from the contradictory of propositions *M*, *N*,..., always with respect to the same ideas. For since *M*, *N*,... are derivable from *A*, all of the ideas that make *A* true must also make *M*, *N*,... true. All ideas, then, that make *M*, *N*,... false, and consequently their contradictories true (and there are some), must make proposition *A* false, and so its contradictory true. The latter is derivable from the former, then.

(25) No proposition not generically false is *incompatible* on both sides with two single propositions that contradict each other. If it is not compatible with the one, the other is compatible with it, indeed derivable from it, always with respect to the same ideas. Suppose a proposition *X* is not compatible with one of two propositions that contradict each other, *A* and *M*, for example *A*. Then every idea that makes *X* true (and there are such) must make *A* false, and consequently make *M* true. So *M* is not only compatible with *X*, but even derivable from *X*.

(26) There are definitely not only single propositions, but even whole sets of propositions, however, which are *compatible* with two mutually exclusive *single* propositions as well as with whole sets of propositions, whether they are merely contrary to each other or even contradictory. They are compatible with both of them at the same time, indeed even derivable from them, always with respect to the same variable parts. Thus in the example in (4), the propositions, $a=m$, $b=m$, are not only

compatible with each of the sets of propositions mentioned there, which stand in the relationship of contradiction to each other, but are even derivable from them. In the same way, to give still another example, the proposition, "Caius deserves blame," is derivable from the proposition, "Caius is stingy," if we regard only the idea of Caius as the variable. Nevertheless it is compatible both with the proposition that is contradictory to the latter proposition, "It is false that Caius is stingy," and the proposition that is merely contrary to it, "Caius is a spendthrift." It is even derivable from the latter.

(27) If propositions *A*, *B*, *C*,... and *M*, *N*, *O*,... are contradictory to each other, and *A*, *B*, *C*,... are compatible with certain propositions *X*, *Y*, *Z*,..., then *M*, *N*, *O*,... are certainly not derivable from the latter propositions; and if *M*, *N*, *O*,... are not derivable from them, at least one of the propositions *A*, *B*, *C*,... must be compatible with them, always with respect to the same variable parts. For if propositions *M*, *N*, *O*,... were *derivable* from *X*, *Y*, *Z*,..., every set of ideas that makes *X*, *Y*, *Z*,... all true also makes *M*, *N*, *O*,... all true, and so *A*, *B*, *C*,... all false, on account of the contradiction. Consequently *X*, *Y*, *Z*,... and *A*, *B*, *C*, would be incompatible. So if *A*, *B*, *C*,... are compatible with *X*, *Y*, *Z*,..., *M*, *N*, *O*,... can not be derivable from them. If, on the other hand, *M*, *N*, *O*,... are *not derivable* from *X*, *Y*, *Z*,... there are ideas that make *X*, *Y*, *Z*,..., all true without making *M*, *N*, *O*,... all true. *A*, *B*, *C*,..., however will not all become false with such ideas either. So there are ideas that make *X*, *Y*, *Z*,... true and one or several of *A*, *B*, *C*,... true as well. And so at least one of the propositions *A*, *B*, *C*,... is compatible with *X*, *Y*, *Z*,....

(28) Every proposition of the form, *X* has *y* (I), stands in contradiction, with respect to ideas *X* and *y*, to the following proposition: "The set of two propositions, idea *X* does not have denotation and the idea of an *X* which does not have *y* has denotation, is not a set of false propositions only." (II) For no doubt ideas can be substituted for *X* and *y* which make proposition I true sometimes and false sometimes. But whenever it is true, idea *X* has denotation and the idea of an *X* which does not have *y* does not have denotation. Thus both of the propositions mentioned in II are false, and so the proposition, which denies this, is itself false. On the other hand, whenever I is false, one of two alternatives must hold. Either idea *X* must have no object or there must be some *X* which do not

have the property *y*, i.e. the idea of an *X* which does not have *y* must have denotation. So it is true that the propositions named in II are not both false, i.e. proposition II is itself true. From this it follows, according to 18 that propositions I and II stand in the relationship of contradiction.

(29) The propositions, *X* has *y* and *X* has the property not *y*, are a pair of propositions that do not contradict each other, with respect to the variable ideas *X* and *y*. They are merely contrary to each other. Two ideas substituted for *X* and *y* never make both propositions true, but they may well make them both false at times. The latter happens whenever we substitute for *X* an idea that encompasses a number of objects and for *y* an idea of a property which neither belongs to all *X* nor is absent in all *X*.

§ 160

Relations among Propositions Resulting from Consideration of How Many True or False Propositions There Are in a Set

There was no talk of whether the given propositions were true or false so far as the relations among propositions considered in §§ 154–159 were concerned. All that was considered was what kind of a relation they maintained, disregarding truth or falsity, when certain ideas considered variable in them were replaced by any other ideas one pleased. But it is as plain as day that it is of the greatest importance in discovering new truths to know whether and how many true – or false propositions there are in a certain, set, either in the form in which these propositions are immediately before us or still better in the infinitely many forms they can assume when we modify certain parts of them to be regarded as variable. The most noteworthy cases that can arise here are the following.

(1) In the first place, all of the propositions in the given set can be true – or all false. This can be so either just in the form they have or in any form they assume when we modify certain ideas in them entirely at will or only in such a way that the subject of these propositions always remains a denotative idea. We shall call propositions that stand in such a relationship to each other simply *true* or *false*. Those that have the second sort of relationship we shall call generically true or false, however. (§ 147)

(2) It may be, to go on, that all we know of a given set of propositions

is that not all of the propositions included in it are false – or true – , without knowing whether just a single proposition, or several of them, or even all of them are true – or false. Consider the four propositions: "This work in the fields will be undertaken in the spring, it will be undertaken in the summer, it will be undertaken in the fall, it will be undertaken in the winter." Of them we know that they are surely not all false, but that either one, or two, or three or perhaps all of them are true, no matter what kind of an idea is substituted for the variable *this*, as long as it is such that the propositions themselves retain denotation with it. It is important to learn to know this relationship where it is present. For even if we do not learn immediately which are the true propositions in a given set when we are only told that some of them are certainly true, still setting before us such a small number of propositions among which the truth is to be sought after usually makes its final discovery much easier. Since I know of no title that has been used for this relationship heretofore, I will permit myself to call propositions of which we know that they are not entirely false *complementary* or *auxiliary* propositions. *Complementary* to each other, because together they exhaust the entire range of suppositions we have with respect to the nature of a certain object (in the above case, for example, concerning the season in which certain work in the fields will be undertaken); *auxiliary*, because we use them as a sort of aid when we do not know which case really holds. But this relationship of complementarity can subsist among given propositions *M*, *N*, *O*,... either only in the specific form in which they are given to us or it can subsist no matter what ideas we substitute for certain ideas *i j*,... regarded as variable in them, either unconditionally or insofar as we do not violate some condition laid down for this substitution of ideas. The former could be called a *material*, the second a *formal* auxiliary relationship or complementarity.

(3) A new relationship enters in when we know that there is only a single true – or false – proposition present in the given set of propositions *M*, *N*, *O*,..., as in the example: "The Gospel according to Matthew either was originally written Greek or was originally written in Hebrew." This may be called a relationship of *one-membered complementarity*. It is ordinarily called a *disjunction*. If the given propositions have the relationship mentioned only insofar as we leave them as given, it is a *material* disjunction, as in the preceding example. But if they maintain

this relationship whatever ideas we substitute for certain ideas taken to be variable in them, then it is a *formal* disjunction, as in the example: This conic section is either a parabola or an ellipse or a hyperbola, when the idea *this* is the only one regarded as variable. If certain propositions M, N, O,... stand in the relationship of formal disjunction to each other with respect to the variable ideas i, j,..., then every set of ideas that makes one of these propositions, e.g. M, true when it is substituted for i, j,... must turn all the rest of them into false propositions. We can say, then, that every one of the propositions M, N, O,... if only it is not generically false, stands in the relationship of exclusion to all the rest (§ 159).

(4) Since propositions that are generically true or false can usually be found out much more easily than others, I will assume that all generically true or false propositions have been omitted from the list of propositions M, N, O,... which is supposed to contain only a single true proposition with every change in ideas i, j,.... By this adjustment the set concerned is reduced to such a small number of propositions that it would lose the property of always having a true proposition within it as soon as we should choose to leave out one more proposition. For since this proposition is also true for certain ideas and since for precisely these ideas it alone is true, if we should choose to leave it out, then for precisely these ideas no proposition in M, N, O,... would be true. For this reason we could call such a set a set of *complementary propositions* of *minimal number* and we could call the propositions themselves *exactly complementary*. Every proposition in such a set would stand in the relationship of (unilateral) exclusion to all the rest. For each of them becomes true in terms of certain ideas, but only in terms of such as make all the rest false. If, conversely, every one of the propositions M, N, O,... should stand in the relationship of (unilateral) exclusion to all the rest, and if there are so many of them that with every set of ideas substituted for certain parts of them thought of as variable i, j,... some true proposition emerges, then these propositions taken together form a set of complementary propositions of minimal number. For there is always only one of them that is true, and if only one more of them is left out, there would be ideas upon which none of them would be true at all, namely those with which the one that has been rejected becomes true.

(5) There is a distinctive case again, when we learn that more than one

of the given propositions $M, N, O,\ldots$ is true, without knowing how many. Such is the case, for example, with the following propositions: "This word begins with a vowel, it begins with a consonant, it ends with a vowel, it ends with a consonant, it is monosyllabic," where there are at least two, perhaps even three true propositions. The relationship of such propositions could be called that of *complex* or *redundant complementarity*. If it holds only insofar as the given propositions $M, N, O,\ldots$ remain unchanged, it is a *material* relationship; if it also holds when certain ideas $i, j,\ldots$ are arbitrarily changed in them, it is *formal*.

(6) The relationship among propositions $M, N, O,\ldots$ is still more worthy of attention when even the number, how many true – or false – propositions there are among them, is also known to us, and when this number remains the same no matter what substitutions we make for certain ideas we may arbitrarily vary within them. For example, the following six propositions stand in a relationship of this kind: "In triangle acb, $ac=bc$; $a=b$; $ac>bc$; $b>a$; $ac<bc$; $b<a$;" wherein two propositions become true every time, no matter what points are represented by the three letters, a, c, b. It would be most appropriate for us to call such a set of propositions, when the number of true propositions in it is $=n$, a set of *n true* propositions. It is either merely *material* or *formal* depending on whether the relationship is tied to a determinate form of these propositions. – When there is not a single generically false proposition among the set of propositions $M, N, O,\ldots$ of which n are true every time, then there must be $(n-1)$ propositions compatible with every single one of them and when these are combined with it, they must constitute a set of n propositions that stands in the relationship of exclusion to the rest. For if the former were not so, then the number of true propositions would have to be smaller than n at times; and if the second were not so, the number would have to be greater than n at times.

(7) Among the sets in which the number of propositions true at the same time is variable, the ones most worth paying attention to are those for which there are ideas that make all of them true at the same time when substituted for the variables assumed. For example, the system of two propositions, Some X are A and Some X are not A, is of this type when X is the only idea supposed to be regarded as variable, while A is a denotative idea which does not have the widest extension of all.

Under these circumstances, not only does one of these propositions become true for every idea substituted for *X*, but for many of these ideas both will become true at the same time. Such sets must also possess the distinctive feature that the propositions they consist of must be compatible with each other, with respect to the same ideas in terms of which they are complementary to each other. And conversely, whenever the propositions in a given set are all compatible with each other, it has the property just mentioned. We can therefore call it a set of *complementary* and *compatible* propositions.

(8) So far we have let the ideas *i*, *j*,... which are supposed to be considered variable be chosen quite arbitrarily, or at most tied them only to the condition that the choice should always be made of ideas which produce propositions with *denotation*. We can lay down very many other conditions in the same way. But it is especially worthy of attention and very much in place in the theory of relations among propositions to consider the case in which the variability of ideas *i*, *j*,... is restricted by certain *other propositions*, in particular by laying down the condition that only such ideas should be selected as make certain other propositions *A*, *B*, *C*,... true. The complementary or auxiliary relationships that emerge among propositions *M*, *N*, *O*,... only under this condition can bear the same title as those discussed in (2)–(7), except that they might be called *conditional*, whereas the earlier ones can be called *unconditional*, to make the distinction. So if any set of ideas substituted for *i*, *j*,... which makes propositions *A*, *B*, *C*,... all true also produces one or more truths in the set of propositions *M*, *N*, *O*,..., I shall say that propositions *M*, *N*, *O*,... are *complementary under the condition of propositions A*, *B*, *C*,.... Such a relationship holds between the proposition, this key is made of metal, as the condition for the following propositions being complementary to each other: This key is made of gold, silver, iron, etc., presupposed that the idea *this* have been regarded as the only variable in these propositions. For then every idea substituted for *this* which makes the first proposition true, also makes one or more of the following propositions true. If every set of ideas that makes *A*, *B*, *C*,... all true when substituted for *i*, *j*,... makes only a single one of propositions *M*, *N*, *O*,... true, I say that *M*, *N*, *O*,... are *simply complementary* or (as some say) *disjoined*, under the condition *A*, *B*, *C*,.... The proposition, this flower belongs to the 14th class, stands as a condition in this relation-

ship with the two following propositions: it belongs either to the gymnosperms or the angiosperms. The idea *this* is the only variable. If every set of ideas which makes A, B, C,... all true when substituted for i, j,... also makes a certain *number* of M, N, O,... true, then I say that propositions M, N, O,... are multiply complementary to each other under the condition A, B, C,.... And if the number of true propositions among M, N, O,... always remains the same and $=$n, I say that there are n *true* propositions in this set under the condition A, B, C,.... Finally, if there are ideas which when substituted for i, j,... make M, N, O,... all true besides making A, B, C,... all true, I say that M, N, O,... constitute a set, under the condition A, B, C,..., of complementary and compatible propositions. Such a set is formed, for example, by the two propositions: Some X are Y, and Some X are not Y, under the condition that both Y and non-X are denotative ideas.

(9) Since I also applied the term complementary to ideas in § 104, this gives rise to the question, how they are related to the *propositions* I identify in this way. If ideas A, B, C, D,... are complementary to each other, with respect to the widest extension of something in general, they have the characteristic that their domains taken all together exhaust the domain of the idea of something in general, or that every object that can even be conceived falls under one of them. Presupposing, then, only that the idea of X is a denotative idea, there is always at least one of the objects falling under it that also falls under ideas A, B, C, D,.... Consider the propositions, "The idea [X] a has denotation, the idea [X] b has denotation, the idea [X] c has denotation," and so on; or according to the more conventional expression, Some X are A, Some X are B, Some X are C, and so on. At least one of them will be true; consequently they stand in the relationship of formal complementarity to each other. This makes it clear how complementary ideas lead to complementary propositions. If ideas A, B, C, D,... should not exhaust the domain of the idea of something in general, but only that of a lower idea, M, then we would still have to preface the propositions just mentioned with the condition, every X is an M, or in other words, the propositions mentioned would have to belong to the genus of conditionally complementary propositions. Only it is not to be forgotten that the set of complementary propositions revealed in just this way do not belong to either of the two most noteworthy types, namely neither to those in which there is always only a

single true proposition nor to those in which the *number* of true proposition is always the *same*. For if we also suppose that the given ideas A, B, C, D,... exclude each other (§ 103), then, depending on whether we take X to be this or that denotative idea, sometimes more, sometimes fewer true propositions will make their appearance among the propositions, Some X are A, Some X are B, etc. Indeed, if we take idea X to be wide enough, and set the idea of something in its place, for example, all of the propositions just cited will become true. The set of these propositions is thus really of the type discussed in (7).

Note: The relationships I have considered in this section are either entirely neglected in ordinary treatises on logic or given only cursory mention in connection with the theory of disjunctive judgments. It is only the two propositions I adduced in (7) as a special example of a set of complementary and at the same time compatible propositions that are included in every logic under the title of *subcontrary* propositions. The definition given of this relationship rests either on a description of form, which was regarded as essential in these propositions, or on their relation to the truth, in which I too place their essence. For example, in Fries S. d. L. (§ 156), we read: "Subcontrary opposition is such that the truth of the one judgment follows from the falsity of the other, but not conversely, since both can be true together." As can be seen, I have adhered to this definition, only with the expansion that seemed necessary to me in order to be able to apply it to more than two propositions. Moreover, it has usually been overlooked that the two propositions, Some A are B and Some A are not B, only stand in the relationship of contrariety to each other when idea A has an object. For if it is objectless, e.g. imaginary, then neither of the two propositions is true, whatever B may be. Thus it is true neither that some round squares are virtuous, nor that they are not virtuous, for there are no round squares at all. It is a more striking mistake, committed by some logicians, when they say that these two hypothetical propositions, If α is, then β is and If α is not then β is, are also subcontraries. Kiesewetter does so, in his W. A. d. L., p. 280: "If it is false that it will be wet if it does not rain, then it is true that it will be wet if it rains. For if the reason for its being wet is not not raining, then raining is." Not at all; such judgments can both be false. For example, if it rains, it freezes; and if it does not rain, it freezes."

§ 161

The Relationship of Comparative Validity or the Probability of a Proposition with Respect to Other Propositions

(1) We have already become acquainted in § 147 with the concept of the validity of a proposition as the concept of a property that can belong to any particular proposition as soon as we regard certain ideas present in it as variable and investigate how the new propositions that can be formed from it by substituting any others we please for those ideas relate to the truth. If we attend to the relations that can obtain between several propositions, there immediately emerges a relation very much worthy of our attention, one which has such a great similarity to validity that it brings it quite involuntarily to mind. Namely, let us consider certain ideas *i*, *j*,... in a particular proposition *A*, or in several propositions *A*, *B*, *C*, *D*,... as variable, and in the latter case suppose that propositions *A*, *B*, *C*, *D*,... have the relation of compatibility with respect to these ideas. Then it will often be unusually important to know the relationship of the set of cases in which propositions *A*, *B*, *C*, *D*,... all become true to the set of cases in which still another proposition *M* becomes true along with them. For if we hold *A*, *B*, *C*, *D*,... to be true, then the relation just identified, between the set of cases in which *A*, *B*, *C*, *D*,... become true and the set of cases in which *M* also becomes true along with them, tells us whether or not we should assume *M* to be true as well. Namely, if the latter set comes to more than half of the former, we can hold *M* to be true merely on account of the truth of propositions *A*, *B*, *C*, *D*,.... If this is not the case, we can not. So I permit myself to call this relationship between the sets cited the *relative validity* of proposition *M* with respect to propositions *A*, *B*, *C*, *D*,..., or the *probability* proposition *M* attains from *presuppositions* *A*, *B*, *C*, *D*,.... I give this relationship the name *relative validity* because of its similarity to the property I call the validity of a proposition in § 147. For just as the question which arises in connection with the validity of a proposition concerns the relation of the set of all the various propositions that can be formed from it by substitution of certain ideas to the set of those among them that are true, what we are asking about in connection with the *relative* validity of proposition *M* with respect to certain other propositions *A*, *B*, *C*, *D*,... is the relation between the set of cases in which *A*, *B*, *C*, *D*,... become

true and the set of cases in which M is also true along with $A, B, C, D, \ldots$. I call this relation *probability*, however, because it seems to me that according to a linguistic usage becoming the more general as time goes on what we understand by probability is really nothing but such a relationship between given propositions, without presupposing that these propositions must be thought and believed by any thinking being.

(2) As a relation between two sets, the relative validity or probability of a propositions can generally be represented by a fraction, whenever it is definable, in which the numerator and the denominator are related as those two sets.

(3) Since the set of cases in which M also becomes true along with propositions $A, B, C, D, \ldots$ can never be larger than the set of cases in which $A, B, C, D, \ldots$ themselves become true, but can often be smaller, the degree of probability can never be greater than 1. It really only attains this value when all the ideas that make $A, B, C, D, \ldots$ all true also make M true, i.e. when M is derivable from $A, B, C, D, \ldots$. In this particular case we are used to saying that proposition M is *certain* with respect to propositions $A, B, C, D, \ldots$. In the opposite case, when there is not a single set of ideas which also makes M true along with $A, B, C, D, \ldots$ when they are substituted for $i, j, \ldots$, i.e. when M stands in the relationship of incompatibility with $A, B, C, D, \ldots$, the degree of probability of M with respect to propositions $A, B, C, D, \ldots = 0$.

(4) If presuppositions $A, B, C, \ldots$ are equivalent to presuppositions $A', B', C', \ldots$, then the probability of a proposition M with respect to $A, B, C, \ldots$ is the same as its probability with respect to $A', B', C', \ldots$, always understood with reference to the same variable ideas $i, j, \ldots$. And if proposition M is equivalent to M', the probability of M with respect to certain presuppositions $A, B, C, \ldots$ is the same as the probability of M' with respect to the same presuppositions, always with reference to the same ideas $i, j, \ldots$. For the very same ideas which make $A, B, C, \ldots$ true also make $A', B', C', \ldots$ true; and the ideas which make M true also make M' true.

(5) If the probability of proposition M with respect to propositions $A, B, C, D, \ldots$ and ideas $i, j, \ldots = \mu$, then the probability of its negation, of the proposition Neg. $M = 1 - \mu$. For since every set of ideas substituted for $i, j, \ldots$ that makes $A, B, C, D, \ldots$ all true either makes M or Neg. M true, the sum of the cases in which M and Neg. M can become true must

be equal in value to the sum of the cases in which $A, B, C, D, \ldots$ become true.

(6) If the degree of probability of a proposition M is only just as large as the degree of probability of its negation, i.e. $=\frac{1}{2}$, we also say that it is merely *doubtful*. Degrees of probability still lower we are accustomd to calling *improbabilities*, and the smaller the fraction constructed according to (2), the greater we call the improbability.

(7) But whenever there is even one idea which makes $A, B, C, D, \ldots$ and M all true, it can immediately be inferred that there is an infinite set of such ideas, since every equivalent idea accomplishes the same thing. Therefore the sets spoken of in (1), if they both actually exist, are both infinite in every case. And consequently the relationship between them can never be found *directly*, i.e. by a mere enumeration of them. Instead one must attempt to define them by considerations of another kind. In order to understand a general way in which this can come about, we must first note that it is possible, still without knowing the specific degree of probability of two or more propositions k, k', k'', k''',..., at least to recognize that these propositions all have an equal probability with respect to presuppositions $A, B, C, D, \ldots$. This will be the case when the propositions identified all have one and the same relationship to the given propositions $A, B, C, D, \ldots$, so that there is no difference among them except what stems from certain ideas taken to be variable in them, with exchange of those ideas transforming the one proposition into another. Suppose, for example, that the presupposition A, relative to which the probability of a given proposition M is to be determined, consists in the statement that Caius has drawn one ball from an urn containing several balls, among them one marked No. 1 and another marked No. 2. I say, then, that if there is no other presupposition, the two following propositions: Caius drew the ball marked No. 1 and Caius drew the ball marked No. 2, both have the same degree of probability, if the ideas, No. 1 and No. 2, are among those ideas we can regard as variable in connection with this investigation. For if we compare these propositions with the given presupposition A, we see that they both have exactly the same relationship to A, since the only difference between them is that one contains the idea of No. 1 and the other the idea of No. 2 in place of it. The ideas in this pair both occur in the same way in the given proposition A (namely not in a rank order, but in a sum). Presupposing, then, that the ideas of No. 1 and No. 2 are among the ideas

we may regard as variable in this investigation, it is obvious that the same ideas by which one of these propositions can be made true along with A also make the other one true. Specifically, let us suppose that the ball Caius really drew was ball No. 3. In that case, the two propositions will only become true, and will be true every time, whenever we substitute for the variables No. 1 or No. 2 in them the idea of No. 3 or an equivalent. Necessarily, then, we must consider the degree of probability of these two propositions as equal. Now if we are set the task of determining the degree of probability of a proposition M on presuppositions $A, B, C, D,\ldots$, in terms of variables $i, j,\ldots$, we can first of all try to think up a certain number of propositions $k, k', k'',\ldots$ which all have the same degree of probability in connection with given presuppositions $A, B, C, D,\ldots$. Moreover, they are to be so constituted that every set of ideas that makes $A, B, C, D,\ldots$ all true when substituted for $i, j,\ldots$ also makes one and only one of propositions $k, k', k'',\ldots$ true, so that these propositions stand in the relationship of *one-membered complementarity* (§ 160) to $A, B, C, D,\ldots$. If we succeeded in doing this, we could say that we had divided the entire infinite set of cases in which presuppositions $A, B, C, D,\ldots$ become true into the cases in which propositions $k, k', k'',\ldots$ become true, as many parts as there are propositions and all equal to each other. In other words, that the set of cases in which k is true, the set of cases in which k' is true, and so on, are all equal to each other and their sum total represents the entire set of all the cases in which $A, B, C, D,\ldots$ become true. Now if it should also be true that proposition M, whose probability we are supposed to determine, had such a relation to propositions $k, k', k'',\ldots$ that none of them left its truth undecided, that we could derive either M or Neg. M from every one of them, then all we would have to count in order to discover immediately how the set of cases in which presuppositions $A, B, C, D,\ldots$ is related to the set of cases in which M also becomes true along with them is how large is the number of propositions $k, k', k'',\ldots$ all together and how many of them make M true. If the total number of propositions k, $k', k''\ldots = k$, and the number of them from which M is also derivable $= m$, it is clear that the entire infinite set of cases in which presuppositions $A, B, C, D,\ldots$ become true can be divided into k equal parts and that the infinite set of cases in which M also becomes true along with them arises from m of these parts. The desired degree of probability, then, would be $=m/k$. To provide an example, let the presupposition be that Caius has drawn one ball from an urn containing 90 black and ten white balls. We are supposed to determine how great the degree of probability is on this assumption that Caius drew a black ball, if the only ideas supposed to be considered variable are the ideas of Caius, black and white. Let us first of all identify the 100 balls in the urn by the numbers 1, 2,... 100, only for the purpose of making it easier to distinguish one from the other in our thinking. Then form the following 100 propositions: Caius drew ball No. 1, Caius drew ball No. 2, and so on, through the proposition, Caius drew ball No. 100. It is evident that these 100 propositions all have an equal degree of probability, if we count the ideas just now introduced, of No. 1, No. 2, etc., as arbitrary. This must always be permissible since they do not occur in the given propositions at all, and so can not change their relation to each other. The equal probability of these propositions becomes clear from the fact that their relation to the given presupposition is the same throughout, so that each of them becomes true when we substitute for the number in it the number of that ball which Caius in fact drew. It is also evident from the nature of these propositions that every set of ideas substituted for the ideas originally cited as variable, Caius, etc., which succeeds in making the indicated presupposition true also makes one and only one of the 100 propositions true along with it. This reveals that each of these propositions contains $\frac{1}{100}$ of the entire set of cases in which the presupposition becomes true. If we note finally that on this presupposition only 90 of the balls in the urn are black, we understand that the proposition that Caius drew a black ball is made true by 90 of the above propositions, false by the other 10, since only 90 of the numbers assigned can apply to balls that are black. Accordingly, our proposition's probability will be $=\frac{90}{100}=\frac{9}{10}$.

(8) There might well be presuppositions $A, B, C,\ldots$ of such a character that the way of determining the degree of probability of proposition M can not be applied, indeed where it is indefinite in itself. For example, if we were told that the number of black and white balls is unequal without being informed which is the greater or how the two are related, the probability of the proposition that Caius drew a black ball can certainly not be determined from that information alone. We must accordingly distinguish between *definite* and *indefinite* probability.

(9) If a proposition M has degree of probability μ under presuppositions A, B, C,... and with respect to ideas i, j,..., and proposition R is unilaterally derivable from M with respect to the same ideas, then the degree of probability of proposition R with respect to the same presuppositions A, B, C,... can *never be smaller than* μ. For every set of ideas substituted for i, j,... that makes M true also makes R true. All of the ideas, then, that make M true along with A, B, C,... for just this reason make R true as well, and consequently the probability of R can not be smaller than μ. It may well be larger, however, for since in general there are more ideas that make R true than ideas that make M true, it would be possible for some of these to make A, B, C,... true also, in which case the probability of R would have to greater than that of M. Thus, for example, if presuppositions A, B, C,... state that there are 40 balls with blue and yellow stripes, 40 with red and green stripes, and 20 of a single color, the probability of proposition R, that a ball with more than one color will be drawn, is significantly greater than the probability of proposition M, that a ball with blue and yellow stripes will make its appearance, if the ideas of blue, yellow, red, green and of one color count as variables. Under this presupposition, however, R is unilaterally derivable from M.

(10) If a proposition M has the degree of probability μ under presuppositions A, B, C,... and with respect to ideas i, j,...; and this same proposition is unilaterally derivable from another proposition L, with respect to the same ideas, then the probability of proposition L with respect to the same presuppositions A, B, C,... can *never be greater than* μ. For since M is derivable from L, the probability of M can not be smaller than that of L (9).

(11) Suppose the following conditions: (1) Proposition M has degree of probability $=\mu$ on presuppositions A, B, C,... and with respect to ideas i, j,.... (2) A second proposition N has degree of probability $=\nu$ on presuppositions D, E, F,... compatible with A, B, C,... and with respect to the same ideas i, j,..., and so on. (3) Propositions A, B, C,..., D, E, F,... etc. are of such a character that every combination of one of the propositions *of equal probability* offered by A, B, C,..., namely K, K',..., with one of the propositions offered by D, E, F,..., namely L, L',... etc. to form propositions such as: K and L is true, K' and L' is true, etc., produces propositions once again with *equal probabilities*. On these

conditions, the degree of probability with which we can state that on presuppositions *A*, *B*, *C*,..., *D*, *E*, *F*,... taken all together, with respect to the same ideas *i*, *j*,... propositions *M*, *N*,... are all true at the same time is equal to the product of $\mu \times \nu \times \ldots$. Understandably, it will suffice to demonstrate this theorem in terms of only two propositions, *M* and *N*. On the assumption, the following propositions, which can be formed by combining each of propositions *K*, *K'*,... with each of propositions *L*, *L'*,..., namely:

K and *L* are true... (x)
K and *L'* are true... (x')
K' and *L* are true... (x'')
K' and *L'* are true... (x''') etc.

are all supposed to have the same degree of probability with respect to the sum total of presuppositions *A*, *B*, *C*,..., *D*, *E*, *F*,... and ideas *i*, *j*,.... If only one of the propositions *K*, *K'*,... becomes true every time presuppositions *A*, *B*, *C*,... become true; and only one of *L*, *L'*,... becomes true every time presuppositions *D*, *E*, *F*,... become true, it is obvious that whenever presuppositions *A*, *B*, *C*,..., *D*, *E*, *F*,... all become true (which does happen, given their assumed compatibility), one of the propositions x, x', x'', x''',... must become true. It is equally obvious that the assertion that *M* and *N* are both true will only become true so often as the components of propositions x, x', x'', x''',... are of such a kind that the truth of proposition *M* is derivable from one of them (which belongs to propositions *K*, *K'*,...) and the truth of proposition *N* from the other (which is one of *L*, *L'*,...). But if the total number of propositions *K*, *K'*,... $=k$, and the number of them from which *M* is derivable $=m$, the total number of propositions *L*, *L'*... $=l$, and the number of them from which *N* is derivable $=n$; then it will soon be found that the total number of propositions x, x', x'',x''',... must be $=k \cdot l$, and the number of those among them from which *M* and *N* are derivable at the same time must be $m \cdot n$. In view of this, the entire infinite set of cases in which presuppositions *A*, *B*, *C*,... *D*, *E*, *F*,... all become true at the same time can be divided into a number of parts, $k \cdot l$ of them, and the infinite set of cases in which propositions *M* and *N* become true as well encompasses $m \cdot n$ such parts. Consequently the degree of probability of the assertion that *M* and *N* are both true at the same time, with respect to presuppositions A, B, C,..., D, E, F,... and ideas i, j,... is $=mn/kl=$ $=m/k \cdot n/l$. But m/k is the probability of proposition M with respect to presuppositions A, B, C,... and thus $=\mu$, and n/l is the probability of proposition N with respect to presuppositions D, E, F,... and thus $=\nu$. It follows that the probability just calculated, of M and N both being true, is $=\mu \cdot \nu$. So, for example, the probability of someone drawing from 6 balls, of which 4 are black and 5 sweet-smelling, one ball that is both black and sweet-smelling is $=\frac{4}{6} \cdot \frac{5}{6} = \frac{5}{9}$.

(12) If everything remains as in (11), and a proposition R is unilaterally derivable from propositions M, N,... with respect to the same ideas i, j,..., then the probability of R relative to presuppositions A, B, C,..., D, E, F,... can *never be smaller than* the product of $\mu \times \nu \times$ .... For R will not be true less frequently than propositions M, N,... are true at the same time.

(13) Since the product of $\mu \times \nu \times$ ... is smaller than any single one of its factors, μ, ν... and becomes smaller and smaller the more there are of them, it is plain that the quantity which can be assumed to be the limit of probability for a conclusion will always turn out to be less than the probability of any one of its premises, and generally will become smaller the more premises it has with mere probability. Premises that are certain do not diminish the probability of the conclusion. For if M is certain, then we have $\mu = 1$.

(14) If everything remains as in (11) and a proposition R has a relation of probability to propositions M, N,... with respect to the same ideas, i, j,... of degree $=\rho$, then the probability of R with respect to presuppositions A, B, C,... D, E, F,... and ideas i, j,... is surely not smaller than the product of $\rho \times \mu \times \nu \times$ .... For the set of cases in which propositions A, B, C,..., D, E, F,... all become true at the same time is related to the set of cases in which propositions M, N,... all become true at the same time as l: $\mu \times \nu \times$ ..., and the set of cases in which propositions M, N,... all become true at the same time is related to the set of cases in which R becomes true along with them as $1:\rho$. So the set of cases in which R becomes true can surely not have a relation to the set of cases in which A, B, C,..., D, E, F,... become true smaller than $\rho \times \mu \times \nu \times \ldots : 1$

(15) If every thing remains as in (11), and propositions M, N,... are unilaterally derivable from a proposition R, with respect to the same ideas i, j,..., then the probability of R relative to presuppositions A, B,

$C, \ldots, D, E, F, \ldots$ is surely no larger than the product of $\mu \times \nu \times \ldots$. For R will not be true more often than $M, N, \ldots$ are all true together.

(16) If everything remains as in (11), the probability of the claim that one of the various propositions $M, N, \ldots$ is true (or that the idea of a true proposition among $M, N, \ldots$ has denotation) relative to presuppositions $A, B, C, \ldots, D, E, F, \ldots$ and in terms of ideas $i, j, \ldots = 1 - (1-\mu) \times (1-\nu) \times \ldots$. For if the probabilities of propositions M and $N = \mu$ and ν, then according to (5) the probabilities of propositions Neg. M and Neg. $N = (1-\mu)$ and $(1-\nu)$. So the probability of the claim that. Neg. M and Neg. N become true at the same time, relative to presuppositions $A, B, C, \ldots, D, E, F, \ldots$ is, according to (11), $=(1-\mu) \cdot (1-\nu)$. Consequently, the probability that they do not all become true at the same time, i.e. that there will also be a true proposition among $M, N, \ldots$, $=1-(1-\mu)(1-\nu)$. Thus the probability of someone who is drawing from two urns, one containing 40 black balls out of 50, the other 45 black balls out of 60, taking out a black one $=1-(1-\frac{40}{50})(1-\frac{45}{60})=\frac{19}{20}$.

(17) If everything remains as in (11), with the further condition that there are only two possible alternatives, either that propositions $M, N, \ldots$ are all true or that they are all false, then the degree of probability of the claim that they are all true, with respect to presuppositions $A, B, C, \ldots, D, E, F, \ldots$ all together and ideas $i, j, \ldots$,

$$= \frac{\mu \times \nu \times \cdots}{\mu \times \nu \times \cdots + (1-\mu) \times (1-\nu) \times \cdots}.$$

Again, the rule is illustrated if we demonstrate it only for the case of two propositions M and N. Now if we add to the conditions in (11) a further one, that there are only two alternatives, either M and N are both true or they are both false, then the combinations of propositions of the form K and L enumerated in (11), having the form K and L are true, do not all have a place. Only some of them do, namely those with which propositions M and N either both become true or both become false. Now if $\mu = m/k$ and $\nu = b/l$, the number of combinations for which propositions M and N are both true $= m \cdot n$, the number of combinations for which they are both false $=(k-m)(l-n)$. Accordingly the number of equally probable cases that are operative here $=m\,n+(k-m)(l-n)$, and the number of cases in which M and N both become true $=mn$. Thus the degree of probability that the latter will occur =

$$\frac{mn}{mn+(k-m)(l-n)}=\frac{\mu\nu}{\mu\nu+(1-\mu)(1-\nu)}.$$

For example, if we knew that someone had reached into two urns, one containing 30 black and 20 white balls, the other 70 black and 50 white, and had drawn one ball from each of them, of which we were told only that they were of the same color (either both black or both white), then the probability of the proposition that they are both black would be $=\frac{21}{31}$. The theorem under this number has application whenever we want to calculate the probability of a proposition for which we have several presuppositions independent of each other and each of them gives it its own specific degree of probability. If we have several independent witnesses for an event's occurrence, for example. Suppose the degree of probability a certain event receives merely from the testimony of witness A were $=\frac{3}{5}$, and the degree of probability it has from the testimony of witness $B=\frac{7}{12}$. Because both witnesses agree, there are only two alternatives, either both of them are telling the truth or both of them are deceiving us, and so the degree of probability the event receives from the joint testimony of both witnesses $=\frac{21}{31}$. One can apply the above theorem even when one or more propositions speak against the proposition to be established (witnesses, for example, who testify against an event's occurrence). What one does is to imagine that in the place of the presupposition that gave the negation of our proposition the probability π, there is a presupposition which gives its affirmation the probability $1-\pi$. So the probability of an event, when one witness speaks for it with probability $=\frac{4}{5}$ and another witness speaks against it with probability $\frac{3}{4}$, must be just as certain as the probability of an event for which two witness declare themselves, one with probability $\frac{4}{5}$, the other with probability $\frac{1}{4}$. So it $=\frac{4}{7}$.

(18) If the probability of one of the propositions $M, N, \ldots =\frac{1}{2}$, then the probability of the claim that they are all true is neither increased nor diminished by this, for

$$\frac{\frac{1}{2}\nu}{\frac{1}{2}\nu+(1-\frac{1}{2})(1-\nu)}=\nu.$$

Propositions with probabilities still lower diminish the degree of probability that would apply without them.

(19) If the probability of a proposition M relative to presuppositions A, B, C, D... and ideas i, j,...$=\mu$, and the probability of a second proposition N relative to the same presuppositions and the same ideas $=\nu$, etc., and furthermore these propositions have the relationship of incompatibility with respect to the same ideas and the assertion is made that one of them is true, then the degree of probability that this true proposition is M

$$= \frac{\mu}{\mu + \nu + \cdots}.$$

For suppose there are only two of these propositions, M and N, and that $\mu = m/k = ml/kl$ and $\nu = n/l = kn/kl$. Then the entire set of cases in which presuppositions A, B, C,... become true can be divided into kl equal parts, and ml of these parts provide the set of cases in which M, kn the set of cases in which N becomes true. But since these propositions conflict with each other, the two sets just identified have no part in common, and consequently the entire set of cases in which one of them is true must contain $ml + kn$ such parts. The probability of proposition M with respect to presuppositions A, B, C,... and the added condition that only M or N is true can be expressed, therefore, by

$$\frac{ml}{lm + kn} = \frac{\mu}{\mu + \nu}.$$

For example if we know that there are 1000 balls of different colors in an urn, but only 10 black and one white, then the probability that someone drawing a single ball has drawn a black one $= \frac{10}{1000} = \frac{1}{100}$, and that it was a white one $= \frac{1}{1000}$. Now if someone without specifying the color of the ball, only informs us that it was either black or white, then the probability of the former $= \frac{10}{11}$.

(20) Since

$$\frac{\mu}{\mu + \nu + \cdots} = 1 - \frac{\nu + \cdots}{\mu + \nu + \cdots},$$

it is clear that even a proposition with a very low degree of probability can attain a very high degree if circumstances enter in so that we have only to choose between it and others which are still more improbable.

For no matter how small μ may be, if only $\nu + \cdots$ is much smaller still, $1-(\nu+\cdots)/(\mu+\nu+\cdots)$ will come as close to unity as we wish.

Note 2: There is complete agreement with my definition above when Lacroix, Laplace and other mathematicians define probability as the relation between the number of favorable cases and the number of all possible cases. All that seemed necessary to me was to define in somewhat greater detail what we understand by *possible* and *favorable* cases, all the more so because we can slip into really substantial errors here. For when we find a number of mathematicians – and very distinguished ones – adding the remark that we must understand *possible* cases to be cases of *equal possibility*, it is very natural, since only probability admits of more of less, not possibility, that we see "in cases which have equal possibility" to be only another expression for the concept of cases which have equal probability, and then consider the whole definition given here a circular definition. When, on the other hand, we read in other authors or even the same ones that cases of *equal possibility* are supposed to be those cases for which there are *equal* but not *fully* sufficient grounds (*quod aeque facile evenire potest*, as Huyghen said), this must be still more misleading. For if a pair of events are really of such a nature that there are thoroughly equal grounds for their occurrence, for example the two events that a balance with equal weights on both sides will incline to one side or the other, the occurrence of one of these events is not only not probable, but we are certain, rather, that neither of the two will and can occur. For this very reason, the mathematician would really only be saying that the balance will not move at all under the given circumstances, since there is a completely equal reason for movement on each side. If we want to reasonably expect an outcome, then, for example that Caius draws precisely one of a number of balls in an urn, we must presuppose that there are not *completely* the same grounds for one being drawn rather than another, since otherwise it would be a certainty that none would be drawn at all. In my opinion, what we really mean to assert by that mistaken expression is that there are no grounds which favor drawing one ball more than the others among the *presuppositions* relative to which the probability is to be calculated. In this case, for example, in the propositions that there are a number of balls in the urn, etc., since the various propositions, "Ball No. 1 will be drawn, Ball No. 2

will be drawn...," all stand in one and the same relationship to the given presuppositions. This still remains true when we add to these presuppositions the proposition, "The balls in the urn do not all have a like relationship to Caius; there is rather an unlikeness such that it will determine him to choose one out of all of them." For since the ideas of No. 1, No. 2... do not occur at all in this proposition, it is obvious that the propositions above are all related to it in the same way too.

§ 162

The Relation of Ground and Consequence

(1) There is a relationship that prevails among truths, as I hope to show in greater detail in the next chapter, which is very much worthy of attention. By virtue of it some truths are related to others as *grounds* to their *consequences*. Thus the two truths, that the sum of the three angles of a triangle is equal to two right angles and that every quadrangle can be divided into two triangles, with their angles taken together forming the angles of the quadrangle, are the ground of the truth that the four angles of every quadrangle together equal four right angles. Likewise, the truth that it is warmer in the winter than it is in the summer includes the ground of that other truth that the thermometer is higher in the summer than in the winter. This latter truth on the other hand, can be regarded as a consequence of the former.... The examples just cited show that a truth which is related to certain other truths as a consequence to its grounds is frequently also *derivable* from these latter truths, presupposing only that we regard certain specific ideas as variables. The proposition, the thermometer is higher in the summer than it is in the winter, is obviously derivable from the proposition, the heat of summer is greater than that of winter, if we regard only the ideas of summer and winter as variable. For no matter what ideas we substitute for them, only those which make the latter proposition true also make the former true. But since propositions obtained from given true propositions by an arbitrary substitution of ideas do not always have to be true, it is understandable how there can also be a relationship of derivability among propositions that are false, a relationship such that the truths produced by substituting certain other ideas for the variables stand in the relationship of ground and consequence to each other. So it is with the two propositions, "It is

warmer in place X than it is in place Y," and "The thermometer is higher in place X than it is in place Y," if ideas X and Y are regarded as the only variables. For there is no doubt that these propositions can become false if we substitute any ideas we please for X and Y. But whenever we choose two such ideas which make the first one true, the second will also be a truth, and one which is related to the first as a consequence to its ground. It is well to note, nevertheless, that what was said just now does not hold wherever there is a relationship of derivability. The relationship between the two propositions just considered is a reciprocal one. Just as the proposition, The thermometer is higher in X than it is in Y, can be derived from the proposition, It is warmer in X than it is in Y, the converse is also true. It is entirely proper to derive the proposition, It is warmer in X than it is in Y, from the proposition, The thermometer is higher in X than it is in Y. All the same, it could never occur to anyone to consider the latter of these two propositions as a consequence flowing from the former, and that as its ground, even if they are both true. No one will say that the true ground of its being warmer in the summer than in the winter is located in the fact that the thermometer mounts higher in the summer than in the winter. Instead, everyone regards the fact that the thermometer climbs as a consequence of the higher heat level, and not the other way around. Not every relationship of *derivability*, then, is so constituted that it also expresses a relationship of ground to consequent holding between its propositions when they are all true. But a relationship of derivability that does possess this characteristic will doubtless be sufficiently worthy of attention to deserve a designation of its own. I will accordingly call it a *formal ground-consequence* relation, while one that holds between *true* propositions may be called, to make the distinction clearer, a *material ground-consequence* relation. So I say that propositions $M, N, O, \ldots$ stand in a *formal ground-consequence* relation to propositions $A, B, C, \ldots$ with respect to ideas $i, j, \ldots$, or are *formal consequences* of them or *formally follow from* them, if every set of ideas which makes $A, B, C, \ldots$ all true when they are substituted for $i, j, \ldots$ also transforms $M, N, O, \ldots$ all into truths, and such as are related to truths $A, B, C, \ldots$ as a genuine consequence to its ground.

(2) The ground-consequence relation also gives rise to a distinctive division of the relation of *probability*. Namely, if propositions $A, B, C, \ldots$, which give to M probability μ, can be regarded as parts of a set of various

propositions A, B, C, D, E,..., to which M stands in the ground-consequence relationship, then the probability of M on the basis of A, B, C,... is called an *intrinsic* probability or a probability *on intrinsic grounds.* If, in the opposite case, none of propositions A, B, C,... belong to the set mentioned, that probability is said to be *extrinsic* or on *extrinsic grounds.* The cloudy skies, for example, make it intrinsically probable that it will rain. The falling barometer, or the metereologist's prediction makes it extrinsically probable.

§ 167

Propositions which Assert a Relation of Probability

When we specify the degree of *probability* a proposition M has with respect to certain others A, B, C, D,... and the ideas i, j,..., we are making, according to § 161, the judgment: "The relation of the set of all the ideas which make A, B, C,... true when they are substituted for i, j,... to the set of those that also make M true along with A, B, C,... has the property μ." – Ordinarily we merely say: "The probability of proposition M with respect to presuppositions A, B, C,... is $=\mu$," expecting the hearer to guess from the context which ideas we are thinking of as variables. But if we say, as is much more often the case, of a proposition M that it is just *probable* with respect to presuppositions A, B, C,..., without specifying the degree of its probability, all we mean is that the degree of its probability is $>\frac{1}{2}$. If, on the other hand, we declare it to be *improbable*, all we mean is that the degree of its probability $<\frac{1}{2}$. If this is correct, what has gone before teaches us how these propositions must be expressed in order to make their logical components visible. Other linguistic variations in designating some measures of the greater or lesser degree of a proposition's probability are not so noteworthy as to deserve being mentioned here.

§ 168

Propositions Which Assert a Relation of Ground and Consequence

(3) In my opinion the concepts of *cause* and *effect* are intimately related to the concepts of ground and consequence. What I believe is that in their proper sense these two words only refer to objects which possess reality

so that we say of a real object α that it is the *cause* of truth M when the proposition, α has existence, is one of the partial grounds on which truth M depends. And likewise we say that object μ is an *effect* of object α when the proposition, μ has existence, is one of the consequences following from the proposition, α has existence. Thus we say that God is the cause of the world's existence, the world however God's effect, because the ground of the truth that there is a world lies in the truth that God exists. If this is correct, then we must count the propositions asserting a causal relation which occur so frequently, or so-called *causal propositions*, i.e. propositions of the form, X is the cause of Y, or X produces Y, or Y is the effect of X, as propositions which assign a relation of ground-consequence to certain other propositions. For X is cause of Y really means: "The truth that X exists is related to the truth that Y exists as a ground (partial ground) to its consequence (partial consequence)."

§ 174

Propositions of the Form: n A are B

Expressions of the type: Two men died without being born, or generally: $n\,A$ are B, allow of a double interpretation. Sometimes all we want to assert by using such an expression is that there are at least n members of A which are B, without denying that there may be more of them. Sometimes our intention would go so far as to make a decision about the latter point. In my opinion, what we are saying in the first case is only that the idea of $n\,A$ which are B has denotation; the components of our proposition follow from that. In the other case, however, the subject of our proposition is the set of all A which are at the same time B, and we are saying of this set that it has the character of a set consisting of n parts, or that the number of its parts $=n$.

Note: In the very same sense as we say, $n\,A$ are B, we also say: There are nA, which are B. The latter expression is to be interpreted like the former, then.

§ 179

Propositions with If and Then

In all somewhat developed languages a very commonplace way of expressing oneself is by *if* and *then*, as they are used in the following proposition, for example. If Caius is a man, and all men are mortal, then Caius is also mortal. I have already claimed in § 164 that we make use of this expression in order to express the relationship of derivability of a certain proposition from one or more other propositions. But I do not believe this is universal, that in every case in which we use *if* and *then* we are thinking of the presence of certain ideas which may be handled as variables and replaced with any others whatsoever without disturbing the proposition's truth. This may well be so with the proposition I just cited as an example, for in this case the idea forces itself upon us that the claim expressed would remain true no matter what we put in the place of the three ideas, Caius, man and mortal. And everyone feels that the meaning of the proposition only extends so far as to say that in every case in which upon an arbitrary assignment of the ideas referred to the two antecedent propositions are true, the consequent also expresses a truth. But since this way of expressing oneself has a great deal of convenience, we often also make use of it where we are not thinking of any variable ideas in the propositions compared at all, or where it is at least not necessary to think of such ideas. I say we have such a case in the following proposition: "If you rearrange the digits in a given number any way you please and substract the resulting number from the original the remainder is always divisible by 9." There are no ideas at all in this proposition that one would have to regard as variable, and the *if*, *then* has been used here purely for the sake of greater convenience, instead of expressing oneself in another way, say like the following: A number which differs from another number only in that its digits are arranged in a different order has a relationship to the latter such that the difference between the two is always divisible by 9. Generally I say that even in those cases where the propositions connected by *if* and *then* do stand in a relationship of derivability to each other, we still do not think of this relationship, when it is not one of those whose frequent occurrence has already made us very facile in its use, but of a certain other relationship more or less equivalent to it. What it is I will show by a few examples. First, consider the propositions included within the following form: *If A is B, then it is also C*, for example, if Caius remains silent on this occasion, then he is ungrateful. It may still be the case that the two propositions connected here, Caius remains silent on this occasion and

Caius is ungrateful, stand in the relationship of derivability to each other. If the one who is expressing this judgment should clearly think this, he would have to think that there is a certain idea (say the idea of Caius) in those propositions, which can be regarded as variable with the result that every assignment of it which makes the first proposition true also would make the second proposition true. Now is this really the thought we have in mind with those words? I do not believe it is. What is vaguely before us is an essentially different thought. What "If Caius remains silent on this occasion, then he is ungrateful," means to say is that among Caius' circumstances there are some such that they are subject to the general principle that anyone who remains silent under such circumstances is ungrateful. In these terms, the proposition we are expressing would really be the assertion of the denotative character of an idea and it would generally fall under the form: The idea of certain properties of A, of which the principle holds that every object which has *b* along with these properties also must have the property *c*, has denotation. It seems to me that propositions of the following form are also to be interpreted in a similar way: *If A is B, then C is D*. For example, if Caius is dead, then Sempronius is a beggar. With these words, too, all we are saying is that there are certain relationships between Caius and Sempronius of which the general principle holds true that of any two men of whom one (in Caius' circumstances) dies the other (in Sempronius' circumstances) is reduced to beggarhood. It is easy to see that a similar interpretation is appropriate even if there are several antecedent or consequent propositions. For just this reason I believe we always attach this meaning to the locution, *if, then*, apart from cases in which the ideas $i, j, \cdots$ that undergo arbitrary variation in the antecedent and consequent propositions compared present themselves too clearly to be overlooked. In those cases the first form of interpretation enters in.

Note: It belongs more to a grammar than to a logic to remark that we construe the *if* and *then* with the *indicative* form of the verb (If *A is*, then *M is*), insofar as we want to leave it unsettled whether the propositions *A* and *M* being compared are true or false, but that we use the *subjunctive* (If *A were*, then *M would be*) insofar as we want to indicate that we hold propositions *A* and *M*, as they are before us, to be *false*.

§182

Propositions Containing the Concept of Necessity, Possibility or Contingency

Propositions that assert a *necessity*, a *possibility* or a mere *contingency*, so far as their linguistic expression is concerned, form a class of propositions very much worthy of attention. But since the meaning we attach to these words is not always the same, those propositions themselves will have to be understood in various ways.

(1) I believe that generally we use the words, necessity, possibility and contingency, like the related words, *must* and *can*, only with some reference to the concept of *being* or reality, when we are taking them in their strictest sense. My opinion is that it can be said of any *must* in the strictest sense that it is a must *be* or the necessity of being, and of every *can* in the strictest sense that it is a can *be* or the possibility of being. And what we call *necessary*, or even only *contingent*, in its proper sense is always something real, and what is *possible* is never anything but what can take on reality. Now so far as the concept of *necessity* is concerned in particular, it seems to me that we say the being of a certain object is *necessary* or has necessity or *must* exist when there is a pure conceptual truth of the form, *A′ is* (or has existence), in which *A′* is an idea that encompasses object *A*. Thus we say that God is necessary because the proposition that God is is a pure conceptual truth. In the opposite case, when it is not the proposition, *A′* is, but rather the proposition, *A′* is *not*, that is a pure conceptual truth, then we say that object *A*, which falls under the idea *A′*, is *impossible*. Thus we say, for example, that an omnipotent created being is something impossible, because the proposition that there is no such being is a pure conceptual truth. We call the being of an object *possible*, on the other hand, whenever it is not impossible. Thus it is possible for a human being to be mistaken, because there is no conceptual truth that would assert the non-existence of a mistaken human being. Now if an object exists, without being necessary, we call it *contingent*. Thus, for example, mistaken human beings exist only contingently.

(2) On these definitions, every object that can be represented by a pure concept that refers to it alone is either one or the other, either necessary, if it has reality, or impossible, if it does not have reality. For if there is some pure concept *A′* that represents this one object exclusively, then the

proposition, *A′ exists*, is a pure conceptual proposition. And if it is true, then the object is necessary. But if this proposition is false, then (since *A′* is a singular idea) the equally pure conceptual proposition, *A′ does not exist*, is true and the object is consequently impossible. So only such objects as we can not represent except by way of mixed ideas or intuitions (§74) can be real and yet not necessary, i.e. merely contingent.

(3) If neither object *A* nor object *M* are necessary in and of themselves, but it is clear that the proposition, *M* exists, is derivable from the proposition, *A* exists, with respect to some of the ideas in *M* and *A*, then we say that *M* is *relatively necessary* on the presupposition *A*. This necessity of *M* we call a *relative* or *extrinsic* necessity, and in contrast to it, we call the necessity in (1) an *intrinsic* necessity. Thus we say that judgment is only relatively necessary, namely only under the presupposition that someone has sinned, since the proposition, creatures will be brought to judgment, is derivable from the proposition, creatures have sinned. When in the opposite case it is not the proposition, M exists, but the proposition, *M* does not exist, that is derivable from the proposition, *A* exists, we say that *M* is *impossible relative* to *A* or *impossible if A* exists. We call this impossibility of *M* a *relative* or *extrinsic* impossibility, and the impossibility in (1) *intrinsic*, to distinguish between them. If object *M* is neither intrinsically impossible nor impossible relative to another object *A*, then we say that *M* is *possible* with respect to this object *A*. Finally, if an object *M*, which is not necessary relative to another object *A*, nevertheless has existence, then we say that its existence is not only intrinsically but also *extrinsically contingent* or *contingent relative* to *A*.

(4) The words, necessary, possible and contingent are taken in a *second* sense, which I call the *broader* or *loose* sense, although it is very common, when we do not apply them to the existence of things as heretofore (1)–(3) but to truths in themselves. Whenever the proposition, *A* has *b*, is a pure conceptual truth, we are accustomed to saying that property *b* *necessarily* belongs to object *A*, without paying any attention to whether this object itself, and consequently that property too, is something existent or not. Thus we say that every equation with an odd-numbered degree necessarily has a real root, although neither equations nor their roots are anything existent. If it is not the proposition, *A* has *b*, but the proposition, *A* does not have *b*, which is a pure conceptual truth, then we say that property *b* is *impossible* for object *A*, even when it would not occur to

anyone to look for this property or the object among the things that are real. Thus we say it is impossible for a simple idea to be imaginary, because a pure conceptual truth tells us that simple ideas are never imaginary. If it is not impossible for property *b* to belong to object *A*, if there is no pure conceptual truth of the type, No *A* has *b*, then we say that it is *possible* for *A* to have *b*. Finally, if the proposition, *A* has *b*, is true without being a pure conceptual truth, i.e. if property *b* belongs to object *A* without necessarily belonging to it, we say it belongs to it *contingently*. If property *b* does not necessarily belong to *A*, and property *p* does not necessarily belong to *M* in and of itself, but the proposition, *M* has *p*, is derivable from the proposition, *A* has *b*, with respect to some of the ideas in *A* and *M*, then we say that property *p* is *relatively* necessary to *M*, namely under the presupposition that all *A* have property *b*. The concepts of relative *possibility* and *contingency* are defined in a similar way.

(5) The words, *possible* and *can*, but not necessary and contingent, are often taken in yet a *third* sense. We say that something is *possible* or *could* be when all we want to indicate is that *we* do not know it to be impossible, or, which comes to the same thing, that no pure conceptual truth asserting the opposite is *known* to us. Understandably, all that follows from this is that we know no such truth, not that there are none. So we may not confuse what we call *possible* in this sense of the word either with what is called possible in the strict sense (1) nor with what is called possible in the derivative sense (4), as has really happened all too often. In order to guard against this confusion, we could use the expression, *apparently possible* or *problematic*, instead of the adjective *possible*. But I would not know any appropriate word at all to recommend in the place of the verb *can*. And since even the words, apparently possible and problematic, are inconvenient, nothing remains to be done except to allow the continued use of the two words in question in their threefold sense, but to take all the more care to distinguish correctly which sense they are being used in every time they occur. We may suspect from the beginning that the word possible is used in the sense of apparently possible whenever it is applied to (future or past) events. For example, if what we hear is, "it is possible that it will rain today," this is only to be understood as apparent possibility. The only interpretation to be given it is that the speaker knows of nothing from which there follows

the impossibility of rain today. When a historian writes, "It is possible that Attila was assassinated," all he wants to say is that he knows of no fact which proves the falsity of this suspicion set forth by Agnellus. But even when we are speaking of mere conceptual truths, we sometimes use the words could and possible in this third sense. For example, many respond to the question whether the two-hundredth decimal place in the quantity π may not be an even number with the answer, that this is possible. All he intends to say by this is that he has no knowledge in hand that would prove the impossibility of that assumption.

(6) After these definitions, it is possible to make a judgment about how I conceive of the meaning of propositions that assert a necessity, possibility or contingency. Thus I believe, for example, that the claim, "God exists with necessity," has no other meaning but: The proposition that God exists is a pure conceptual truth. "Every effect must necessarily have its cause," to me means nothing other than: It is a pure conceptual truth that every effect has its cause, and so on.

CHAPTER THREE

ON TRUE PROPOSITIONS

§ 198

The Concept of a Ground-Consequence Relationship between Truths

Of all the relations that hold between truths, the one most worthy of attention in my opinion is that of ground and consequence, by virtue of which certain propositions are the *ground* of certain other propositions and the latter their *consequences*. I have already had occasion to speak of this relationship several times (particularly in § 162), but here is the place where I must concern myself with it in detail. Let us first of all properly establish the *concept* of it. When we compare the following three truths:

> "One should never prefer one's own advantage to the greater advantage of another;"
>
> "One is preferring one's own advantage to another's greater advantage when one destroys what is necessary for satisfying the essential requirements necessary to another's life only for the sake of producing an unnecessary physical pleasure for oneself."
>
> "One should never destroy what is necessary to another for satisfying his essential requirements for life only in order to procure an unnecessary physical pleasure."

with each other, we shall soon be aware that the first two have a very distinctive relationship to the third. It asserts its existence right away by the fact that we come to see the last truth with the utmost clarity and distinctness when we have previously acknowledged the first two and brought them before our consciousness.

On further consideration it becomes clear (at least I think so) that the essence of that relationship is still not fully expressed by the *effect* just mentioned, namely that the last truth *can be known* from the first two. For this is also the case with truths of which it is obvious that they do not stand in the relationship of the ones considered above. Thus, for example, it is also appropriate to say of the three truths:

> "If the thermometer is higher, it is warmer;"
>
> "In the summer the thermometer is usually higher than in the winter;"
>
> "In the summer it is usually warmer than in the winter;"

that we *know* the last as soon as we have known the first two and brought them before consciousness. All the same, who would not feel that the relationship between the truths in the first example is quite different from the one in which the truths of the last example stand? If we are supposed to express the distinctive character of that first relationship in *words*, we feel ourselves almost forced to call it a relation between *ground* and *consequence*, to say that the first two truths in that example are the *ground* of the last and that it is the *consequence* of them. We shall not make the same assertion in the second example, if we want to speak precisely. To be sure, we are accustomed to using the locution "The truth that it is warmer in the summer than in the winter is *grounded* in [based upon] the truth that the thermometer is higher in the summer than in the winter," at times. But we soon modify our claim. We had really only been speaking of the *knowledge* of these truths, we had only wanted to indicate that the *knowledge* of the one of these truths *produces* the knowledge of the other. But that the first truth is in and of itself the *ground* of the latter is a claim we are so little inclined to make that instead we achnowledge that precisely the opposite relationship obtains. Now since everyone will see for himself that the above example is not the only one of its kind, I conclude that there are truths which stand in such a relationship to each other that we have no better way of identifying it in terms of linguistic usage than by means of the relationship of a *ground* to its *consequences*.... On the presupposition that, as the example cited reminds us, it is frequently not a single truth, but a whole set of truths that stand in the relationship of ground to consequent with one or a whole set of several truths, it will be permissible to call the individual truths that make up such a set *partial grounds* and *partial consequences,* indeed even simply grounds and consequences themselves, if we only take some care that the double meaning of these words does not give rise to any misunderstanding. For example, where we are speaking of grounds or consequences which are not mere partial grounds or partial consequences, for the sake of greater clarity to call them *complete* grounds and consequences. Finally, because it happens all too often that we think of merely subjective *cognitive grounds* or cognitive consequences, i.e. of truths which as premises produce knowledge or arise from it as conclusion, in connection with the words, *grounds* and *consequences,* we will also call the grounds and consequences of which we are now speaking *objective* grounds and consequences at times, so as to indicate that their relationship exists among truths *in themselves,* independently of our thinking.

§ 199

Can the Inference Rule also Be Counted among the Partial Grounds of a Conclusion?

In the example we used to illustrate the concept of the ground-consequence relationship in the preceding section, the truth we looked upon as a consequence stood in the relationship of *derivability* to the two truths we represented as its ground. Now for every relationship of this latter kind there is a *rule* that describes it, i.e. a proposition which states what properties the premises *A, B, C, D,...* present in this case must have and what property the conclusion following from them must have. And if the rule by which we wanted to derive from propositions *A, B, C, D...* certain other propositions *M, N, O,...* were incorrect, it is obvious that we could not claim that propositions *M, N, O,...* are truths that *follow* from truths *A, B, C, D,....* Taking this into account, someone could get the idea that the rule by which propositions *M, N, O,...* are derivable from propositions *A, B, C, D,...* is to be regarded as a *truth* which must be added on to truths *A, B, C, D,...* in order to obtain the *complete* ground of truths *M, N, O,....* On this view of the matter, for example, the two truths, "Socrates was an Athenian" and "Socrates was a philosopher," would still not constitute the complete ground of this third proposition derivable from them: "Socrates was an Athenian and a philosopher." Instead, the rule by which one proceeds in this inference would also belong to the complete ground, namely the truth: "If two propositions, *A* has *b* and *A* has *c*, are true, then the proposition, *A* has $(b+c)$, is also true." Very much as I might agree with this point of view from one side, there is a strong argument against it in the following consideration.

If one maintains that the inference rule by which truths *M*, *N*, *O*,... are derivable from truths *A*, *B*, *C*, *D*,... must also belong to the complete ground of *M*, *N*, *O*,..., along with *A*, *B*, *C*, *D*,..., one is really claiming that propositions *M*, *N*, *O*,... are only true because this inference rule is correct and propositions *A*, *B*, *C*, *D*,... are true. In fact, one is constructing the following inference:

"If propositions *A*, *B*, *C*, *D*,... are true, then propositions *M*, *N*, *O*,... are also true; now propositions *A*, *B*, *C*, *D*,... are true; therefore propositions *M*, *N*, *O*,... are also true." Just as every inference has its inference rule, so does this one. If we designate the first of the three propositions just stated briefly by *X* and the second by *Y*, then this rule reads: "If propositions *X* and *Y* are true, then propositions *M*, *N*, *O*,... are true." – Now if one required to start with that the rule of derivation be counted in the complete ground of truths *M*, *N*, *O*,... along with truths *A*, *B*, *C*, *D*,..., the same consideration forces one to require that the *second* inference rule just stated also be counted in with that ground, for one can also say of this rule with the same right as before that if it were not correct, truths *M*, *N*, *O*,... could not follow. One can see for oneself that this sort of reasoning could be continued *ad infinitum*, and consequently that if one were justified in counting the inference rule within the ground of truths *M*, *N*, *O*,... even the very first time, there could be an *infinite* set of them belonging to this ground. But this would seem absurd.

§ 200

Is the Relation of Ground and Consequence Subordinate to that of Derivability?

If we are not permitted to count the inference rule by which the truth considered the conclusion in the next to the last section can be derived from the two others as part of its ground, then it will surely be agreed that the two truths combined already contain the *complete ground* of the third one. And so we would immediately have an example here of some truths that, in entirely the same way as they stand in a relationship of derivability to each other, are also related as *ground* and *consequence* The two truths that constitute the *premises* in the former relationship, constitute the *ground* in the latter. And the truth that appears as a mere *conclusion* there, appears as a genuine *consequence* here. Now the question arises, whether these two relationships always have to coincide in this way, or what sort of difference there is between them?

That the relationship of ground and consequence is not *entirely the same as* that of derivability, also that the difference between them by no means consists in the fact that the latter is encountered among *propositions generally*, the former only among *truths*, has already been settled by what was said in § 198. For the propositions I introduced there, as an example of propositions in which the relation of ground and consequence was not supposed to have a place, also stood in the relationship of derivability to each other and were all true. But someone who did not presuppose a relation of ground and consequence wherever he encountered a relation of derivability between true propositions, could still suspect the converse, that wherever the former prevails, the latter must also be present. If truths are supposed to be related to each other as ground and consequence, they must always, one might believe, be *derivable* from one another as well. The relation of ground and consequence would then be such as to be considered a particular species of the relation of derivability; the first concept would be subordinate to the second. Probable as this seems to me, I know of no proof that would justify me in looking upon it as settled. As you know, I call certain propositions derivable from certain others when we find ideas in them which can be exchanged for any others you please with the result that whenever the one of these propositions is true, so is the other. So if one wanted to claim that the relation of ground and consequence is subordinate to that of derivability, then one would have to prove that a relationship of ground and consequence could only hold between certain truths $A, B, C, D, \ldots$ on the one hand and $M, N, O, \ldots$ on the other if there are certain ideas in them which can be exchanged for any others you please so that propositions $M, N, O, \ldots$ remain true whenever propositions $A, B, C, D, \ldots$ are true by themselves. Now how to prove that this is and must be universally the case?

Furthermore, on this view of the matter it would be self-evident that we would also have to ascribe the same relationship of ground and consequence assumed between truths *A*, *B*, *C*, *D*,... and *M*, *N*, *O*,... to all of the truths produced from them by varying the arbitrary ideas *i*, *j*,.... But it would follow from this that for every set of truths *A*, *B*, *C*, *D*,... *M*, *N*, *O*,... that are related as ground and consequence, there would be an infinite number of other sets standing in the same relationship, if infinitely many other truths can be produced by varying the ideas to be considered as arbitrary in them. Now is it probable that for every set of truths from which another follows as consequence from them as ground, there is an infinite number of other sets of truths from which other truths follow in the same way, so that the distinctive character of the very ideas of which these sets of truths are composed never has any influence on the type of ground-consequence relation between them? – The following example seems to me to prove the opposite. Anyone who does not deny the existence of a relationship of ground and consequence in general will be inclined to grant that there is a certain practical truth of the form. One ought to do (or will) *A*, which is so constituted that all of the other practical truths, e.g. one should not lie, etc., can be derived from it as a consequence from its ground by the addition of some theoretical proposition of the form, For *A* to happen, *X* is necessary. But that first truth (the so-called highest moral law) also appears to have a ground. For if *A*, were impossible, there could be no obligation to will it. Consequently the obligation to do *A* is grounded either wholly or partly in the truth that *A* is possible. We may assume, however, that the truth just stated by itself, or a set of truths in which it is only one part, is the complete ground of that supreme moral law; then it is clear that it does not follow from its ground by any of the usual rules of inference. "One should do *A*" flows as a consequence from a set of truths in combination, but none of them can already include the concept of should within it (state an obligation, because otherwise it would be a practical truth). Therefore we can see distinctly that none of the usual rules of inference would justify us in accepting it as part of the conclusion.

§ 203

Only Truths Are Related as Ground and Consequence

(1) If we are not in a position, whether because of ignorance or because

it is impossible in itself, to analyze the concept of the relation of *ground and consequence* into other simpler concepts, it becomes all the more necessary for its correct interpretation for us to describe the distinctive features of this relation in a series of special theorems. And since it is the relation of derivability above all that could be confused with that of ground and consequence, because of the similarity between them, it will be useful to define the distinctive features of the latter concept by way of a comparison with those of the former. As we know, the relation of derivability holds only between *propositions*, but it holds between them without regard to whether they are true or false. The question is, then, how does it stand with the relation of ground and consequence in this respect? To this question I believe I must reply that the relation of ground and consequence has in common with that of derivability the the fact that it, too, applies only to propositions (hence to nothing that would possess reality). It differs from the latter, however, in that it can never hold between false propositions, only between *true* propositions or truths. What I maintain is that the designations of ground and of consequence, if taken in their proper sense, may be exclusively ascribed to truths only, single truths or whole sets of them. Whatever is supposed to be called a ground, or a consequence, must be a truth, some particular one or a whole set of several truths.

§ 204

Can Something Be Ground and Consequence of Itself?

According to the definition of the relationship of derivability given in § 155, whole sets of propositions, like single propositions, can stand in the relationship of derivability to one another. Indeed, we can even say to a certain extent that a proposition, likewise a whole set of propositions has this relationship *to itself.* In my opinion the same does not hold true of the relationship of *ground* and *consequence*. Neither of a single truth *A* nor of a set of truths *A*, *B*, *C*, *D*,... can we maintain that they stand in the ground-consequence relation to themselves, i.e. they are the ground and the consequence of themselves. That is already implied in the concepts I connect with the words ground and consequence. If I think something is a ground, I think of something else *of* which it is the ground; if I think something is a consequence, I think of something else along with it of which it is supposed to be a consequence.

§ 205

Are Ground and Consequence in Each Case only a Single Truth or a Set of Several Truths?

With the relationship of *derivability*, as we have already said, it is sometimes only a single proposition, sometimes a set of several propositions from which others can be considered to be derivable. Likewise, it is sometimes only a single proposition, sometimes a set of several which can be regarded as derived. Let us investigate whether this is also the case with the relationship of *ground and consequence*, then.

(1) To begin with, so far as the *ground* is concerned, it seems clear to me that it can be only a single truth at times and at times a set of several truths. Whatever sort of a truth *A* may be, the truth "that proposition *A* is true," is a genuine consequence of it; and this consequence surely does not require any second truth at all besides *A* alone to be grounded. Consequently *A* constitutes its complete ground. There are, therefore, grounds that consist of a *single* proposition. But very many of the examples already introduced teach us that there are also grounds consisting in the union of *several* truths. To be sure one could reply that in all of the cases in which we had regarded several truths, *A*, *B*, *C*, *D*,... as the ground of one or more other truths, it is really not these several truths as such, but the single truth produced by combining them, namely the truth, "Every one of the propositions *A*, *B*, *C*, *D*,... is true," that is the real ground of the consequences we cited. But my response to this is that the truth, "Every one of the propositions *A*, *B*, *C*, *D*,... is true," must already be regarded as a consequence flowing from the several truths *A*, *B*, *C*, *D*.... One obviously can not say of it that its ground is only a single truth. It is certain, then, that there are also grounds which consist of a whole set of several truths.

(2) So far as *consequences* are concerned: I might almost doubt that the *complete* consequence ever consists of only a single proposition. Instead I might claim that even in the cases in which the ground is only a single truth, and so much the more in others, there are always several truths that can together be regarded as its consequence. If someone said that

each of these truths already deserves to be called a *consequence* by itself, not only the whole set of them, I would reply that this seems to me to be a confusion of the concept of consequence, namely complete consequence, with that of a partial consequence. For if we do not wish to use the words consequence and partial consequence pointlessly, we must understand by consequence, which more precisely we also call the complete consequence, the set of all of the truths that stand in the relation of consequent to ground to certain others $A, B, C, D, \ldots$. And such an interpretation of this concept could be justified, because the concept of a sum of all the truths which stand in the relationship of consequent to ground to given other truths $A, B, C, D, \ldots$ is surely worthy of our attention, and so deserves to be designated by a term of its own.

§ 206

Can One Ground Have a Variety of Consequences or One Consequence a Variety of Grounds?

According to the concept of the relationship of *derivability* we set forth, diverse conclusions can flow from the same premises and the same conclusion can be derived from diverse premises. It is not quite the same with the relationship of *ground* and *consequence.* For it is already clear from the preceding section that there is not a variety of consequences belonging to the one given ground, unless we understand these to be merely partial consequences. Nothing short of all the truths having the ground-consequence relation to certain given truths constitute their complete consequence. It is not so evident that there is only one consequence from the same grounds and that different grounds consequently always have different consequences. On the contrary, one might believe that examples can be given in which the same consequence comes from different grounds, like effects from unlike causes. Thus the commandment, Thou shouldst not lie, can be derived from the highest moral law in a large variety of ways, namely by citing each of the manifold kinds of damage lying does to the general welfare, and each one of these derivations could deserve to be titled a relationship of ground and consequence. It is well known that there are also an infinite number of different combinations of two or more mechanical forces that are called equivalent to each other because as causes they bring about completely identical movements as their effects. And in terms of the connection between grounds and causes that was assumed above (§ 201), it follows that equivalent causes pre-suppose equivalent grounds as well. Nevertheless, all that these examples prove, when examined more closely, is that different grounds can sometimes have *the same partial consequence*. But they do not show that their *entire consequences* are the same. The complete consequence which certain truths *A*, *B*, *C*, *D*,... have includes, among others, the truth "that each of propositions *A*, *B*, *C*, *D*,... is true." This, however, is a consequence (a partial consequence) which obviously no other set of truths has but precisely this one. And so it is clear that every distinct ground has a consequence that is at least in some parts distinctly its own.

§ 207

Can One Regard the Consequence of a Part as the Consequence of the Whole?

With the relationship of derivability it was permissible, according to the definition given in § 155, to consider propositions already following from one part of several mutually compatible propositions *A*, *B*, *C*, *D*,..., e.g. from *A*, *B*, to be derivable from the whole set of them as well. We shall not proceed in the same way with the relationship of *ground and consequence*. We shall not be permitted to look upon a consequence that follows from several truths *A*, *B*, *C*,... as a consequence of the whole set of truths *A*, *B*, *C*, *D*, *E*, *F*,.... This is already clear from what has gone before, for if we could consider one truth *M*, or even one set of truths *M*, *N*, *O*,... as a consequence of *A*, *B*, *C*,... and then also as a consequence of *A*, *B*, *C*, *D*, *E*, *F*,..., it would be false that the same consequence always comes from the same grounds. The error in the opposite point of view manifests itself in a particularly striking manner in connection with the kind of grounds that relate to something *real*. For if we choose to consider a consequence of the truth, *a* exists, as a consequence of several truths, *a* exists, *b* exists, etc., as well, we would involve ourselves in out-and-out contradictions. For example, if it is true that a certain object *A* is cause of another *B*, but equally true that there is yet a third object (itself, for example, a mere effect of *B*), then it would have to be possible to say that the sum of *A* and *C* can also be regarded as the cause of *B*. And so on.

§ 209

Can a Truth or a Whole Set of Truths Be both Ground and Consequence in One and the Same Relation?

(1) There is no doubt about the relationship of *derivability* that there are propositions *A*, *B*, *C*,... from which certain other propositions *M*, *N*, *O*,... are derivable in such a way that the latter are also derivable from the former. We called propositions of that kind equivalent propositions in § 156. And just as there are propositions generally that have this relation of equivalence to each other, there are also *true* propositions in particular that are equivalent to each other, i.e. truths each of which can be derived from the other. Consequently the question arises whether the relationship of *ground and consequence* can not be just as reciprocal as that of derivability, i.e. whether there are not pairs of truths or entire sets of truths so constituted that we could regard each of the two parts compared both as ground and as consequence of the other with equal right? – I am inclined to give a *negative* answer to this question, both when ground and consequence are supposed to be understood as complete and when they are supposed to be understood as partial ground and partial consequent.

(2) It is already clear that truths *A*, *B*, *C*,..., which form the complete ground of truths *M*, *N*, *O*,... can not at the same time be regarded as the consequence of the latter, because the proposition that *A*, *B*, *C*,... are all true also belongs in the complete consequence *M*, *N*, *O*,.... So it would also have to be possible to regard *it* as a partial ground of truths *A*, *B*, *C*,.... But this is absurd.

(3) But even when we are speaking of partial grounds and partial consequences, it seems to me that their relationship can not be reciprocal, that what we call ground of something else can never be called its consequence as well. To be sure, we often speak of objects in the relationship of *reciprocal causation* to each other and explain this as a relationship such that by virtue of it each of the various objects involved in it is at the same time cause and effect. Now since according to my explication of the concepts an object is only the cause of another when the truth that states its existence is the ground of the truth that states the other's existence, it comes to seem as if there were truths of which one can

reciprocally be regarded as the ground of the other. On some consideration it becomes clear, however, that strictly speaking it is not the various objects involved in a reciprocal causation themselves which we have to regard as the really efficient causes in this case, but only certain *forces* immanent in them. Once again, it is not the things themselves, but only certain *changes* going on in them that we must look upon as the effects produced. And once this distinction is made all of the illusion as if the same thing that is cause of something else were also its effect collapses all by itself. Thus, for example, a pair of balls colliding with each other might both undergo a certain change in form, a sort of flattening, but the flattening of the one is not a cause of the flattening of the other, so that precisely what is the cause would also be an effect. Instead, the flattening of both of them is caused by the quantity of the motion with which both balls approach each other.

(4) I admit, nevertheless, that there are examples which look very much as if the same truth which is a ground of another could also be regarded as its consequence. One of the most plausible is the relationship between two truths of the form: "What is *A* is also *B*" and "What is not *B* is also not *A*." For it seems beyond doubt that one can regard the latter as a consequence of the former. But since the proposition, "What is not not *A* is also not not *B*," follows from the proposition, "What is not *B* is also not *A*," in the very same way that the proposition, "What is not *B* is also not *A*," follows from the proposition, "What is *A* is also *B*;" we must, if the second proposition is a consequence of the first, also say that the third is a consequence of the second. But since the ideas, not not *A* and not not *B*, are equivalent to the ideas *A* and *B*, we can easily translate the third proposition into the first, and consequently consider the first also to be a consequence of the second. The deceptive look of this objection disappears at once when we notice that the third proposition, What is not not *A* is not not *B*, is at *most* to be considered as equivalent to the original, What is *A* is also *B*, and not as identical with it. We may always look upon the third, more complicated truth as a consequence of the simpler ones, then, but who will choose to look upon the simpler truth as a consequence of the more complicated one? – We shall see, moreover, in the theory of inference that the double derivability assumed here is by no means completely correct.

§ 210

Can a Set of Several Grounds Be Regarded as the Ground of a Set of Several Consequences?

One distinctive feature of the relationship of *derivability*, with which I continue to compare that of ground and consequence, was this: when propositions *M*, *N*, *O*,... were derivable from propositions *A*, *B*, *C*, *D*,... and propositions *P*, *Q*, *R*,... from propositions *E*, *F*, *G*,..., the set of propositions *M*, *N*, *O*, *P*, *Q*, *R*,... could be looked upon as derivable from the set of propositions *A*, *B*, *C*, *D*, *E*, *F*, *G*,..., if only the ideas relative to which this derivability holds were the same on both sides and propositions *A*, *B*, *C*, *D*, *E*, *F*, *G*,... were compatible with each other. (§ 155 (22)) The question is, whether the like is also admissible in connection with the relationship of *ground and consequence?* In my judgment this is also to be denied. For anyone will see that one cannot call the set of several truths *A*, *B*, *C*, *D*, *E*, *F*, *G*,... the ground of several truths *M*, *N*, *O*, *P*, *Q*, *R*,..., at least not generally and without limitation, just because part *A*, *B*, *C*, *D*,... is the ground of part *M*, *N*, *O*,... and part *E*, *F*, *G*,... is the ground of part *P*, *Q*, *R*,..., if he takes into consideration the absurdities that would follow from that. Since the same truth or the same set of several truths can be a consequence in some respect and a ground or partial ground in another, it is possible that truth *E* or the several truths *E*, *F*,... are the consequence of *A*, *B*,..., further that the single truth *G* or the several truths *G*, *H*,... are the consequence of *C*, *D*,..., and finally that *M*, *N*, *O*,... follow from *E* or *E*, *F*, *G*, and *P*, *Q*, *R* from *H*. Now if it were permitted to combine grounds and consequences, of whatever sort they are, into sets, then we would also have to look upon the set of truths *A*, *B*, *C*, *D*, *E*, *F*, *G*, *H*,... as the ground of the set of truths *E*, *F*, *G*, *H*, *M*, *N*, *O*, *P*, *Q*, *R*,.... That is, the same truths would have to be capable of being part of the ground and part of the consequence as well. No one will admit that. To be sure, this absurdity would not come into the picture if we were only permitted to combine such grounds as do not stand in the relationship of ground to consequence with each other. But apart from the fact that we have nothing at all to justify us in asserting what we would have to presuppose for this purpose, that a truth's partial grounds could not also have the relationship of ground to consequent to each other,

I appeal to what everyone conceives by the words, ground and consequence, and indeed by the related words, cause and effect, in order to decide whether one could rightly say of a set of several grounds that it is a single ground, of a set of several consequences that it is a single consequence? Who does not feel that the connection between ground and consequence is much *more intimate* than it would be if the mere fact that some of the grounds and consequences are combined in thought were supposed to make only one ground and one consequence out of them? Nevertheless, an example may make this still clearer. The truth that God is omnipotent contains the ground of the truth that this world is the best. The truth that an isosceles triangle has two equal angles contains the ground of the truth that all of the angles of an equilateral triangle are equal. Now who would say that the ground (the simple ground) of the two truths, that this is the best world and that the angles of an equilateral triangle are all equal, lies in the set of two truths, God is omnipotent and an isosceles triangle has two equal angles? Who does not say, rather, that in this case there are two grounds and two consequences?

§ 211

Is there a Rank Order among the Parts of the Ground or of the Consequence?

As far as the relationship of *derivability* is concerned, there is neither a specific *rank order* among the various premises that belong to a conclusion nor among the various conclusions that follow from a single premise or from a set of several premises. That is to say, there is no rule that requires us to look on one of these propositions as the first, the second, etc. in order to find them to be in the relation of derivability mentioned. It is the same with the relationship of *ground and consequence*, when either the ground or the consequence or even both of them consist of a set of *several* truths. Here too it is thoroughly unnecessary to think of these particular propositions in a certain *rank order*, one of them as the first, another as the second, etc. They constitute a mere *set*. (§ 84) Scarcely anyone will dispute this unless he generally rejects the existence of an objective connection among truths, independent of our knowledge, such as I have previously described the relationship of ground and consequence to be. For if the truths which I call the *ground* of certain others are considered merely as items of *knowledge*, which are supposed to bring about the knowledge of the latter truth, there can always be controversy over whether the order in which we bring them before our consciousness is always entirely indifferent, whether the knowledge of the consequential truth can not at times be produced more easily by one order than by another. And it can be shown in the same way, on occasion, that one of the consequential truths will be more appropriately brought to the fore, the other left back. But who would maintain this of truths in themselves as well and say, for example, that the two truths I described in § 198 as the ground of the truth, that one should not be stingy with the necessities of life, constitute this ground only if we regard one of them as the first, the other as the second? – Even with the concepts of cause and effect, in which order we may think of the individual parts that combine to make up one or the other is a decidedly indifferent matter. We can define the nature of the formula for the resultant force equivalent to given external forces by observing that the result of this formula could not change no matter how we transposed the external forces.

§ 213

Can the Consequence of the Consequence Be Considered a Consequence of the Ground?

It is an important feature of the relationship of *derivability* that propositions derivable from certain conclusions can also be looked upon as conclusions from the premises from which the former are derived. (§ 155 (24)) Here there should be some discussion of whether this is equally true of the relationship of *ground and consequence*. Anyone who answers this question in the affirmative permits himself to call *C* the consequence or partial consequence of *A*, *A* on the other hand the ground or partial ground of *C*, if *B* is the consequence or partial consequence of *A* and *C* the consequence or partial consequence of *B*. I must confess that prevailing linguistic usage favors this. For we do not usually make any objection to calling one truth the consequence of another when it is really only a consequence of its consequence; and likewise we say that many a phenomenon is the effect of an object although we know that it is really only an effect of its effect. Thus we call all of the theorems established by the geometer consequences of those few basic propositions from which they seem to us to be derived, even in a very indirect way. Likewise, we call a consumptive's ultimate death an effect of the cold drink by which he contracted this illness once upon a time. Much as this point of view has in favor of it, however, I can still not refrain from doubting its correctness. It seems to me that the relationship of ground and consequence is of such a kind that one cannot say of a consequence of a consequence, just because it is a consequence of a *consequence*, that is the consequence of the *ground* of its ground, without altering the concept.

If we abide by the usual point of view, we must say, contradicting § 206, that the same consequence has several grounds. For we could now declare not only what really makes up its ground, but also the ground of this ground, etc., to be *grounds* of it. And so the same *effect* would have to have not only several *partial causes* but several *complete causes* as well. For suppose that the several realities *a*, *b*, *c*,... together contain the complete cause of realities *m*, *n*, *o*,... and that these contain the complete cause of reality *r*. Then the complete ground of the truth, *r exists*, must lie within the truths, *m* exists, *n* exists, *o* exists,... and the complete ground of these truths in the truths, *a* exists, *b* exists, *c* exists,.... Now if the ground of a ground is also the ground of its consequence, then the complete ground of the truth, *r* exists, also lies within the set of truths, *a* exists, *b* exists, *c* exists,.... Consequently, the existence of things *a*, *b*, *c*,... already includes a complete cause of the existence of thing *r*. Thus we have a complete cause of *r* both in *m*, *n*, *o*,... *a*, *b*, *c*,.... One and the same object, then, has more than one complete cause of its existence. Should we definitely grant this? Should we rather not be justified in saying that *several* causes (presupposing that they are not partial causes, but complete causes) must also produce several *effects?*

Finally let us take the case where the ground on which some individual truth *M* depends (as a partial consequence, say) is composed of *several* truths *A*, *B*, *C*, *D*,..., so constituted that each of them has a ground of its own, so that *A* (as partial consequence) is grounded in truths a, a', a'',..., *B* in truths b, b', b'',..., and so on. The first example I set up in § 198 was an example of this sort, in that the truth of the practical principle, one should not prefer one's own advantage, etc., certainly depends upon a ground entirely different from what supports the theoretical principle: one does not derive one's own advantage, etc. Now presupposing that in § 210 I was correct in maintaining that one can not immediately regard the sum of the grounds as a ground of the sum of their individual consequences, for that reason one can not say that the ground of the ground is always also a ground of the latter's consequence. There are cases where there is no such (single) *ground* of the *ground*, since the ground really consists of a *variety* of consequences, which also each have their *various* grounds. As a single truth, the truth M can also have only a single ground. If it should now be permitted to consider the total set of truths a, a', a'', b, b', b'',... as this *single* ground of M, then we would have to be permitted to regard this set as the single ground of the set of truths A, B, C,..., which is not the case.

§ 214

Can Every Truth Be Regarded not only as Ground but also as Consequence of Others?

There is no doubt about the relationship of derivability that any proposition can be regarded in both ways, as a premise *of* other propositions, and as a conclusion *from* others. So we raise the question, whether something similar also holds true of the relationship of *ground and consequence*, i.e. whether every truth can be regarded both as ground of other truths and as consequence of others?

The foregoing has really already settled the first part of this question, whether for every truth there are certain other truths that *follow* from it. For if what I have assumed several times is correct, that the proposition, "*A* is true," can be considered a genuine consequence of the truth of *A*, it is plain as day that for every given truth it can be proved that there is at least one truth which follows from it.

The other part is the only controversial part of our question: whether every truth can also be regarded as *consequence* (partial consequence, at least) of others, whether another single truth or a whole set of several can be cited that may be regarded as its ground? I suspect that this must be denied, or that there are and must be truths that have no further ground of their truth. But I must confess that I do not yet know of any proof for this which would satisfy me.

My suspicion is prompted primarily by some examples that seem to me to belong to the class of truths that have no ground. One such that strikes me in particular is the proposition, that *there is something, generally speaking*, or (as one would have to express this according to my view) that the idea of a something has denotation. Any other truth one might choose to give as ground for this one is rather itself to be considered a consequence of it or of others that follow from it, and so on.

But if one admits the existence of truths that have no further ground, no doubt one will find them so worthy of note as to require that they have a name of their own. I would call them *basic truths*, since they are only the grounds or bases of others, never consequences themselves. All the rest, on the other hand, I would call *consequential truths*.

§ 215

Is there More than One Basic Truth?

If whether there is even a single truth with no further ground is already a controversial question, it is all the more difficult to decide whether more than one of them can be assumed. I take the position, however, that this is also to be affirmed, because I do not comprehend how all the rest of the truths there are are supposed to follow from one single truth.

Note: Since Leibniz (*op cit.*) speaks of *vérités primitives* in the plural number, he must have been of the same opinion. On the other hand, when we read in a large number of logicians that the so-called *basic principle of identity*, or the principle that *what is, is*, is the highest truth of all, it appears that many of them looked upon only this truth as having no further ground.

§ 216

Does the Process of Mounting up from Consequence to Its Grounds Have to Come to an End for Every Given Truth?

When someone beginning with a given truth *M* asks for its ground, and having found it in the single truth *L* or in several truths *I*, *K*, *L*,..., now asks once again for the one ground or several this one truth or several truths have, and continues this as long as grounds can be given, I call this process the process of *mounting up from the consequence* to its ground. Insofar as there are truths that have no further ground of their truth, as was assumed in the preceding section, it is clear that this process of mounting from the consequence to its ground can not be applied with success at all in the case of certain truths (namely basic truths). And it is equally true that there are other truths (namely those that follow from such basic truths by way of a merely finite set of ground and consequence relationships) where we come to an end with this process after a definite number of repetitions. But it does not yet follow from what I have said so far that this has to be the case with every given truth, i.e. that we will always reach an end to the process of mounting from the consequence to its ground with every truth we apply it to. For even if there are truths that depend on no further ground, there could still be other truths constituted in such a way that they not only have a further ground themselves, but that the truths making up their ground once again have their own grounds and that this goes on to infinity. If I am not mistaken, every proposition that describes one of the changeable states or properties of a created substance provides an example of such a truth. Since every such state has a cause that lies at least in part in the preceding state, there is a series of causes here that goes on to infinity. But if there is an infinite series of causes, there must also be a series of grounds that goes on to infinity, since reality *M* can only be called a cause of reality *N*, if the truth that *M* exists contains a ground or partial ground of the truth that *N* exists.

§ 217

What the Author Understands by Subsidiary Truths

The truths we discover in the process of mounting up from the consequence to its ground, beginning with a certain truth *A* have such a noteworthy relationship to *A* by virtue of this very fact that it is worth the trouble of giving them a designation of their own. If it were permissible to regard the ground of a ground as the ground of the consequence (§ 213), we could call all said truths *grounds* or at least *partial grounds* of *A*. This is what they have generally been called heretofore, the first one discovered in this way the *proximate* or *immediate* ground, those following *more remote* or *mediate* grounds. But if one takes the view I presented as the more probable one in § 213, the most one can permit oneself is to use these terms with the explicit comment added that they are only being used *loosely*. It might be more to the purpose to use the term *subsidiary* truths, which leaves undetermined whether the truth we have before us is a mere partial ground or a complete ground, whether it is the proximate (proper) ground or a so-called more remote ground (consequently a ground in a loose sense). Truths related as a subsidiary truth is related to those truths with reference to which it is a subsidiary truth could be called *dependent* on each other or *subordinated* to each other, the latter in particular *dependent* on the former or *subordinated* to the former. Truths that have no relationship of dependence to each other could be called *independent*. I will not conceal the fact that this designation also has some inconvenience to it, however, since we shall find ourselves forced to take the term, *subsidiary truth*, in quite a different sense also.

§ 218

No Truth Can Be a Subsidiary Truth of Itself

No matter how numerous and varied the truths we discover in the process of mounting up from the consequence to its ground, beginning with a given truth *A*, may be, it is nevertheless certain that *A* itself will not reappear among these truths, or what comes to the same thing, that no truth can be a mere subsidiary truth of itself. I call this certain, even though the only proof I am capable of giving for it is the following. The relationship of ground and consequence must be of such a sort that knowing all of the subsidiary truths that belong to some truth *A* can produce the knowledge of this truth itself as well. But if truth *A* itself could be among the truths that belong to it as subsidiary truths, this would result in the absurdity that truth *A* itself belongs among the truths by knowing which we come to realize truth *A*.

§ 220

What Kind of Pictorial Representation Can Be Given for the Relationship that Prevails between Truths with Respect to Ground and Consequence?

From all I have said so far about the relationship of ground and consequence between truths, it is clearly to be seen that it is a very complicated one. So much so that it is very much to be wished that someone might provide us with a sensible representation, a spatial one, to ease the understanding of it. What I can say in this respect is not very much.

The *terms* introduced by previous linguistic usage are already pictorial. They indicate certain relationships that are to be encountered partly in time, partly in space. The images borrowed from *time* consist merely in the fact that we picture the various truths that make up the parts of a ground or the parts of a consequence in terms of the relationship between several things that are *contemporary* with each other. The relationship of the ground to its consequence, on the other hand, we commonly picture by way of the relationship of something earlier to something later. The partial grounds are imagined to be *contemporary*, the consequence *later*. In attempting to illustrate the relationship of ground and consequence in space, it was taken for granted that the parts of a ground of a consequence have to be represented as things which are *next* to each other (i.e. in the same horizontal plane) and grounds and consequences themselves as things which are one *above* the other (i.e. in different horizontal planes). But the same opinion does not always seem to have prevailed as to whether the ground should be represented as the one below and the consequence as the one above; or *vice versa*. For we encounter in the same language (e.g. in our [English] language) locutions of which one is in accord with one of these points of view and others in accord with the other. In itself, the word *ground* already means, in space, an object in the *lowest* position, on which another (the consequence) is *erected*. From this point of view, it also came about, then, that there was talk of whole scientific systems *supported* by some basic truth, or *built up* on it, or that they would immediately *fall down* when it was shaken. It is equally obvious that the other point of view is being expressed when certain propositions, rightly or wrongly thought to permit the derivation of all of a science's theorems from them, were given the name of the *highest* basic propositions in this science, when the transition from a consequence to its ground was called *mounting up* and the transition from the ground to its consequence *going down*, etc. Actually it does not appear that either of these two points of view has any discernible advantage over the other. All the same, since it is customary to read from top to bottom, while mentioning grounds before consequences is most advantageous to knowledge, to that extent it could suit our proposes better to assign the higher place to the grounds, the lower to the consequences. All the more so since in representing the relationship of *derivability*, which consequences so often have to their grounds, we are used to proceeding in a similar fashion, writing the premises above, the conclusion below. And just as we are used to separating the former from the latter by a horizontal line, this may also be done with grounds and consequences. Whatever else still needs to be noted in order to avoid an entirely unsuitable representation can best be shown by an example. In the diagram immediately following, *A* may represent the truth with which the process of mounting up from the consequence to its grounds began. *B, C, D, E, F, G, H, J*,... are supposed to be the subsidiary truths that belong to it.

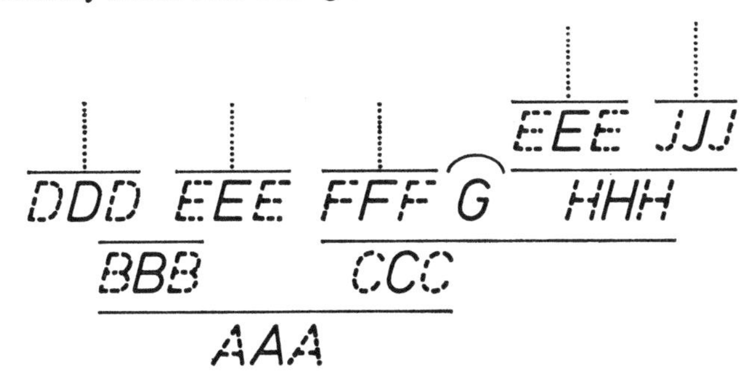

The proximate or proper ground of *A* is supposed to lie in the combination of the two truths *B*, and *C*, the ground of *B* in the combination of *D* and *E*, the ground of *C* in the combination of *F*, *G*, and *H*; the ground of *H* in the combination of *E* and *J*. The arc over *G* is supposed to indicate that the process of mounting ends here, i.e. that it is a basic truth. But since every truth that is a consequence could have some truths alongside it which follow from the same ground (§ 205) as it does, yet which are not always necessary to derive the consequence that is supposed to be drawn from the truth itself, the existence of such truths is supposed to be indicated by the letters written with dots. These are therefore missing with *G*. The horizontal lines above *D, E, F, E, J*,..., along with the dotted lines rising from them are supposed to show that the process of mounting has not come to an end with these truths. Since the same truth can occur as a subsidiary truth more than once, this kind of representation will require us to write it down more than once. I wished to remind us of this case by writing the letter *B* twice. But it is just this point that is also this representation's most important drawback. One and the same truth should reasonably appear in only a single place as well, not in several places, as if it were present more than once. If we wished to avoid this, we would have to make the following adjustment, say, in the diagram:

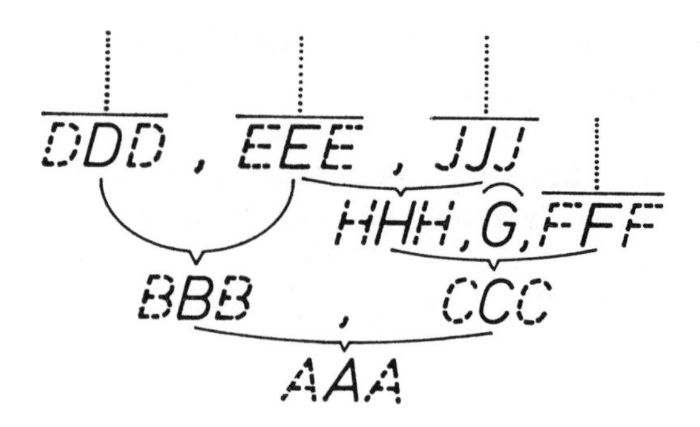

But who does not see that this would present a very complicated representation in cases where the same truth is repeated often? Let others invent a better one, then!

§ 221

Some Criteria for Determining whether Certain Truths Have the Relationship of Dependence

Nothing would be more desirable for logical purposes than the discovery of some universally applicable criteria for determining whether or not given truths have the relationship of ground and consequence or at least of dependence, i.e. whether or not we must regard one of them as the ground or partial ground of the other or as one of its subsidiary truths. But so far I have nothing but a few conjectures on this matter.

(1) In the first place it seems to me that the distinction between conceptual and intuitive propositions noted in § 133 admits of an important application here. Namely truths that contain nothing but pure concepts (conceptual truths) appear to be distinguished by the fact that they are dependent only on other pure conceptual truths, never on intuitive propositions. It may well be that intuitive truths (experiences) are very often useful to us in attaining *knowledge* of a pure conceptual truth, but the *objective* ground of such a truth can never reside in them. Insofar as there is any ground for it at all, it must always lie in other pure conceptual truths.

(2) It is my further belief that any pure conceptual truth on which another depends may certainly never be more *complex* than the latter, even if it may be no more simple. None of the particular propositions constituting the objective ground of a truth which is a pure conceptual truth may contain more simple parts than it has. For example, I would

not permit myself to acknowledge a relationship of ground and consequence in the very commonplace argument form:

Whatever has *a*, has *b*;
Whatever has *b*, has *c*;

Whatever has *a*, has *c*

if either concept *b* is more complex than *a* or *b* more complex than *c*. For in the first case the second premise would be more complex than the conclusion, and in the second case, the first premise would be. I would not choose to make this claim still more general, extending it even to *intuitive propositions*, since I believe I discern cases there that contradict it. For if *J* is some simple idea, referring only to a single real object, i.e. an intuition, then without a doubt there is an infinite number of truths that have idea *J* for their subject and some one of the properties belonging to this object for their predicate. Many of these properties (particularly those of which the idea is very complex) will be grounded in others. For example, if *J* is a *cry* I have just heard, the truth that this cry obliges me to look about for the person in distress it may have come from surely has its ground in certain other truths, e.g. that this cry has a resemblance to a call for help. It is easy to suppose, however, that in such a case the complete ground of a truth like *J* is *B* can not be situated in a single truth of the form, *J* is *A*, but only in the combination of this truth still another of the form, every *A* is a *B*. Now no matter how simple idea *a* (the abstractum belonging to concretum *A*) may be, the proposition, *J* is *B*, is always somewhat simpler than *A* is *B*, which nevertheless is prior to it as partial ground.

(3) If principles (1) and (2) are both granted to me and the conjecture already expressed in § 78 accepted, that there is only a finite number of simple concepts, it then becomes a simple matter to give a strict proof, with respect to conceptual propositions, of the claim asserted in § 214, that there are truths that have no ground. For if the number of simple concepts is finite, then the number of pure conceptual truths that do not exceed a specified level of complexity, e.g. do not contain more parts than a given *M*, is also only finite. But anyone who wanted to claim that every pure conceptual truth has a ground would also have to claim that every specific truth, like *M*, depends on an infinite number of sub-

sidiary truths. For if every ground once again required a ground, and so on to infinity, the number of propositions on which truth *M* depends is infinite. But if these subsidiary truths are all pure conceptual propositions and are not supposed to be more complex than *M* itself, this contradicts what has just been asserted. And of course that would settle it that there are *genuine basic truths*, at least in the domain of pure conceptual truths. If all of this is conceded to me, I will venture the following further claim. If we conceive of all pure conceptual truths with a complexity that does not exceed a specified limit (e.g. all truths that do not contain over 100 parts) combined in one collection, then it follows from what has been said so far that every truth there present which is not itself a basic truth finds its complete ground within this stock of truths, and not only its proximate ground, but all of its more remote grounds, right up to the ultimate grounds that have no further ground. We also already know that there will be truths *A*, *B*, *C*,... here which are completely without a ground Besides them, there will supposedly be some truths *D*, *E*, *F*,... that have a ground, to be sure, but do not have any relationship of derivability to this ground. (§ 200.) All of the rest of the truths *M*, *N*,... *Z* in this collection, finally, appear as consequences which are at the same time derivable from their grounds. Obviously, if we should attempt to change the order of these truths at will it would be an easy matter to assemble them into a great many groups such that when one part has been conceded all of the rest can be derived from it. And now I make the claim: if only we always observe the rule of never deriving a simpler truth from premises individually more complex than it is, the precise number of propositions we must assume (i.e. without having first derived them from others) will always turn out to be greater for every combination we might try than it is when we arrange the propositions according to their objective connections. So the relationship of ground and consequence is distinguished by the fact that it provides for the largest number of conclusions from the *smallest* number of premises, given only that the conclusions are not simpler than their premises.

(4) Another feature of this relationship could consist in the fact that truths *A*, *B*, *C*, *D*,... which form the ground of a truth *M*, are always the simplest of all those equivalent to it considered individually. Likewise, any truth to be regarded as consequence of several truths *A*, *B*, *C*,... is always the simplest of the individual truths equivalent to it. Consequently,

all of the truths equivalent to a particular *M* and yet more complex than it is can never be anything but consequences of it, and so on.

(5) Generally, it seems to me, truths *A, B, C*,..., which contain the ground of a truth *M*, at the same time derivable from them, must always be the *simplest set* of truths from which *M* can be derived, if the same ideas are always supposed to be regarded as variable and none of the premises is supposed to be more complex than the conclusion.

(6) And not only must propositions *A, B, C*,... be the simplest from which *M* is derivable, they must also be the most *general*. So, for example, I do not at all consider the proposition, Caius has obligations to God, a consequence of the two propositions, Caius is a man and All men have obligations to God. For these two propositions are not the most general ones from which the former can be derived, for not only men, but all rational and finite beings have obligations to God. So I believe I can assume with certainty that there is a relationship of ground and consequence in the type of argument in (2) only if *B* and *C* are *equivalent ideas*. In that case, the two propositions, All *A* are *B*, All *B* are *C*, are surely the most general propositions from which the conclusion, All *A* are *C*, is derivable.

(7) If proposition *M* has the relationship of *exact* derivability (§ 155 no (26)) to propositions *A, B, C*,..., with respect to ideas *i, j*,...; and if besides this propositions *A, B, C*,... and *M* are the simplest among their equivalents and no member of *A, B, C*,... is more complex than *M*; then we may well suspect that *M* has a genuine relationship of *ground and consequence* to *A, B, C*,..., and one such that whenever ideas are substituted for *i, j*,... which not only make *A, B, C*,... true, but also free of all redundancy, truth *M* is to be regarded as a genuine consequence of truths *A, B, C*,.... So I do not hesitate to declare the relationship of the propositions:

Every *A* is *B*,
Every *A* is *C*,

Every *A* is both *B* and *C*,

to be a relationship of ground and consequence, since not only is the latter proposition derivable from the first two, but they are also derivable from it and moreover are as simple as what has just been said requires.

Note: Some of what I have just asserted is surely almost universally accepted. For if it is said that a priori truths can not be proved by experience, what is meant is scarcely different from saying that the objective ground of a pure conceptual truth can not be situated in intuitive truths. And when it is said that the genuinely scientific presentation proceeds from the general to the particular, from the simpler to the more complex, this is surely advanced only because the more general, simpler truths are conceived to be the ground of the particular and more complex. Besides, this might be the most appropriate place to confess to my reader that at times the doubt occurs to me that the concept of the relationship of ground and consequence, which I have asserted to be simple in the foregoing discussion, is not perhaps complex, and in the end nothing but the concept of such an arrangement of truths that by virtue of it the largest possible number of conclusions can be derived from the least number of simple premises.

CHAPTER FOUR

ON INFERENCES

§ 223

Content and Purpose of this Chapter

Since as soon as we know that from certain propositions of the form *A*, *B*, *C*,... certain others of the form *M*, *N*, *O*,... are *derivable*, the truth of the latter can immediately be seen from the known truth of the former (§ 155), it is understandable that in the business of discovering new truths it is unusally important to know what are the various conclusions that can be derived from any given particular proposition or set of several propositions, when we may regard these ideas or those ideas as variable in them. It is definitely incumbent on logic, then, to make known the most general rules there are in this connection. But since it is possible in itself to derive from any given proposition (if only it is not generically false) an infinite number of others, and all the more so from any set of several (mutually compatible) propositions, it would be absurd to demand that all of the propositions derivable from one or more given propositions be accounted for. Besides, many of these conclusions are not worth enough attention to deserve being mentioned separately. It must be perfectly satisfactory to us, then, if logic makes known to us only those rules of derivation by which conclusions follow that are worth paying attention to in some respect. But even this would still be too much to demand. For in terms of the broad conceptual sense in which I have taken the word, *derivability* (§ 155), there are also derivations of such a kind that to judge their correctness or incorrectness it is necessary to have knowledge completely different from knowledge of logic. Thus the proposition, this is a figure with an angle sum of two right angles, is derivable from the proposition, this is a triangle, with respect to the idea, this. And the proposition, Caius has an immortal soul, is derivable from the proposition, Caius is a human being, with respect to the idea, Caius. For whenever other ideas are substituted for the ideas in these two pairs of propositions cited as variable that make the first of them true, the second will surely also become true. In order to see this, however, one most know the two truths, that the angle sum of a triangle is two right angles and that all human souls are immortal. Now since these truths do not concern logical objects at all, since they concern something quite different from the nature of concepts and propositions and the rules by which intellectual expositions are to proceed, no one will demand that logic teach derivations of this sort. All we can expect here, then, is the description of such types of derivation as can be seen to be correct from merely logical concepts; or, which comes to the same thing, in which nothing but concepts, propositions and other logical objects are spoken of. We would have such an example in the way in which the proposition, Caius has an immortal soul, can be derived from the combination of the two premises, that he is a human being, and that all human beings have immortal souls, in which the two ideas, human being and immortal soul, can be regarded as variables along with the idea of Caius. For in order to know that this derivation is correct, nothing else is necessary but the knowledge of the general truth that from any two propositions of the form, *A* is *B*, and *B* is *C*, a third can be derived, of the form, *A* is *C*. But this can be seen without knowing anything about the nature of human beings, of what is called dieing, or the like. Nevertheless, since it is understandable that different conclusions can be derived from different *kinds of propositions*, or, what means the same thing here, according to the different characters of the parts in given propositions supposed to be regarded as *invariable*; if we wanted to claim some completeness in this presentation, we would have to proceed in order through all of the kinds of propositions with which we became acquainted in the second chapter. We would have to consider them not only individually, but in all of the combinations that emerge when we bring together two or three or more of them in this way or that, and in each case investigate what kind of conclusions can be derived from them. But this would be such a huge undertaking, that I am even less capable of encompassing it here, when I do not yet know whether the different kinds of propositions I have set forth above are also recognized by others, or how else the list of them might be corrected or completed. Enough, then, indeed for many perhaps already too wearisome, that I should consider only the most important forms and the most noteworthy combinations, primarily of any two of them. Now as I see it, among these most important forms are those in which assertions are made about ideas or propositions themselves, if it is just these that are regarded as their variable parts. Before all, however, the form that makes its appearance when we regard everything in a proposition as arbitrary which can change only when it is supposed to remain a proposition. It would be completely superfluous and counter to our purposes, on the other hand, if we chose to consider anything here except *exact* inferences (§ 155 (26)). For supposing that we begin from the consideration of individual propositions and do not make the transition to a more complex combination without first having considered all of the simpler ones; the guide to the discovery of conclusions that would not require all of the premises assumed, but only part of them, has always already been provided by what went before. It is not equally necessary that we never mention anything but *simple* inferences (§ 155 (33)), for it can also be useful to consider complex inferences, if they occur very often, and imprint them on the memory. In any case, the error of bypassing completely some simple inference that occurred frequently and required nothing but logical knowledge to be understood would be more important. This would be an error, I say, not because of concern that someone who had not read of such inference would not know how to make them in cases coming before him, but because apart from all of the immediate advantages how many and what kind of simple inferences there are remains something worth knowing. Nevertheless it still seems to me to be too venturesome to reach a definite decision on this matter and so I will be satisfied to show that there are *more* of them than has been believed heretofore. Anyone who finds such investigations too dry can leave this chapter out, anyway.

Moreover, there can be talk of the relationship of derivability between given propositions $A, B, C, D, \ldots$ on the one hand and $M, N, O, \ldots$ on the other only if certain ideas in these propositions are thought of as variable. And in order to make a proper judgment as to whether certain propositions are really derivable from certain other ones or not, it is necessary to know which of the ideas in them are being regarded as variable. Therefore, I say once and for all here that I will regard as variable in the various propositions of which I am going to speak in the future only those ideas, but also all of those ideas in each case that I indicate by general signs (letters). For example, if I shall claim that from the two propositions, whatever a has, b has, and whatever b has, c has, the third proposition, whatever a has, c has also, can be derived, it is to be understood that the relationship of derivability holds between these propositions if all of the ideas indicated by the letters *a*, *b*, *c*, and no others, are the ones regarded as variable in this case. What I shall be representing in this way, by the words used, then, is really not the propositions themselves, which stand in the relationship of derivability to one another, but only the *form* these propositions must have, and consequently not the arguments themselves, either, but only their *forms* (the rules according to which they must be constructed). To indicate in abbreviated form that certain propositions *M*, *N*, *O*,... are derivable from certain others *A*, *B*, *C*, *D*,... I write the latter above and the former, separated from them by a horizontal line, below them. If I bring in several arguments one right after the other, in which some of the premises are supposed to be shared, for the sake of abbreviation I shall sometimes place them only in the first argument and indicate their presence in the ones following by a mere asterisk (*).

Although there is no rank order among the various premises in one argument, it will nevertheless be permissble to call one or another of them, which has greater *generality* or some other feature that distinguishes it above the rest or is only the first one to be cited, the *major* premise and the other one, where there are only two, the *minor* premise.

Note: Even if the distinction I have made here between two kinds of derivations, one requiring merely logical knowledge to be correctly judged, the other requiring many other items of knowledge, is not encountered in previous treatises on logic, it is nevertheless so well known that it is customary to call only the former kind of inferences, in contrast with the latter kind, *logical inferences*. And because of the complete lack of knowledge from any other source with which one is in a position to make these inferences they have often been looked down upon with depreciation.

§224

Some Rules by which Conclusions to Given Premises Can Be Sought out

Before I move on to the development of particular conclusions which can be derived from any given combination of premises, it could be useful to put forward some generally applicable rules concerning how conclusions to given premises can be sought out.

(1) If we have already derived certain conclusions *M*, *N*, *O*,... from the given premises *A*, *B*, *C*, *D*,...and we know how to derive some new conclusions *R*, *S*, *T*,... from one or more of them, singly or in combination, always with respect to the same variable ideas *i*, *j*,..., it will be permissible according to § 155 (24) for us to look upon these latter conclusions as conclusions also belonging to the premises laid before us. Only this sort of derivation by itself will not tell us, to be sure, whether conclusions *R*, *S*, *T*,... discovered in this way also have a relationship of exact derivability to the given premises *A*, *B*, *C*, *D*,..., even if the inferences by which *M*, *N*, *O*,... follow from *A*, *B*, *C*, *D*,..., and *R*, *S*, *T*,... from *M*, *N*, *O*,... should be exact inferences. (*Ibid.*, (34))

(2) If we have already considered propositions *A*, *B*, *C*, *D*,..., for which we are supposed to find conclusions, combined on another occasion with certain other propositions *E*, *F*, *G*,..., and we have derived conclusions *M*, *N*, *O*,... from that set, with respect to ideas *i*, *j*,..., we may say that propositions *M*, *N*, *O*,... become true whenever the truth of *E*, *F*, *G*,... is added on to the truth of propositions *A*, *B*, *C*, *D*,.... And so we may set up the following hypothetical judgment as a conclusion following from propositions *A*, *B*, *C*, *D*,... alone: If *E*, *F*, *G*,... are true, then *M*, *N*, *O*,... are also true. That the inference we arrive at in this way, namely:

$$\frac{A,\ B,\ C,\ D,\ldots}{\text{If } E,\ F,\ G,\ldots \text{ are true, } M,\ N,\ O,\ldots \text{ are also true,}}$$

is in fact different from the original inference, namely:

$$\frac{A,\ B,\ C,\ D,\ E,\ F,\ G,\ldots}{M,\quad N,\quad O,\ldots,}$$

becomes clear when we express both of them in the way we learned to express the meaning of any inference in § 164. The last reads: Every set of ideas substituted for *i*, *j*,... that makes propositions *A*, *B*, *C*, *D*, *E*, *F*, *G*,... all true also makes *M*, *N*, *O*,... all true. But the other one: Every set of ideas substituted for *i*, *j*,... that makes *A*, *B*, *C*, *D*,... all true also makes the proposition true that every set of ideas substituted for *i*, *j*,... that makes the further propositions *E*, *F*, *G*,... true will make propositions *M*, *N*, *O*,... true at the same time. Now in this way from any given inference with *n* premises others can be derived that have only $n-1$, $n-2,\ldots$, or indeed only a single premise. Thus, for example, from the two premises, *A* is *B*, *B* is *C*, there follows the conclusion, *A* is *C*. We shall be justified, then, in also deriving from the single premise, *A* is *B*, the conclusion. If *B* is *C*, then *A* is also *C*. And in this case if the inference by which propositions *M*, *N*, *O*,... follow from the set of propositions *A*, *B*, *C*, *D*, *E*, *F*, *G*,... is exact it is obvious that the new inference is too.

(3) If we have discovered that the negation of one or more of the premises laid before us, *A*, *B*, *C*, *D*,..., or even just the negation of a conclusion derivable from them, *M* or *N*, follows from certain propositions *H*, *J*, *K*,...; we will be able to infer that propositions *H*, *J*, *K*,... are never all true together whenever *A*, *B*, *C*, *D*,... are. So we attain the conclusion: the set of propositions *H*, *J*, *K*,... is not a set of true propositions only. But whether this conclusion has the relationship of exact derivability to its premises can not be determined from this way of deriving it. For even if it was an exact form of inference by means of which we derived from propositions *H*, *J*, *K*,... the negation of one of *A*, *B*, *C*, *D*,... or the negation of a proposition following from them by an exact inference, it still could be the case that the falsity of one of propositions *H*, *J*, *K*,... and consequently the correctness of the proposition that the set of these propositions is not a set of true propositions only is already derivable from a mere part of propositions *A*, *B*, *C*, *D*,.... Thus it is by means of an exact inference that from the two propositions:

All men are mortal (*A*)
Caius is a man (*B*)

the proposition:

Caius is mortal (*M*)

follows, and from the two premises, all men are immortal (*H*), Caius is a man (*I*) there follows in the same way a proposition from which the negation of the previous conclusion *M* can be derived. It is definitely correct, therefore, that we can derive from the first two premises the conclusion: "The propositions, all men are immortal, and Caius is a man, are not both true." But we may not represent this derivation to be an exact inference, for since the falsity of the proposition, all men are immortal, already follows from the first premise by itself, the whole conclusion requires no more than the first premise for it to be established.

§ 243

Continuation [Assertions About Numbers]

Finally if we combine several assertions of a singular idea with each other, we obtain the following inferences:

(1) Idea A is a singular idea,
Idea B is a singular idea;
____
If ideas A and B exclude each other, the set of objects represented by A and B numbers 2.

(2) Anyone can see for himself that this sort of inference can be extended to a larger number of premises. Thus, for example, from n premises of the form:
Idea A is a singular idea,
Idea B is a singular idea
Idea C is a singular idea, etc.

one can derive the conclusion: "If every pair of ideas in A, B, C, D,... exclude each other, the sum of the objects represented by them $=n$."

But since the condition that every pair of ideas in A, B, C, D,... exclude each other will be fulfilled if only every idea which can be formed by substituting for x and y any pair from a, b, c, d,... in the form, [Something] $(x+y)$, is objectless, our conclusion can also be expressed this way: "If every one of the ideas produced by substituting any pair from a, b, c, d,... for x and y in the form, [Something] $(x+y)$, is objectless, the sum of the objects represented by n ideas A, B, C, D,... $=n$." It is also easy to conclude that if the condition assumed here it drawn into the premises, the resulting conclusion (the consequent of the above conclusion) has the relation of equivalence to the premises.

(3) Idea A is a singular idea,
Idea [Not A] b is a singular idea also;
____
The sum of objects A and [Not A] b is $=2$.

(4) The idea of [Something] $(a+b)$ is a singular idea, and:
The idea of [Something] (non $a+b$) is a singular idea;
____
The set of all B is two.

Example. B = a human being who appeared on earth without being born of another; A = a being of the male sex.

(5) It is worth noting how this sort of inference can also be extended to several premises, and then serves to count objects. If we have n premises:

$[M]$ a is a singular idea,
$[M]$ $(b+n\, a)$ is a singular idea,
$[M]$ $(c+n\, b+n\, a)$ is a singular idea, and so on:

the conclusion follows from them: "The sum of objects $[M]$ a, $[M]$ $(b+n\, a)$, $[M]$ $(\mathrm{c}+n\, b+n\, a)$,... is $=n$."

VOLUME THREE

PART THREE

Theory of Knowledge

CHAPTER ONE

ON IDEAS

§ 270

Concept of an Idea in the Subjective Sense

I have already deemed it necessary in § 48, when I was attempting to arrive at an understanding of the concept of an *idea in itself*, to set alongside of it the concept of an idea in the *subjective* sense. I have already explained there what I understand by the expressions, a *subjective* idea, an idea *thought of* or an idea *had.* And so, after everything that was mentioned there, it could scarcely be necessary to remind anyone that I do not take the word, *idea,* by any means in that narrower sense in which the object of an idea is always something *absent* and must be of a more or less *different* nature from what the idea depicts it to be. This is the sense in which we say: "My brother, who is standing in front of me, I *see*; but my sister, who is absent, I can merely *have an idea of*," or "It is not a mere *idea* of mine, but the real thing." – It may very well be important in many respects to distinguish ideas of this sort, to which either no object corresponds whatsoever or at least no object immediately present, from other ideas, which have such an object and are produced in us just by its action on our capacity for ideation. Nevertheless, the purposes of logic require us to take the word, idea, in the broader sense that embraces both of these types and generally anything which can be a component of a judgment without constituting a complete judgment itself. Just as what was supposed to be understood under *ideas* in the preceding chapters was only *ideas in themselves*, *objective* ideas, even when the suffix, *in itself* or *in themselves*, was not added; in this chapter I am asking [the reader] to conceive the word, idea, always to mean a subjective idea, i.e. an *appearance* or *apprehension* of what I previously called an idea in itself in and by the mind of an intellectual being. It will also be opportune to remark here that we are used to calling our soul's ability to produce subjective ideas under circumstances its *capacity for ideation* or its *power of ideation.*

§ 271

There Is an Idea in Itself Attached to Every Subjective Idea

We must now consider some properties shared by all subjective ideas. One of the first is that for every such idea there is an idea in itself attached to it. The subjective idea consists in its appearance. This is already implicit in the concept we just gave of a subjective idea. But the following observation also makes this clear. If a certain subjective idea *A* is really supposed to be present in the mind of a thinking being, then the proposition, "this being has the idea *A*," is a truth, even if there is no one thinking this truth, i.e. it is a truth in itself. The components that make up this truth, then, are just so many ideas in themselves. But it is as plain as day that idea *A* must also be present in this proposition, not the subjective idea, however, but the idea in itself that corresponds to it. – The objective idea that corresponds to a subjective idea, i.e. the idea whose appearance in the mind *is* the subjective idea, could be called its *material*, as I called it in §48 and on several other occasions. But then neither the object to which the idea refers nor the parts of which it may be composed must be confused with this material. What I understand by *object*, namely, is completely the same in the case of subjective ideas as it is with what I called objective ideas, the something represented by these ideas. There is such an object, as we know (§66), for many ideas, but not for all of them; and with subjective ideas it is always the same as the one the objective idea attached to them has. But the subjective idea can never lack a material, while the objective idea has no such material, but is this material itself. Thus the subjective idea of a round square definitely has a material, namely the objective idea of a round square, but there is no object to which either of these ideas, the subjective or the objective, refers. It is even less permissible to mix up the material of a subjective idea with its parts, which are once again other ideas.

Note: Since the distinction between subjective and objective ideas has been touched upon only infrequently heretofore, there could be no felt need for a designation of the concept I have here indicated by the word, *material*. Moreover, I admit from the very beginning that this word is not convenient, and I would wish a more appropriate one would be found.

For since material is basically the same as what is called *matter* or *content*, these latter words have already been used, even by myself, to designate a completely different concept, so it is something to be concerned about that what was thought of in connection with the first would be what one is accustomed to thinking of in connection with the latter two.

§ 272

Every Subjective Idea Is Something Real, but only as an Attribute of a Substance

(1) A second property of all subjective ideas is that they are already as such something *real*, i.e. that they claim existence at a specific time and in the mind of a specific being. To be sure, it appears as if this is denied sometimes, when the statement is made, "that this or that is not real, but a mere idea." On closer examination, however, it becomes clear that for the most part all such an expression is intended to assert is that there is no external object corresponding to the idea. But one does not intend to say that this idea itself is never real of its kind. What we are doing in this case is making a kind of contrast between the mere idea and something real insofar as the idea is not *that* reality its object would be, if it should have one.

(2) Obvious as it is, at least to me, that every subjective idea as such already deserves to be called a *real thing*, still it surely only belongs to the class of realities that exist merely as properties* of other real things. For just this reason, we do not call them substances, but only *inherences*. Subjective ideas can only exist if there is an entity that has them, and they exist only as phenomena in that entity's mind. We may well call them properties which exist in such an entity; we may not apply to them, however, the term, substance, something real which exists (as we say) *in and of itself* (i.e. not just in something else). This seems to me to hold true so generally that I would take it to extend even to God's ideas; and so I can by no means agree with those who ascribe a kind of substantiality to the ideas God has, by maintaining that the substances in the universe are nothing but these divine ideas. In my opinion, the substances in the world

* The word being taken in such a broad sense that even very transitory *states* are properties (of the entity in which they occur, *at that time*).

have God's creative power to thank for their existence, not His ideas. His power of ideation embraces far more than what exists in reality. God has ideas of the merely possible as well, which it would be contrary to His purposes to elevate to reality. He even has ideas of the impossible. And these ideas in Him are the only things that already possess reality merely by virtue of His thinking of them, not their objects.

(3) Since we distinguish between two kinds of properties, intrinsic and extrinsic, or attributes and relations (§80), the question arises, to which of the two do a being's ideas belong? I believe they are *intrinsic* properties, for what could be more internal to a being than his ideas? – If it nonetheless sounds somewhat odd (as I admit) that the ideas a being has are *properties*, indeed *attributes* of it besides, the reason for this is probably only because ordinary linguistic usage attaches a certain accessory idea to both of these words, one which we have left out of their meaning as broadened for the purposes of science. Not everything an object has at this or that particular moment is ordinarily called a *property* or *attribute* belonging to it, only what it has over a longer period of time, what is enduring in it. Now since ideas for the most part have a very short duration and are then displaced by other ideas, it is understandable why they have customarily been denied the name of a property or attribute.

§ 285

Naming Our Ideas

(1) The characteristic of our minds considered in §283, that ideas once associated will mutually revive each other, permits us to use ideas that are easily evoked to produce others that would be more difficult to arouse but which are associated with them, sometimes in ourselves, sometimes in other people as well. We call an object we use for such a purpose, i.e. with the intention of reviving an idea associated with it in a thinking being, a *sign*. The objective idea corresponding to the subjective idea that is supposed to be aroused by the idea of the sign is called the idea *signified*, also the *meaning* of the sign. If the idea signified is a denotative idea, we are also accustomed to calling its *object* itself *what is signified* or the *meaning* of the sign. On occasion we also use the words, *sense* and *significance* as synonyms of the word, *meaning*. Yet a distinction can be drawn between the former words and the latter. The *meaning* of a sign means only

the idea it has been designed to arouse, which it in fact ordinarily arouses. Its *sense* or *significance*, however, means the idea it is intended to arouse in a particular case. Conceivably, someone can take a sign, out of ignorance, say, in a completely different sense or significance from its real meaning. Using certain signs to stimulate ideas in someone is called *speaking* or *talking* to him, in the widest meaning of these words. Attending to certain signs in order to learn what ideas their originator wants to awaken in us is called *reading*, in the broadest meaning. Really learning from them which ideas their originator intended to evoke is called *understanding* them. Mistakenly imagining that they had this or that sense, when they had another, is called misunderstanding them. Finally, a proposition in which it is asserted that the sense of certain signs is thus and so is called an *interpretation* of them. It, once again, can be merely thought or it can be expressed by symbols of another kind, when it is called an interpretation *in words*. If a certain object is used by all men to designate an idea, it is said to be a *universally applicable* sign. If it is used only by some men, e.g. only by a certain people, it is said to be a conventional sign with *those* men only. The set of all signs conventional to a man, especially those he uses to communicate his thoughts to others, is called that man's *language*. When the reason why an object is useful for the designation of a certain idea lies in human nature, i.e. in certain properties common to all of us, it is called a *natural* sign of that idea. If the reason lies in a merely accidental circumstance, i.e. one which does not apply universally, the sign is called an *accidental* sign If the accidental sign originated through an act of will i.e. if an object A serves to designate an idea B because we consciously and intentionally associated the idea of A with B and were determined to do so by nothing but an accidental circumstance (not true in all cases), the sign is called *arbitrary*. Thus, for example, wringing the hands could be considered a natural sign, threatening with the finger only an accidental one, and the way it was indicated that the vanquished gladiator should live or die at the Roman contests a merely arbitrary sign. Insofar as the sign is an external object, the thing signified something internal to us, e.g. a thought, a sensation, the sign is usually called an *expression*. So clapping the hands is looked upon as an expression (and a natural one) of joy. Most of the signs we use, especially for arousing ideas in other men, consist of phenomena which can be apprehended by the sense of hearing or of sight. The former might be called *audible* or *auditory* signs, the latter *visible* or *visual* signs. The former are usually produced by the organs of the mouth, i.e. by *voice*.

(2) Ideas that are compounded from others, e.g. ideas of entire propositions, we usually express only by employing the signs of the individual ideas that combine to make them up in a certain order. If we use audible signs, we have them follow one another in temporal succession approximately in the same way the simple ideas must follow each other in order to produce the compound idea. For example, if we want to evoke the idea of an entire proposition, we first bring forth the sign that serves to produce its subject idea, then we let follow the sign that means the connective concept, and finally the sign that is supposed to excite the predicate idea. If we use visual signs, on the other hand, for example certain figures, we place them alongside each other in a definite direction (e.g. from left to right, from top to bottom) and stipulate once and for all the temporal sequence in which they are supposed to be apprehended by the eye.

CHAPTER TWO

ON JUDGMENTS

§ 290

The Concept of a Judgment

The reader has already become acquainted with the concept I attach to the word judgment from §19 and 34. It follows very clearly from what was said there, especially in comparison with §48 that a judgment is something entirely different from a mere idea (even a subjective one) of a proposition. Someone who is merely thinking of the proposition, the sun is a ball of fire, does not therefore have to holding it to be true at all. If he does not do so, one can not say that he is making the judgment: "The sun is a ball of fire." Now we surely do make judgments, and just as surely we can ascribe to ourselves a capacity, or what means the same thing here, a power to do so. We give it the name, the *power* of *judgment.*

§ 291

Some Properties that Belong to All Judgments

(1) Just as we assumed that every subjective idea had some objective idea which constituted its material, we shall also have to assume that there is a proposition for every judgment. Precisely what the judgment consists of is the appearance of this proposition. Consequently we call it the judgment's *material.*

(2) Just as reality is ascribed to every subjective idea, as such, we must also ascribe reality to every judgment, as such, namely reality in the mind of that being who makes it and during the time he is making it. For since judgments consist of ideas, how could the parts have reality if the whole did not have it? But just as the idea is not a real thing existing in and of itself, but merely a thing that exists in something else, namely the soul, as an *inherence*, this holds true of judgments as well. It exists only in the being who makes it.

(3) If every proposition is composed of parts that are ultimately anal-

yzed into ideas, so must every judgment, as the appearance of a proposition, be composed of parts. And the judgment must contain as many corresponding subjective ideas as there are objective ideas that can be discriminated in the proposition that is the judgment's material. On the other hand, it certainly does not suffice, in order to be able to say that someone is making a judgment, merely that all of the ideas which belong to this judgment are present in his soul. For what a variety of judgments can arise from different combinations of the same parts!

(4) For ideas *a, b, c, d,...*, of which judgment *M* is composed, actually to be able to contribute to its formation, they must all appear at the same time, at least to the extent that before one of them has disappeared entirely, the other has always already begun. For if the opposite were the case, for example if we had already forgotten the subject idea of a proposition before we came to consider its predicate idea, it would surely not be possible for us to make the judgment. But just as the mere simultaneous existence of certain ideas in our souls still does not suffice for us to be able to say that they constitute parts of a single idea, this can not be sufficient for them to be considered as parts of a single judgment, either. Instead what is necessary for that is that they be combined with one another in a certain quite distinctive way. If the question is asked, in what does this combination consist, all I can say is that it must be a kind of mutual influence these ideas have on one another. But what the nature of this influence must be in order to produce a judgment, I am unable to define more precisely. Perhaps the only definition it is capable of is by way of the concept of the effect it is supposed to produce.

(5) No matter how great the influence our *will* exerts on the generation and nature of our judgments may be, that we voluntarily direct our attention to certain ideas and withdraw it from others, and so can change the entire course of our thoughts; whether we make a judgment or not still never depends directly on our will alone. This is a result, rather, merely of the sum total of the ideas immediately present in our souls, according to a certain law of necessity.

§ 292

What We Call a Single Judgment, and when We Say of Several Judgments that They Are Like or Unlike

It becomes clear from considerations similar to those in §273 that it would be most appropriate to regard the whole time during which one and the same objective proposition appears before our mind without any interruption, even if it is in various ways, as the duration of only a single judgment. And then to speak of more than one judgment only when either the same proposition appears before the same mind at separated times, or when it is making its apperance in the minds of different beings or when we are speaking of the appearance of different propositions. It is equally clear that we must also regard judgments we distinguish by virtue of the definition of number just hit upon as *unlike* if they are appearances of different propositions. If I speak of *like* judgments in the future, then, the least that has to be understood by that is that they are judgments which contain the same proposition. Depending on the circumstances, however, one will be able to extend this likeness to cover some other properties as well.

§ 294

Classifications of Judgments that Arise from Classifications of Propositions with the Same Names

Since for every judgment there is a proposition it brings to appear before the mind of a thinking being, it is understandable that there must also be a distinct class of judgments corresponding to every distinct class of propositions and that it will be permissible to name them after the latter. If we have distinguished simple and compound propositions, true and false, analytic and synthetic conceptual and intuitive propositions in the theory of propositions, and several other such classes, we can make precisely such distinctions among judgments as well. And anyone will know for himself what we understand by a simple or a compound, true or false, analytic or synthetic conceptual or intutitive judgment. Let it be noted of judgments containing an intuition that they are usually, especially if they are thought to be true, called *experiences* or *empirical judgments*.

§ 298

Does Every Judgment Leave a Trace of Itself behind after It Has Passed away?

If I am correct in maintaining (§ 283) that when any idea has ceased to exist itself, it still leaves a certain trace of itself, which is something real, it will be permissible for us to assume this to be true of all judgments as well.

(1) So far as judgments which have become clear to us are concerned, i.e. judgments of which we acquired some intuition, it already follows from this intuition, which has to leave a trace of itself behind after it has passed away, being a kind of idea, that we also retain a kind of trace of the judgment itself, namely in this surviving trace of its intuition. For since the nature of this latter trace must be such that the intuition from which it stems can be known from it, the judgment to which this intuition referred as its object must be capable of being known from it as well.

(2) But judgments that did not become clear to us could also leave behind a kind of trace of themselves after they have disappeared. Since every judgment as such consists of ideas in a specific combination and interaction, it can not cease to be unless its ideas either cease to be themselves or at least exit from the combination that previously existed. But since this combination is something real, it seems to follow from the very reasons for which I made a similar claim corcerning ideas (§ 283 (2)) that that combination could not cease to be without a certain trace of it remaining behind in our soul.

(3) We also proved above that every idea leaves a trace of itself behind from the remarkable capacity we have for remembering our former ideas under certain circumstances. But we also have such a capacity with respect to our judgments. We are also used to remembering them at times. This can only come about by virtue of our becoming aware, not of the judgments themselves, but of a trace of them that remained behind after they had disappeared.

(4) Surely, then, there are such traces, and not only some but all of our judgments leave certain traces behind them upon ceasing to be. According to their various characters, the greater or lesser liveliness and confidence we make them with, they may be stronger or weaker, of this kind or that.

§ 300

Mediation of a Judgment by Other Judgments

(1) It is now time to begin investigating the important question of how our judgments *come to be*. That they do not always originate in the same way is clear at the first glance. Surely the judgment that the ratio of the diameter to its circumference is irrational arises within us in an entirely

different way from the judgment, say, that "I am now feeling a pain," or one like it.

(2) The way a judgment comes to be is particularly worth paying attention to when it is *caused* by one or more other judgments, or as we can also say, *mediated* by them. I say that a judgment *M* is caused or mediated by one or more other judgments *A*, *B*, *C*, *D*,..., if what causes us to make judgment *M* lies in the fact that we have just made judgments *A*, *B*, *C*, *D*,..., too. In such a case, I call judgment *M* the judgment *caused*, *produced* or *mediated*, judgments *A*, *B*, *C*, *D*,... *causative*, *productive* or *mediative* judgments. The act of the mind by which it makes the transition from judgments *A*, *B*, *C*, *D*,... to judgment *M* is ordinarily called an *inference* or *inferring*, also *drawing a conclusion*, For this reason, judgment *M* is also said to be *inferred* or *derived* or a conclusion *drawn*, judgments *A*, *B*, *C*, *D*,... its *premises*, frequently (although quite improperly) its *grounds* also. Finally, our mind's capacity for finding the cause of affirming proposition *M* in the affirmation of propositions *A*, *B*, *C*, *D*,... is ordinarily called its *capacity for inference* or *faculty of inference*. Judgments that are not mediated I shall call *unmediated* or *immediate* judgments, no matter what may bring them about. So in order to say truly that there is a relationship of *mediation* in the sense just defined between judgments *A*, *B*, *C*, *D*,... on the one hand and judgment *M* on the other, it is not enough for judgments *A*, *B*, *C*, *D*,... merely to provide a *motive* for forming judgment *M*. Instead, the *complete* cause of judgment *M* coming to be must lie in the fact that they are present. Yet I do not understand this *complete* cause, to define it still more precisely, to be exactly that nothing else whatsoever would be necessary for judgment *M* to arise besides the existence of judgments *A*, *B*, *C*, *D*,..., not even the activity of one of the capacities of our own mind. All that I mean is that no further judgment besides the given judgments *A*, *B*, *C*, *D*,... would be necessary to bring about judgment *M*. For example, if we have made the judgment that Caius is a learned man, and it happens that very soon afterwards there comes into our head the judgment, "Learned men are often inclined to be vain," we may not yet consider this latter judgment to have been mediated by the former. It may well have been motivated by it, in that the concept of a learned man that occurs here led to that of vanity as one associated with it and thereby finally evoked the judgment "that learned men are often inclined to be vain." But the causes that lead us to

make this judgment (that lead us to suspect learned men of vanity) lie in experiences and judgments with quite a different content from the judgment that Caius also belongs to the class of learned men. This latter judgment is absolutely unnecessary for the former. On the other hand, if we express the judgment that Caius is a man and soon afterwards the second judgment that Caius is fallible, to be sure one can not say that the former was indispensable to the latter, but did not suffice all by itself to produce the second judgment. Instead, still another judgment, say that *all* men are fallible, would have to work along with it for that. So if we want to speak precisely, we shall not assert a relationship of mediation between the single judgment, that Caius is a man, taken by itself, and the inferred judgment, that Caius is then fallible, We shall assert it, rather, between the latter judgment and the combination of the two judgments, that Caius is a man and that all men are fallible.

(3) It becomes clear right away, from these and similar examples, that there is in fact such a relationship of mediation among some of our judgments and consequectly that there are judgment which can be said to be mediated. It is equally certain, however, that there are *unmediated* judgments. For in the last analysis the existence of mediated judgments can only be understood through the existence of unmediated judgments.

(4) A judgment M, which comes to be through the mediation of judgments A, B, C, D,... follows these judgments in time, yet only in such a way that they could not yet have completely disappeared when it comes into being. The rightness of this claim is confirmed in part by the observation each of us can make of our own judgments. It also follows in part from the relationship of contemporaneity which must hold between an effect and its causes.

(5) If certain judgments A, B, C, D,... have mediated judgment M in us on *one* occasion, it by no means follows that they have to precede it as its mediating causes on every occasion, that we are not able to form it except by mediation and by the mediation of precisely A, B, C, D,.... Still less can it be claimed that just those judgments that have mediated a certain other judgment M in *us* are also necessary for mediating it in every other thinking being. For example, on one occasion I can derive the judgment that Caius is a learned man from the fact that someone who knows him better assures me of it, and on another occasion I can reach that conclusion from a conversation with him myself, and the like.

(6) That it would have to be a very welcome thing if we could specify for each of our judgments whether it belongs to the class of unmediated or mediated judgments, and if the latter specify which judgments produced it in us, is an opinion we arrive at readily. But this is no easy matter. For not all of the judgments we actually make attain to clarity within us, i.e. we do not intuit them ourselves. Rather, the most of them remain obscure and we are therefore not in a position to remember them, or become conscious of them, still less to state them in words. Nevertheless they bring forth many effects in our mind and in particular they can mediate the origination of other judgments. That is why it often happens that we have derived a judgment from other judgments, and yet we do not know how to specify the judgments from which we have derived it. On the contrary, it seems to us as if it were an unmediated judgment, while in fact it has been mediated by many others. So we may not conclude merely from the fact that we do not know how to specify which judgments a judgment we have just made has come from that it is unmediated. At the utmost we could do this when closer consideration shows us that the judgment in question either could not be mediated, by its very nature, or could only be mediated if we were aware of this mediation.

(7) But to be able to judge this matter, we must first acquire a more precise knowledge of the various *ways* in which judgments can be mediated by other judgments. Now I believe that a judgment M can only be *mediated* by certain other judgments $A, B, C, D, \ldots$ if one of the following cases holds: (a) either propositions $A, B, C, D, \ldots$ must all be true and stand to proposition M in the relationship of the objective ground to its consequence, in the sense of § 198; or (b) proposition M must be *derivable* from propositions $A, B, C, D, \ldots$ in the sense of § 155, even if it is not precisely a consequence of them; or, finally (c) proposition M must have a definite degree of *probability* relative to propositions A, B, C, $D, \cdots$, if not complete certainty. If the correctness of this claim is supposed to become clear, I must show first of all that each of the three cases enumerated here does occur sometimes, and then that there is no other way to form a judgment through the mediation of other judgments.

(8) Now no one will deny that there are cases that belong to the way judgments arise described under (a), if he only grants the general point that truths do stand in the relationship of ground and consequence to each

other and that we humans are capable at least sometimes of seeing that there is such a relationship. For if the ground of truth *M* lies in truths *A*, *B*, *C*, *D*,... how should the knowledge of *M* be produced any more surely and completely than through the knowledge of *A*, *B*, *C*, *D*,..., which make up its ground?

(9) But there are also cases that belong under (b); i.e. believing certain propositions *A*, *B*, *C*, *D*,..., from which another proposition *M* is derivable with respect to certain ideas *i*, *j*,..., is sufficient, I say, not always, but sometimes to produce belief in proposition *M*. To be sure, it does appear as if arriving at this belief in proposition *M* required something more besides believing propositions *A*, *B*, *C*, *D*,..., namely the knowledge that *M* is derivable from *A*, *B*, *C*, *D*,.... And I willingly concede that this is actually so in many cases. For example, not everyone who believes the two propositions:

All *P* are *M*, and
Some *S* are not *M*,

thereupon goes on to the judgment: Some *S* are not *P*. Perhaps no one will do so unless the inference rule known in the schools as *Baroco* has occurred to him. But at least sometimes we can proceed immediately from believing certain propositions *A*, *B*, *C*, *D*,... to believing a proposition *M* derivable from them, without first having recalled that a proposition of the form *M* is derivable from propositions of the form *A*, *B*, *C*, *D*,... or even knowing it. That can be proved, I believe, from the fact that otherwise not a single inference could take place, i.e. not a single judgment grounded in other judgments as its cause. For if I could not know the truth of proposition *M* immediately from the known truth of propositions *A*, *B*, *C*, *D*,..., but needed first to have insight into the truth that a proposition like *M* can derived from propositions like *A*, *B*, *C*, *D*,.... then it is really the following two items of knowledge which must precede the knowledge of proposition *M*: (a) "Every set of ideas substituted for *i*, *j*,... which makes propositions *A*, *B*, *C*, *D*,... true also makes proposition *M* true and (b) proposition *A*, *B*, *C*, *D*,... are true as they stand before us, or ideas i, *j*,... originally present in propositions *A*, *B*, *C*, *D*,... are a set of ideas which make them true." – Now who does not see that the way in which judgment *M* arises from these two judgments is once again only a matter of knowing the conclusion from its premises? So if no conclusion is supposed to be capable of being known to be true from its premises without the rule on which the inference is made being known itself, the two judgments cited still do not suffice to produce knowledge of proposition *M*. This not only contradicts the presupposition that had just been made, but also shows that it would generally be impossible ever to arrive at one judgment from other judgments in this way, since basically one would need an infinite set of judgments. For what I just said about the necessity for adding a third judgment to the one before us states that the proposition we are supposed to derive from them as a conclusion really is derivable from them as a conclusion. Now a fourth judgment will also be required, which shows us that the three judgments just revealed really do relate to the one to be derived as premises to their conclusions, and so on to infinity.

(10) Finally, it also can not be denied that the third way of one judgment arising from others as listed in (7) does take place, i.e. that believing certain propositions *A*, *B*, *C*, *D*,..., which give another proposition *M* a mere probability can at times determine us to believe the latter. There are so many examples which demonstrate that when we believe the presuppositions *A*, *B*, *C*, *D*,... and they are present to our mind, we ordinarily not only make the judgment that *M* is probable (which can be derived from them by way of a perfect inference) but make the judgment *M* itself. So we do not say merely that it is probable that the surface of the earth will be covered with plants and flowers next spring, because this has happened so often, but we actually expect it, i.e. we make the judgment itself, that this is going to happen (with a greater or lesser degree of confidence).

(11) Now it remains to be proved that there is no fourth way in which one judgment can be produced by way of certain other judgments besides these three. The first thing that could occur to us here is this: do we not sometimes also proceed to a *new* judgment, *M*, from certain judgments *A*, *B*, *C*, *D*, ... we have made, just because we *imagine* that it has one of the three relationships mentioned to those judgments, even though it does not have it at all in reality? Does it not happen, and only too often, that we make *false* inferences, too, and so derive propositions from other propositions that do not follow from them at all in reality? – I reply that we do sometimes take a false rule of derivation to be correct and by means of it we do go on from certain judgments *A*, *B*, *C*, *D*,... to a new judgment *M*, which really does not follow from them. This is not to be denied. But if

our progression to the new judgment *M* does not result *immediately* from considering propositions *A, B, C, D,...*, but ensues only because we regard a false rule of inference as correct, namely because we mistakenly imagine that a proposition like *M* is derivable from propositions like *A, B, C, D,...*, then judgment *M* is not produced by way of judgments *A, B, C, D,...* alone, but only by way of them and the judgment (even if only implicitly added in thought) that a proposition like *M* can be derived from propositions like *A, B, C, D,...* . In other words judgment *M* makes its appearance because we make both of the following two judgments: "If propositions *A, B, C, D,...* are true, then proposition *M* is also true," and' "propositions *A, B, C, D,...* are true." But from these two propositions, proposition *M* is not just apparently derivable, but truly so. Thus the way we have arrived at judgment *M* in this case is no exception to the claim made in (7), but is included under the way a judgment comes to be as described *sub lit.* (b). In order to find a real exception, one must only maintain that we produce judgment *M* so directly from the consideration of judgments *A, B, C, D,...* that we do not need implicitly to bring in the false rule of inference at all. For this it would obviously be necessary for there to be a peculiar inclination in our mind for us to produce a judgment of the form *M* out of judgments of the form *A, B, C, D,...* although it does not stand in the relationship of ground and consequence to them, nor that of derivability, nor even probability. Now who would believe there is such an inclination in our mind? Its existence could only be maintained if it were shown that some judgments we make as carefully as possible are decidedly contradictory to other judgments formed with equal care. But no one has yet proved this. But if we were to fear the existence of such a fallacious inclination in our mind, without any definite proof of it, for a like reason we could not trust a single one of our judgments, i.e. we would have to make no judgments at all. That, since it would itself only be the consequence of some trust we placed in the judgment which moved us to make this resolve, would be a contradiction.

(12) If after what has gone before I should cite some examples of judgments I hold to be unmediated, I would say this: not all of the particular judgments falling under the following two forms have to be unmediated, but each of them includes *some* that are unmediated. (a) The one of these forms: is *I – have – the phenomenon A*; i.e. the subject of all of these judgments is the speaker himself (I), while their predicate represents the pos-

session of some phenomenon going on in him at that very time, e.g. an idea immediately present, a judgment being made, a sensation immediately present, an act of will, and the like. (b) The other form reads: *This* (*what I am intuiting right now*) – *is an A*; i.e. here the subject idea is an intuition immediately present in the being making the judgment which falls under the concept *A*, for example, this (what I am intuiting right now) is something red, or a sweet smell, and the like. It is true, to be sure, that we can also derive such judgments inferentially at times, as when we conclude from perceiving an action we have ourselves performed that we must have had an idea of it, or when we judge of an intuition present within us at the time that it is the effect of an object of such and such a sort. But it is impossible for all of the judgments with these two forms to be mediated, since every judgment of the kind presupposes another judgment of the same kind. In order to conclude from a judgment I have just performed that I must have had a certain idea neccessary for its execution, I must first make the judgment: I have accomplished action *A*. And if I am not supposed to make the judgment, Intuition *X* is an *A*, immediately, I have to derive it from a pair of other judgments: Intuition *X* is a *B*, and all *B* are *A*. – Let me now be permitted to call all of the judgments falling under one or the other of these two forms, insofar as they are made immediately, *perceptual judgments*. All mediated judgments, insofar as they contain an intuition, may be called *empirical judgments*, however, in the narrower sense of the word (§ 294).

(13) On some reflection it will be seen that there must be a great many more unmediated judgments than we have just described. In particular, there must also be various unmediated judgments among so-called *pure conceptual judgments*. For if the judgments in (12) were the only immediate judgments, then all the rest of the judgments we make would have to have arisen from them, directly or indirectly. Now it is totally inconceivable how this could come about. So, for example, from a proposition of the form, I have intuition *A*, we can at most draw the immediate conclusion, there must be some real thing that produces this intuition in me. But the judgment, "Every intuition that arises in me or any other finite being presupposes the existence of a real thing that produces it," can not result, mediately or immediately, from the judgment given, nor from any of the types of judgment cited in (12), nor from any combination of several of them.

(14) As was already noted in (6), among the judgments we derive from other judgments a considerable number of them are such that we are not

aware of the manner of their derivation. Now this may come from the fact that the inferences necessary have become so facile, by frequent repetition from earliest childhood, that at the present time we carry them out too quickly for our powers of intuition to be able to apprehend them in such flight. Or there may be other causes. Whatever they are, judgments of this sort deserve always to be distinguished from those we arrive at consciously, often with a good deal of effort. Now since the former are regarded as immediate judgments for the most part, because we are not aware of the way they are derived, for lack of another term, I will permit myself to call them by this name, too, but only in cases where there is no possibility of misunderstanding or with the added clarification, that this is in a *broader* or *looser* sense of the word. The judgments in (12) should then be called *immediate* in the *preferred* or *proper* or *strictest* sense. Thus I hold (to give an example) that we can not know immediately of the existence of a single (external) object, much less its properties or changes, e.g. whether it is in motion right now or not in motion, and the like. For example, if we make the judgment, there is a bird flying, I hold it to be an inferred judgment. But because we are not ordinarily aware of the premises on which it rests, for the most part we consider it to be an immediate perception and therefore claim that we did not infer the bird's flight but saw it directly. Certainly there is a great difference between the way we know the flight of the bird and the way we know of the movement of the hour hand on a pocket watch, for example. We know of the latter because we notice that the hand is at a different position now from the position we had perceived it to be in some minutes earlier, i.e. we are aware of the grounds from which we infer its movement. In order to give a linguistic expression to this difference, we say that we did not see hand's movement directly, but only inferred it. Now although these expressions are not quite precise, we may make use of them nevertheless, if only in cases where there is no possibility of misunderstanding or with qualifications which prevent it from arising.

§ 303

How We Do Arrive at Our Most General Empirical Judgments and how We Can Arrive at Them

It is a different matter with mediated judgments than with unmediated judgments. The way they come to be must of course be specified more precisely. Nevertheless, so as not to be too lengthy, I shall only take up a single class of them, namely those *empirical judgments* we usually hold to be *unmediated* because we rarely manage to have a distinct awareness of the inferences that mediate them. But we should take note that what we are concerned with here is not just to state the mere inference forms we go by in producing those judgments (which have already been discussed adequately elsewhere), but to describe these inferences themselves, i.e. to state the propositions of which they are composed that most deserve our attention. Finally, I call attention to the fact that we do not merely wish to describe how such judgments are actually formed, but how they *could* be formed and would *have* to be formed if the rational man wished to account for his judgments of this type as completely as possible.

(1) The greatest difficulty is to be found right where in my opinion the beginning should be made, however, in the explication of how the judgments arise in which we specify the *temporal relations* that obtain between certain phenomena going on in us, for example the judgment, "Idea *A* arose before idea *B* did," and the like. It seems to me that we could not say such judgments are entirely unmediated. We do not see immediately which of our ideas, sensations, etc. is earlier or later, and the like. We have to infer this from noticing certain of their properties or from other circumstances. It is still more obvious that we do not always infer the temporal relationships between the changes going on in our own minds from perceiving the temporal relationships that obtain among certain *external* changes. For true as it is that sometimes we determine the temporal relationships in which certain ideas or sensations succeed each other in our minds from the temporal sequence of certain external things we have noticed, this can not be universally the case. Instead, as a rule we have to infer the temporal sequence of the external changes from the temporal sequence inside of us. We could not possibly know what happens earlier or later in the external world if we did not know beforehand what takes place earlier or later in ourselves. For example, I may not know how to answer the question as to which of two objects in my vacation spot, this chapel or that house, I have seen for a longer period of time until I hear that the latter was built some years later than the former; but generally it is the other way around. I can only know the temporal sequence of phenomena in the external

world from the perceived temporal sequence of events in me. But from this it also follows that in order to explain this knowledge of temporal sequence within ourselves we can not have recourse to our knowledge of *spatial* relations. For since the locations of things are nothing but those features of them we must assume if we are to be able to explain how (with the powers we have allowed to them) they affect each other in precisely such and such temporal relationships, it is evident that spatial relationships can never be known unless certain temporal relationships are known first. This is not contradicted by the fact that we make use of space (certain motions perceived in space) to measure time. For any closer examination will show that the only way we finally decide whether a change in space belonging to a specific time has taken place or not is from a certain easily recognizable temporal relationship, namely the relationship of simultaneity. It is certain, then, that at least sometimes we make judgments about the temporal relationships the particular phenomena in our mind (ideas, sensations, etc.) have to each other on the basis of their intrinsic character alone. And the question is, how does this come about? I say that sometimes we must be able to know *immediately* whether an idea or sensation or phenomenon in general is *present* in our mind *right now*. For if we were never supposed to know even this immediately, there is no way to see how we could find it out by making inferences. For it is plain as day that no moment in time can be defined conceptually if one moment is not assumed as given in the first place. But it is equally obvious that if we do not want to regard the present moment as given, any past or future moment could be considered still less so. But a moment is only given for us by virtue of the fact that we designate it as the one at which a certain phenomenon is going on in our mind. And so we must know at least sometimes that an idea or a sensation is present in us right now. Yet no doubt this will only be able to be the case with phenomena in our mind that we have elevated to clarity. (§280)

(2) It is obvious, furthermore, that at least sometimes we must be in a position to know immediately that certain ideas *A*, *B*, *C*, *D*,... occur as parts of an idea we form for ourselves or a judgment we make. But if this is the case, we can conclude that these several ideas *A*, *B*, *C*, *D*,... have found a place in our soul at the same time. Here, then, is revealed a means by which we are capable of knowing the *simultaneity* of several ideas within ourselves, and it is self-evident that the like holds true of judgments,

sensations and other phenomena within our mind when they occur as parts of a whole that could only come to be as a consequence of their mutual influence (and so of their simultaneous existence).

(3) It is equally certain that sometimes we must know, immediately in part, partly from grounds quite independent of the present investigation, that it would be a contradiction to say that two phenomena *A* and *B* are simultaneously present in our mind. Thus, for example, it would obviously be a contradiction to want to say that we had made both a judgment *A* and the genuine negation of it at the same time. Therefore, if we are ever forced to assume the existence of certain contradictory phenomena in our mind, we conclude that they took place at different times. Here, then, is a means by which we know that certain phenomena did not exist in our mind at the same time, but at different times.

(4) If we know besides this that a phenomenon *A* is present right now, we know that the other phenomenon *B*, contradictory to it, which we also encounter within ourselves took place in the past. For example, if we know that we are making judgment Neg. *A* right now, we know that in case we also made judgment *A*, we must have made it in the past. In this way we are capable of determining the relationship by which one time is *past* with respect to another.

(5) Yet there could be more resources of this kind. For there is a variety of phenomena in our mind of which we can be sure that one of them must be later or earlier in time than the other. A wish must surely be present in our mind earlier than the feeling of its being fulfilled. We must have thought of the individual ideas that compose a compound idea earlier than we did it. We must surely also have made the individual judgments from which, considering them together, we have derived another judgment before we made it, and so on. Now suppose we know on the basis of the characteristic cited in (2) or any other characteristic, that a certain phenomenon in our mind was simultaneous with the wish, another with the feeling of fulfillment. We could then make a judgment straightway about the temporal relationship between these two other phenomena.

(6) If phenomenon *A* exists in our mind as a component of several phenomena *M*, *N*, *O*,..., or if we know in any other way that it took place simultaneously with each of them and if we also know that in some part phenomena *M*, *N*, *O*,... are contradictory, and so surely did not take place at the same time, or we draw this conclusion on other grounds, we

discover that phenomenon A existed at *several* times and so (depending on the circumstances) it must have either existed without interruption throughout these times or have been revived several times in our soul.

(7) Since what has been said also holds true when phenomenon A is composed of several individual phenomena, e.g. ideas, sensations and the like, we can know in this manner that we have had several ideas (or phenomena) *at the same time on several occasions*. For example, I can become aware in this way that I have already had the idea of certain colors (namely those of a rose) and the sensation of a certain odor (that of a rose) at the same time on several occasions.

(8) Now we know some ways that judgments in which we assert that an idea or phenomenon in our mind belongs to an earlier time can make their appearance. There are still more of them. Thus we can conclude at least with probability that a certain idea must have been present in our mind earlier and probably on several occasions if we produce it more easily and with less concentration than is the case with ideas of a similar composition. So far as judgments are concerned in particular: we can suspect that we must already have made the judgment, A is B, if it can come to mind without our giving the reasons for our having made it, even though it obviously does not belong to the class of judgments that can be known immediately. Thus I can correctly draw the conclusion that I must already have made the judgment $\sqrt{2}=1.414...$ if it presents itself to me without having been calculated.

(9) We need a definite period of time to produce a certain sum of ideas, judgments, or whatever other phenomena in our mind; and to form the same sum of ideas or judgments a second time, we consume about the same time or a somewhat shorter time. The same thing holds true when these phenomena constitute a certain sequence, i.e. when there is a definite order in which they follow each other in time, as is the case, for example, with the ideas, one, two, three, four, etc. in the business of counting. Now let $a, b, c, d, e,... y, z$ be such a sequence of phenomena in our mind. And let a certain phenomenon A be simultaneous with the first of them, a, and a certain phenomenon B be simultaneous with the last of them, z. We know then that the two phenomena A and B are separated by a space of time which was necessary for producing the whole set of ideas $b, c,... y$. If after the conclusion of the series we begin it again, and so that we have idea a a second time along with its final idea, z, and when we ar-

rive at its end, *z*, a second time, phenomenon *C* enters into our mind, then we know that phenomena *A* and *C* are approximately twice as far apart as *A* and *B*. And so on. Here we would have a means of evaluating the relationship of different temporal intervals to some extent. We actually use this means when in order to measure the space of time between phenomena *A*, *B*, *C*, *D*,... we count between them, and the like.

(10 If we have an intuition *A*, we must know whether it belongs to the class of outer or of inner phenomena (§286), i.e., in the first case, that the change in our soul we are intuiting has to be thought of as the result of the effect of certain external objects on us. We learn in this way that there are also *external* objects which have the power of producing intuitions in us.

(11) If we have *like* intuitions, i.e. if the same concept the one falls under also applies to the other (and we must be capable of knowing this immediately on at least some occasions, §300), we conclude, not with complete certainty, but with probability that the objects which produced them could be alike. This conclusion is not certain, since unlike causes can also produce like effects on occasion. For example, the intuition of yellow can be produced by the action of a yellow object on our eye on one occasion and by the humour in our eye itself, when it is unnaturally colored, on another occasion, and the like. As long as no special circumstance enters in which permits us to suspect an exception, however, it is more probable that like effects arise from like causes.

(12) Now if it comes to pass that we have certain intuitions which fall under concepts *A*, *B*, *C*, *D*,... simultaneously on several occasions, and none of them ever in the absence of the other, or very infrequently, we conclude with a great probability that it is one and the same real object that produces these various intuitions in us. We may therefore ascribe several powers to it at the same time, namely a power to produce intuition *A* and another power to produce intuition *B*, and so on. By means of this sort of inference we gradually discover what various kinds of *powers* the things around us possess. For example, if we have already had the intuition of a certain rosy color and of a certain very pleasant smell at the same time on several occasions, we conclude that the very same object that produced the one intuition in us was also the cause of the other, or, what comes to the same thing, that the rose has an odor.

(13) We still have to explain, however, why the object we have just assumed to be present, which arouses said intuitions *A*, *B*, *C*, *D*,... in us at

times, does not affect us continually. This can only be explained by the fact that this object changes, or that there are other objects that interfere with its effect upon us, or that its spatial relationship to us changes.

(14) If we learn that in most cases something happens when we have previously *willed* it, we conclude that our will is the cause of this result and that consequently we have the power to bring about such results. By this means we become acquainted with many kinds of powers inherent in *ourselves*. In particular, we become aware that under certain circumstances we have the power to procure for ourselves the intuition of a certain object or to suspend it at will. Thus, for example, when we have a rose in our hands, we can sometimes produce the sensation of its smell, sometimes bring it to an end. Since this presupposes (13) that there is a change either in those objects themselves or at least in their spatial relationships to us, or in some other thing around us, we discover that we have the power to produce at will a great many changes in the external objects round about us or in their spatial relationships to us.

(15) We could not immediately affect all the external objects we are capable of affecting by our will as a consequence of this experience. We can affect many of them only by way of changes we have first produced in others. There must nevertheless always be some external objects on which we operate directly. The sum total of these is ordinarily called the *organic soul*, and if everything which has an *organic* connection (§394) with it is counted in, it is called our *body*. It will soon become plain how we become more and more informed about the existence and properties of this body.

(16) If we *never* have certain intuitions $M, N, O, \ldots$ (i.e. intuitions which fall under the concepts $M, N, O, \ldots$) without certain other intuitions $A, B, C, D, \ldots$ preceding them in time, we conclude that the things or changes necessary to bring about intuitions $A, B, C, D, \ldots$ are a *condition* of those things or changes which pertain to bringing about intuitions $M, N, O, \ldots$. If intuitions $M, N, O, \ldots$ *follow* directly upon the precedent intuitions $A, B, C, D, \ldots$ every time, we conclude that the things or changes that serve to bring about intuitions $A, B, C, D, \ldots$ contain the *sufficient ground* or *complete cause* of those things or changes which lead to the production of intuitions $M, N, O, \ldots$ coming to be. If, on the contrary, intuitions $M, N, O, \ldots$ never occur without $A, B, C, D, \ldots$, but these do sometimes occur without being followed by the former, we know that the things or changes which cause $A, B, C, D, \ldots$ are not yet the complete ground, but only a

partial condition of the things or changes by which intuitions *M*, *N*, *O*,... come to be. In this way we can make judgments about the *means* and *conditions* which lead to the occurence of many an object or change in it.

(17) If phenomena *M*, *N*, *O*,..., which I want to bring forth in my mind, do not always make their appearance, but only when I have first succeeded in preceding them with certain others *A*, *B*, *C*, *D*,..., I conclude that my soul does not have the power to cause the changes involved in bringing about phenomena *M*, *N*, *O*,... directly, but only by *means* of those changes signaled by phenomena *A*, *B*, *C*, *D*,... . Consequently it could only cause them when I had the opportunity to bring about *A*, *B*, *C*, *D*,... as well. For example, if I want to secure for myself the taste of an apple, but I accomplish this only when I have first had before me all of the intuitions caused by looking at an apple, bringing it into my mouth, and the like, I conclude that I can not produce that taste directly, only indirectly. If, on the other hand, there are certain phenomena in my soul, *M*, *N*, *O*,..., which ensue whenever, or almost whenever, I will it, I conclude that the objects necessary for bringing them forth can be directly affected by my soul or that they are most intimately connected with objects it affects directly. I therefore count them to my own *body*.

(18) I have intuitions of colors and I am not exactly able to produce them whenever I will to, but I am able to eliminate them (namely by just closing my eyes). From this I infer that these intuitions do not occur in me directly, but by the mediation of an object dependent on my volition, i.e. a part of my body. I call this part my eyes. In a similar way I learn that I also have instruments by which I receive odors, others through which I receive intuitions of taste, still others for intuitions of sound, and so on.

(19) If a rose stands before me, the sight of it arouses, by the well-known laws of association of ideas, a memory of that pleasant fragrance I have often had at the same sight; perhaps a desire for renewing this sensation will also be aroused. But this desire will not be satisfied until (by chance, say) I should pluck it with my hand and bring it to my nose. In doing this, the change taking place with my hand will be *seen* by me, i.e. it brings forth in me various intuitions I must explain by assuming a change going on with this part of my body. Now because the satisfaction of that desire (to smell the rose) follows upon this change, I conclude that that part of my body which has just brought forth the various intuitions mentioned is equally the cause (partial cause) by the change in it

(its movement) of the satisfaction that followed. In this way I become more and more familiar with the services my hands and the other members of my body can perform for me.

(20) I experience daily very many sensations, some pleasant, some unpleasant, which are simultaneous with some changed look of this or that part of my body and become livelier or weaker because of influences affecting this part, and the like. I conclude from this that the cause of those sensations lies in the alteration of this part. For example, I can have a very bad toothache without knowing in which tooth the cause of the pain lies until I finally know it from the change in this pain that results from touching one of them.

(21) All of these pieces of knowledge become incomparably more specific when we begin to make judgments about the *spatial relations* that external objects have to ourselves, partly, and partly to each other. I imagine this comes about approximately in the following way, or could come about so, if we would proceed quite strictly. For every change in my intuitions referring to external objects, I must presuppose as cause a change either in me or in some external objects, either in their intrinsic properties or their spatial relationships to me, or (which is more correct) in all of these things at once, only in varying degrees. For since my intuitions, when they refer to an external object, represent changes produced in me by the action of an external object, their nature is determined by my nature and the nature of the external object and the spatial relation between us. Without anything being changed in one of these circumstances, nothing can change in the intuitions, either. If one of the circumstances identified is changed, however, because of the mutual influence all things in the world have on each other, really all of these circumstances are changed, only the change in one or the other can be so slight that we do not need to take it into consideration. Now if the same set of intuitions A, B, C, D,... is produced in me from time to time, I conclude (as mentioned in (11)) that it is one and the same object that brings them forth. So far I do not know whether in the meantime, while I did not have intuitions A, B, C, D,..., it had changed, perhaps ceased to be, and did not return to existence until the very moment I had them, or whether the way intuitions of it come and go depend more on changes in its spatial relationships to me. But if I now venture upon something different, which usually causes a certain sum of intuitions A, B, C, D,... to disappear without returning

i.e. something different which otherwise usually causes objects to change, without this occurring here, then I suspect that the disappearance and reappearance in this case does not come from an intrinsic change in the objects concerned itself, but from a change in its spatial relationships to me, or from another object's having come between. But to it I must now ascribe a certain power to persist in its state (or in the combination of its parts), a certain solidity. For example, when I take a piece of wood in my hands and handle it in various ways, from time to time (namely, whenever I chance to bring the object back into the same position between my fingers) I obtain the same intuitions, if it is solid; while in the case of a soft or fluid object the sensations it causes in me change continually without the same ones I have already had ever returning. (We become convinced in a similar fashion that even our hands, or rather their particular members, are such solid objects.)

(22) If I obtain two sets of intuitions, $A, B, C, D, \ldots$ and $M, N, O, P, \ldots$, taking one to be the effect of object X and the other as the effect of object Y, either simultaneously or one however quickly after the other, sometimes $A, B, C, D, \ldots$ after $M, N, O, P, \ldots$, sometimes the reverse, I suspect that the locations at which these two objects exist are very close to each other, or that the objects are in *contact*. This inference is not certain; no matter how simultaneously a pair of stars may strike our eye, they may nevertheless be very far apart. But so long as there is no situation pointing to an exception, that presupposition has probability. This is especially true when the objects appear almost simultaneously not only to the eye but to other senses as well, e.g. to feeling, hearing, smell, etc.

(23) If a certain phenomenon I ascribe to the action of an external object becomes stronger every moment, or weaker, in the opposite case, I may suspect that in the first case the object is coming *closer* to me, in the second that it is moving *farther away*.

(24) If I have found objects $M, N, O, P, Q, \ldots$ and A all to be solid, and I have perceived two different parts in object A, a and α, and now notice that at one time a is in contact with M, and α with N, at a following time a is in contact with N, and α with O, at a third time, a with O and α with P, etc., I conclude that the distances between things M and N, N and O, O and P, etc., are approximately equal. The object A I can use for this purpose can be my own hand, for example, and parts a and α a pair of prominent points on it (thumb and little finger, say).

(25) If object *A* passes out of proximity to *M* into proximity to *N*, and from there to proximity to *O*, and so on, and distances *MN*, *NO* are equal, and the times that pass in between are also equal, I know that objects *A* and *M*, *N*, *O*,... are *moving uniformly relative to each other.*

(26) If an object *B* had not yet changed its relationship to certain other objects *M*, *N*, *O*, *P*,... (among them my own body, for example), another object *A* changed its relationship, on the other hand, and thereby gradually came into contact with *B*, and since this happened *B* has also changed its relationships to those other objects *M*, *N*, *O*,..., I conclude with probability that *A*'s contact with object *B* (i.e. the impact transferred to it) is the cause of *B*'s motion that has now started. In this way I learn that the members of my own body, e.g. my hands, can produce motion by impact.

(27) I also often notice that so long as an object stands in a certain relationship with another object, e.g. my hand (is supported or held by it) it does not change its spatial relation to me. I sense at the same time a certain strain, which becomes the more burdensome the longer it lasts. And so I can conclude that this strain is precisely the cause that maintains the object in that stable state, from which it further follows that there must be a tendency in it to change its place in a certain way. So I learn that certain objects *attract* or *repel* each other and the general law of gravity.

(28) Sometimes I notice that a certain object *A* (e.g. a ball) can be in contact with another object *B* (e.g. a flat table) in the most diverse ways, so that it always remains at rest even though the parts in which the two objects touch, at least on *B*'s side, are always different. I learn furthermore that it apparently requires the same exertion and time to bring object *A* from the place on object *B* where it initially rested to any other place equally distant from it. Now if object *A* is at rest and I give it a push with any other object (e.g. my hand) so that in moving it remains continually in contact with *B* (it is not bounced over *B*), I can conclude that this motion, insofar as it is caused by the push alone, lies on a plane. I can also apprehend the motion described by *A* in this case with my eyes. And if I now find that an identical line is described, from whatever point and in whatever direction I initiate the motion, it follows that the surface of object *B* on which these motions all take place can be either the surface of a sphere or of a plane, the lines either arcs of a circle or straight lines. But my eye

really offers me an easy way of distinguishing between a straight line and any other (which means in this case distinguishing a series of small objects differentiated by their color that lie in a straight line from such as lie in a curved line). The former has the property that from whatever standpoint I look at it it causes an image on the retina which coincides with the one brought about by a mere piece of it when the distance is changed. So the ideas the entire line brings forth and the ideas produced by a mere part of it at a different distance must be identical with each other. And I know every line in which I observe this property to be a straight line. But once I have seen a number of straight lines, the distinguishing feature of their appearance soon becomes so well known that I can then decide without much effort from the mere look of it whether a line can be counted as straight or curved. And so I find that in the case that was mentioned at first the line is straight and the surface of object *B* a plane. Soon I also learn to judge whether a surface is a plane surface or a curved one merely from the light and shadow on it. Once I know how to distinguish between straight and curved, I also know how to measure distances. For if lines *mn*, *no* are straight lines and are put together in such a way that *mo* is also a straight line, it is well known that distance *mo* is the sum of distances *mn* and *no*. Finally, since all spatial relationships can be defined in terms of distances, as soon as I known how to estimate them there is no relationship in space it would be impossible for me to define.

(29) Yet one must not believe that the way of estimating these relationships described above is the only one possible and that the sense of sight is therefore indispensable for it. We could define such relationships even in its absence, although with greater difficulty. It can be seen from (21) how mere feeling (the sense of touch) can instruct us as to whether a body is solid or not. We can learn by way of this same feeling that there are also bodies in which the parts in contact resist complete separation and dissolution, but still can be altered in their spatial relationships in a great many ways without much effort, and in particular brought closer to each other. We customarily call such objects *flexible*. Now if we grasp two parts of it not immediately in contact with each other (with our two hands, say) and strive to move one of them in a direction it resists, we know that the distance between the two parts is now as great as it could be without the connection between them being broken. Now if the individual parts of the object are small, e.g. so that we can grasp them comfortably, but

the entire object significantly larger (longer), then in the situation into which we have brought it it represents a straight line (strictly speaking a filament). By handling it we can learn how such a line feels to the touch. If some other object we handle offers the same sensations at every place, it can be concluded that its surface is either plane or cylindrical or spherical. Which of these three is the case can be decided in various ways. For example, if we have an object with parts that remain in whatever disposition we choose to impose on them, we could only press the surface under investigation into it. If the surface produced by the impression is like the given surface (produces the same sensations upon being handled), the latter would be a plane surface. In the contrary case, it would be either cylindrical or spherical, the first if we do come upon parts somewhere that cause a different sensation when handled; the second if this is not the case. If two plane surfaces meet in an object (as at the edge of a prism), the parts where this happens (where the changed sensation enters) lie in a straight line. By handling it frequently and attentively we can learn the distinctive way straight lines feel, and so on.

(30) Once we have become acquainted with a few objects which apparently maintain an invariable shape and size, it will become easier day by day for us to judge of the shape, size and distance of other objects with their help, particularly if we have the sense of sight. We can do so by making use of the following rules in particular, that (a) every object concealed from our eyes by another object must be farther away from us than the one that conceals it; (b) every object that does not change in size must be the farther away from us the smaller the visual angle under which it appears to us. (We measure the size of this visual angle, however, by comparison with another object that changes in neither size nor distance), that (c) every object that does not change in size and shape, must be the farther away the more obscure it appears to us under otherwise similar circumstances, the less we are able to discriminate its parts. And so on.

(31) If someone should ask how the child, how even animals acquire knowledge of these relationships in space, since the latter surely can not manage the inferences I have alluded to here, my answer is that the comprehension of spatial relationships we encounter in children and animals scarcely requires more for its explanation than the well known law of association of ideas, or the expectation of like cases under like circumstances that arises from it, especially if we take some of the effects of instinct to

be assisting, too. We say that an animal apprehends relations in space when we observe that it avoids the stone, for example, it finds in its way, or seeks out something it wants to eat to the right or to the left depending on whether it actually is to its right or its left. But all of this is sufficiently explained by mere association of ideas. It avoids the stone because it has already experienced a hindrance in its course many times when it has had intuitions of the very kind it now has of this stone. It turns to the right in order to capture its prey because it has had to turn to the right to get hold of it before when it had the very same intuitions as it does now, and so on. The child goes a step further, when it is old enough to make some of its own ideas clear to itself, by elevating the expectations generated by association of ideas that are successful into rules for future cases and consciously acts in this way.

§ 306

Survey of the Most Noteworthy Activities and States of Our Mind that Concern the Business of Making Judgments

The end of this chapter may once again be the most appropriate place for bringing together into a concise summary the most noteworthy activities and states of our mind relating to the business of making judgments. (1) In the first place it may happen to be the case that a proposition M has neither been maintained by us now or at any earlier time in our life. In that case, we may say that judgment M is entirely *missing* or *absent* from our mind. It is by no means necessary that M not even be encountered in our mind as a mere idea, or not have been there earlier as such. We can have the idea of a proposition, e.g. the idea of the proposition that there are creatures capable of transporting themselves from one heavenly body to another, without having decided anything about its truth or falsity. In that case, we lack the judgment, then, without lacking the ideas necessary for it. Again, we can lack a judgment by way of never having had an idea of it or thought of it at all. (2) It can also be the case that we are not making judgment M right now, but have made it at some earlier time. It is then called a judgment *made previously*. There are various cases of this sort, in that the trace it leaves behind after it has disappeared can be such that sometimes we can remember it *easily*, sometimes only with *effort*, sometimes only if the idea of it is generated by an *external object*, sometimes *not at all*, The meaning of these expressions is to be understood just as they were in connection with ideas in §289 (2). (3) It does not always have to be the case that when we remember a judgment as one we have made previously, we make the same judgment again. But if we make the judgment, *S* is *P*, whenever the relevant ideas, *S* and *P*, come to mind or at least whenever the question arises in our mind whether the predicate *P* belongs to the subject *S* or not; then it is said that we are *consistently committed* to judgment *M*, committed namely throughout the entire time within which this relationship between us and *M* exists. Since there is no word to describe such propositions without also specifying whether they are true or false and with what degree of confidence we are committed to them, I am accustomed to using the words, *opinion* and *point of view*. I use them in such a way that what I understand by a being's opinions are propositions he holds to be true, no matter whether they are or not and no matter whether the degree of confidence with which he holds them is slight or the strongest possible. (4) The more quickly the judgment *S* is *P* follows upon the mere arousal of the ideas of *S* and *P*, or the question whether *P* belongs to *S*, the more familiar we say it is. (5) On the other hand, if we *raise* the question as to whether proposition *M* is true or not, i.e. if we wish to make one of the two judgments, *M* or Neg. *M*, and despite all of the attention we have therefore directed to ideas *S* and *P* make neither *M* nor Neg. *M*, this state of mind is called *doubt* with regard to proposition *M*.* (6) Not infrequently what confronts us is that as we turn our attention sometimes to this circumstance, sometimes to that, we alternately make a certain judgment *M* and then again the judgment Neg. *M* with some degree of confidence, even if only a slight degree. This relationship between our mind and proposition *M* is called *vacillation*. (7) It may well happen occasionally that we adhere to a judgment *M* for some moments even though it contradicts a great many other judgments *R*, *S*,... to which we are consistently committed, and when we notice this, it is given up. This may be called a passing *fancy*, a fleeting *thought*. (8) If we turn our attention to everything we suspect it will be useful to take into consideration in order to know whether a proposition *M* is true or false, we are said to be *testing M*. It is understandable that we can decide upon such a test even if we already hold *M* or Neg. *M* to be true, with

* This doubt, which is a state of mind, must be distinguished from a *doubt* which is a thought. It will be discussed later.

greater or lesser confidence. Now if we test a proposition while we are already more or less confident that *M* or Neg. *M* is true, our judgment *M* or Neg. *M* called *provisional* with respect to that testing. (9) If we make a judgment *M* without having previously carried out the test the nature of the proposition demands, this judgment was *hasty*. And if this hastily adopted judgment is one we use to derive a number of others, it is ordinarily said to be a *prejudice*. (10) If we make a judgment we have derived from other judgments without being in a position to cite the propositions from which we derived it (because we have not attained any clear idea of them), it can be called a *judgment from obscure premises*. (11) If the ground that determines us to make a judgment is of such a character that we can prevent this judgment from arising in us merely by arbitrarily turning our attention to other objects, I call it a *voluntary* or *discretionary* judgment. In the opposite case I call it *forced* judgment. Thus, for example, the judgment that I am feeling a certain warmth right now is forced upon me. On the other hand, the judgment that the square on the hypotenuse, etc. and most mathematical and scientific items of knowledge are voluntary. (12) if the propositions from which we derive a proposition *M* are pure conceptual propositions, and those from which we have derived them are equally so, and so on back to immediate judgments, we can call *M* a judgment *based on pure concepts*, or *pure*, *a priori*. In any other case it may be called a judgment *drawn from experience*, or a judgment *a posteriori*. (13) If we make a judgment merely because we remember that we have already made such a judgment before, I will call it a judgment based merely on *recollection* or from *memory*. Such judgments from memory necessarily come into play when we calculate and in every case of deriving a judgment from a long series of inference, where the earliest inferences are already lost to consciousness before we have arrived at the last ones. But since any proposition expressing a memory is by nature merely an empirical proposition, all the judgments we take from memory and consequently all of the judgments we further derive from them must be counted as empirical judgments. Judgments we need certain other perceptions to arrive at, in particular such perceptions as do not refer to phenomena in our mind but to *external objects*, are said to be *drawn from experience* in a narrower sense. On the other side, judgments that require no experience but memory to be derived are still counted among judgments from *pure concepts*. (14) If the cause of our having adopted a certain judgment partly lies in the fact that someone who was anxious to have us think that way took great pains to direct our attention to all of the grounds that could make proposition *M* probable or certain in our eyes, and to divert us from all of the grounds that could bring us around to the opposite opinion, being indifferent to whether or not the former were correct and the latter incorrect, then I say that we have been *persuaded* to adopt the judgment. And if it was ourselves, say, who applied these means, it has been produced by *self-persuasion.* (15) Just as it is often sufficient for making a judgment that we just remember having made it at an earlier time ourselves (13), it is also sufficient sometimes just to find that some other being is making it or has made it. Such a case is before us, for example, when we judge that a sick man's condition is very dangerous because we see that the doctor is at a loss. It is a special case of this when we make the judgment *M* because we conclude from what we perceive that it is the will of a thinking being that we accept proposition *M* because he himself holds it to be true. An action someone performs with the specific intent that we, following our best insight, will conclude from perceiving it that it is his will that we accept proposition *M* because he himself also holds it to be true, we ordinarily call giving *evidence* or *testifying* in favor of this proposition. It can be said, therefore, of judgment *M*, which we make because of such testimony, that it is a judgment *supported by testimony.*

CHAPTER THREE

THE RELATIONSHIP OF OUR JUDGMENTS TO THE TRUTH

§ 307

More Precise Definition of the Concepts: Knowledge, Ignorance and Error

In this chapter the relationship in which our judgments stand to the truth (§ 269) is supposed to be examined more closely. We already know from § 43 that there are both true and false judgments among those we make. Now it is necessary to investigate how they both happen, i.e. the ways in which both our true and our false judgments come to be. The latter in particular, because it can be be useful to us in learning how to avoid error to the greatest possible extent. With respect to true judgments, something will also have to be said about the various objects they range over and about the limits of our knowledge. Conceivably, however, a false judgment could never arise if we had knowledge of every truth, i.e. if we were not *ignorant* with respect to a great many objects. The way in which ignorance can arise must also be explained, then, and prior to the explanation of how errors come to be. First of all, though, I shall have to establish more precisely the three *concepts* of *knowledge*, *ignorance* and *error* themselves.

(1) The first was already defined in § 36, yet only insofar as it was necessary and feasible there. All that was said was that I apply the term *knowledge* to any judgment that is in accord with the truth. Taken as a definition, this would give us too narrow a concept, for now we would have to deny of anyone who is not making a certain judment in accord with the truth right now that he has knowledge of this truth, at least at this time. This is not only completely contrary to linguistic usage but would not have any use for science either. According to linguistic usage we ascribe knowledge and continuing knowledge of a truth *A* to someone if only he has made it at some time in the past, and at present is not only in a position to remember it under the same circumstances which are sufficient for remembering other judgments that have

not been forgotten, but would still adhere to it as well. So we say that a boy already knows 100 Latin words, not when he has all of their 100 meanings in his consciousness at one and the same time, but as soon as we find that when we identify one or the other of these 100 words to him he gives its meaning correctly. We choose to abide by this usage also; but we will use *knowledge* and *cognition* as equivalents, since the distinction between the two words is of no importance for our purposes. Accordingly we will understand both of them to mean that state of our mind in which we have not only already made a judgment in accord with the truth (its material is to be called the material of our knowledge), but also remember it and are consistently committed to it. (§ 306 (2), (3))

(2) Now so far as the two concepts of *ignorance* and *error* are concerned, what I believe is that both of them have to refer not to mere ideas but to entire judgments. This will readily be granted me with respect to the word *error*. For no matter how often we speak of mistakes in concepts and ideas, it immediately becomes clear nevertheless, that what we have in mind is nothing but actual judgments, and judgments that deviate from the truth. We say that someone has mistaken ideas about some object if we want to indicate that he makes erroneous judgments about that object. But does this hold true of the word *ignorance* as well? Must it always be understood in terms of certain truths when we say of someone that he does not know certain concepts or is ignorant of them? Could we not also call the lack of a certain concept ignorance of it? I admit that this does happen and is permissible, whenever we have no cause to be very strict in choosing our expressions. But the situation in which someone has not yet acquired a certain concept can already be described clearly and conveniently enough by using the expressions, *lack* of that concept, its *absence*, and others similar. Therefore I hold that we have to relate the word *ignorance*... not to mere ideas but to genuine judgments. And now we shall be justified, I believe, in accusing someone of ignorance of a specific truth *A* if we can not ascribe to him any knowledge of it. Thus according to (1), he is ignorant of *A* if he has either never made the judgment *A*, or at least no longer remembers it, or in any case when we put the question whether *A* is true before him he does not subscribe to judgment *A* immediately, as if he were merely remembering it. The concept of error is to be defined in a similar way. Namely, an error is any false proposition to which someone is committed in the sense of § 306 (3). That is, he must not only actually

have made the false judgment, but he must also still be ready to make the same judgment now, if he should be asked about it.

(3) On these definitions, it is not yet sufficient in order to be able to ascribe knowledge of a truth *A* to someone that he should hold certain propositions to be true from which *A* is derivable, if it can not be presupposed for certain that he has actually undertaken that derivation. It is still less permissible to say that someone is entertaining an error, *A*, when all that has been proved is that he entertains other errors from which *A* follows. And so on.

§ 309

What Is the Basis of the Possibility of Error and what Circumstances Promote Our Errors' Occurrence?

(1) It is incomparably more difficult to explain how *error* makes its appearance, to such an extent that we can regard the trouble this costs us as a notable proof of the magnitude of our ignorance. We make mistakes so often, and yet we not only do not know how to explain how these mistakes came about in each particular case, but we do not even know how to make sufficiently clear to ourselves how it is that an error steals upon us in the general case. We make all of our judgments either immediately or we derive them from others. The question is, into which of these two types of judgment can errors insinuate themselves? We can not very well suspect error among judgments we make without the mediation of others. For if we chose to doubt one of them, we would have to doubt them all, since their origins are all the same. But if we chose to doubt all of our immediate judgments, then we would also have to doubt all derivative judgments. Consequently, we would have to doubt all of our judgments whatsoever, which is absurd. Reasonably, then, we can look for errors only among the class of judgments that we derive from other judgments. But how do they make their appearance even here? – It can not be assumed that the *form of inference itself*, by which we proceed in deriving some of our judgments from others is incorrect, insofar as it does not consist in a *rule* we have just constructed ourselves, i.e. in a judgment. This is based on what has already been said on the subject in § 300. But if all of our immediate judgments are correct and the forms of inference by which we derive mediate judgments from them are also valid, then how do erroneous judgments come to be?

(2) I can conceive of only one way this can come about. According to the discussion in §§ 300 and 301, there are also *inferences with mere probability*, i.e. inferences which really say no more in their conclusions than that a specific proposition M has a greater or a lesser degree of probability. By virtue of its finitude, our mind is forced to affirm any proposition that appears probable to it with a degree of confidence commensurate with this degree of probability, i.e. to make the judgment M. Now since a proposition that is only probable does not always have to be true, it becomes clear how the mind is capable of making a mistake at times merely because it elevates any proposition with a higher degree of probability to a judgment. We make mistakes, that is to say, whenever it happens to be the case that precisely what we know to be probable according to the most correct rules of inference, and so expect or hold to be true, is in fact not true.

(3) Hopefully, no one will raise it as an objection against this explanation that merely expecting something probably to occur, having proceeded according to correct rules, is not to be termed an error at all, even though it finally does not come about. Where would be the error in the mathematician saying that the degree of probability that upon randomly reaching into an urn that contains 99 black balls and one white one, a black ball will be drawn $=\frac{99}{100}$? – My reply to this is that the error does not lie in the proposition that assigns the degree of probability of such a judgment, but in the judgment itself. We are not in error in estimating the degree of probability of a black ball making its appearance as $=\frac{99}{100}$. Our only mistake is in *expecting* (as a consequence of this estimate or for any other reason) the emergence of a black ball. This is not a voluntary mistake. It is forced upon us, since it is necessary for us, under the circumstances that obtain in this case, to expect the emergence not of a white ball, but of a black one as the far more probable outcome.

(4) Now since (as I have said) I can not conceive of any other way for mistakes to insinuate themselves into our judgments, it seems to me that the single source of errors lies in the feature of our faculty of judgment that has just been mentioned, of elevating to actual judgments propositions that have a mere probability for us. If this is a correct explanation, it follows, (a) that only beings who are forced to accept probable conclusions are subject to the danger of making mistakes and that all of them are; (b) that mistakes can occur in that portion of our judgments we

derive by probable inferences, and for that very reason make with only a greater or lesser degree of confidence, never with complete certainty; (c) that finally every error is a proposition that has some probability in connection with the rest of the judgments the being who is making the mistake assumes to be true. This probability every mistake has is ordinarily called its *apparent* truth or *plausibility*. This plausibility can arise out of propositions that are all true or out of propositions that are already incorrect themselves. In the first case, the error may be called an *original* mistake, in the second a *derivative* one.

(5) It is well known that the judgment we initially and inevitably make use of mere probable inferences to derive are so-called *empirical judgments*. Accordingly, it is the domain of experience wherein error initially and unavoidably dwells. But an error can also insinuate itself into the picture in connection with pure *conceptual judgments* as well if somehow we are forced or induced to make use of merely probable inferences in arriving at them. In particular, this is the case with all of the judgments we produce from memory (§ 306 (13)). For this reason the danger of making a mistake finally hovers over every judgment that depends on a lengthy series of inferences, of whatever type they may be. For we never arrive at such judgments except by means of premises of which some are drawn merely from memory.

(6) From what has been said, the origin of error becomes understandable only in general. But in order to obtain rules for the avoidance of error, what has been said is as good as useless. For this purpose it is necessary to seek knowledge of the most important circumstances most favorable to the origination of a mistake. For it is precisely these circumstances we often have the power to avoid, even if not always. I believe the following are the principal cases: (a) *Ignorance* is without a doubt the most fertile mother of error. For apart from the fact that it is the inevitable condition of all error, since there can be no mistake in connection with what is already known to us, it also contains the real reason for a proposition that is false in itself attaining a high degree of probability in our eyes, because of which we assert it with a great deal of confidence. Out of ignorance we often do not know of all the cases that have an identical relative probability under a given set of circumstances. We therefore estimate the degree of probability of accepting one of them as much higher than it ought to be in terms of the real situation. Thus, for example,

before the discovery of Saturn's rings, we had to suppose with a great deal of confidence that all of the heavenly bodies have a globular form. (b) Another occasion of error lies in the unsuitable *direction* of our *attention*. For as we direct our attention to these ideas or those ideas just now occurring to us, this series of ideas or that series will arise within us. And it can very well be understood how this situation has the greatest influence over the nature of the new judgments we make, so that it can induce us to maintain a lot of things of which we perhaps would have assumed the exact opposite under different circumstances. A truth we are not thinking of at the moment (because our attention is directed on something else) has almost as little effect at that time as if it were completely unknown to us. For example, someone who is not thinking of the fact that two straight lines do not always have to lie in the same plane, not at this moment, will straightway maintain that any two straight lines that do not meet must be parallel to each other. (c) Similarly the *accidental association* our ideas can obtain by way of the law of simultaneity can also provide the occasion for thousands of mistakes. For this association is what causes this idea or that, this or that judgment we have made before to come to our mind, that our attention is directed to this or that object, and so on. (d) It follows from this alone that an *unreliable memory*, as well as an *imagination* that is either too *weak* or too *lively* can also be responsible for a great many errors. For a memory that lacks reliability permits us to forget truths we should have taken account of. And an imagination that is too weak does not give the ideas it brings up the strength necessary to combine them into a judgment. Finally, an all too lively imagination gives some of our ideas so much liveliness that other ideas not in harmony with them can not enter into our consciousness at all. (e) Our own will has a great share in the origination of perhaps most of the mistakes we human beings fall into, in that sometimes with the intent to deceive ourselves and sometimes without it we direct our attention only to the grounds that *favor* a certain proposition and withdraw it from those that are opposed. This is the source of the familiar phenomenon that for the most part we human beings believe what suits our passions and deny what would disturb us in satisfying them, that the views of one and the same man change just as often as his inclinations and his needs do. (f) If we have made some mistake for the first time, we are very much in danger of repeating it more often in the future. For what has already appeared to be thus and so once has in its favor the assumption that it would not come out any different if we should go over the grounds that moved us the first time anew on another occasion. Therefore we are used to omitting this new test. But even if we do not omit it, it is still very much to be feared that the same association of ideas that made us unaware of the error when we committed it the first time will return to mind once again and induce us to make the same error. The more often this happens and generally the longer we have attached ourselves to a certain opinion, the stronger our faith in it becomes, the more difficult it is for us to give it up. (g) But an error once committed does not only have the effect of making us repeat it still more often. It commonly leads to many *other* mistakes as well. For as we know, from one false proposition a number, indeed an infinite number of other false propositions can be derived. So it is no wonder if accepting one incorrect proposition as true inclines us to accept more incorrect propositions. For example, if we hold the mistaken opinion that making our own selves happy is the supreme moral law, how many other errors there are that this basic mistake will plunge us into! – (h) The association of our ideas with *symbols*, which we either imagine or actually set forth (e.g. in writing) is generally one of the richest means of facilitating thought. In particular cases, however, this very way of going about things is also a distinctive occasion of error itself. It is so in the first place because every sign, being a distinct idea itself tied up with various other ideas, brings up these latter ideas and so exerts its own influence on the sequence of ideas that make their appearance in our reflection. In the second place, because it is not infrequently our experience to confuse a pair of similar symbols, which do not mean the same ideas at all, with each other. The consequence of this is that we replace a judgment we made earlier, now only to be repeated, with an essentially different judgment. Finally, because the same symbol often has more than one meaning and without noticing it we go from one meaning over to another. Any proper name that carries a secondary meaning with it offers an example of the influence the name of something often has on the series of ideas aroused in us and through that on the judgments we make about it. Every error in calculation caused by exchanging a pair of similar numbers offers an example of an error that arises from confusing similar symbols. Finally, it is an example of an error produced by the ambiguity of certain words when the claim is made that divine revelation is impossible, if there is an attempt to prove this impossibility by saying on the one hand that such a revelation has to be information imparted *immediately* to us by God and indicating on the other hand that we have no criteria at all for distinguishing between God's *immediate* and mediate actions. What is overlooked in this connection is that we take the word *immediate* with two different meanings in those two sentences. The first meaning is one according to which only means of a certain type are not supposed to be applied, not all means whatsoever. The second is the strict meaning which excludes any instrument, no matter what it may be. (i) It is most useful to the development of our understanding and the enlargement of our knowledge for us to grow up not all alone but in the society of other rational beings, who share their views with us from an early age. True as that is, it is nevertheless not to be denied that this is the way in which many errors are brought about. For without testing it, indeed without even being able to test it, we accept almost everything that others tell us, particularly when we are young. In doing so, then, we accept a great deal they are mistaken about themselves or that they intentionally want to put over on us. Now if in later years we attain the capacity for correcting a number of these mistaken views, such correction does not take place, sometimes from inertia, sometimes because several of these views are welcome to us, sometimes too because we have already become too accustomed to them to be capable of becoming aware of their incorrectness so easily. That is why pupils commonly remain committed to their teacher's opinions and why one and the same mistake often spreads throughout entire ages and peoples; and so on.

§ 314

Are there Definite Limits to Our Capacity for Knowledge?

(1) The thought that there are infinitely many truths, but that we as finite beings are never in a position to know more than a finite number of them, leads very naturally to the question *whether there are definite limits to our capacity for knowledge* and whether we are not possibly in a position to specify them? Before we are able to answer this question, we must first establish more precisely what we understand by such limits to our capacity for knowledge and by determining then. In general, what we ordinarily mean by specifying the limits of a faculty is nothing but certain propositions which state truly what this power is capable of producing and not produc-

ing. It will be permissible to call specifications of this sort *complete* when everything this faculty can produce and not produce can be determined from them. They will be called *incomplete* or *partial* when they only determine some of the things this faculty can produce or not produce. In these terms, what we should understand by the specification of certain limits on our capacity for knowledge should be nothing but certain propositions which state with truth what we are in a position to know or not. On reflection, however, it becomes clear that this concept has to be defined more precisely. For since every judgment that agrees with the truth is called a piece of knowledge, even if that agreement is only accidental and had come about only by way of previous errors, it can very well be seen that the limits of our capacity for knowledge, if we were to abide by such a broad definition, would fluctuate everywhere, since mere chance and even a mistake could contribute to its enlargement. So, for example, we could not deny even to the most ignorant of men the capacity for learning this very day what the inhabitants of Mars, of Venus, and so on are like. For could it not happen that by some circumstance (like Ennemoser's clairvoyant woman, for example) he came to have the illusion of being able to make a judgment about such matters, and then by chance hit upon something true? – Truths of this sort known merely as a consequence of error surely should not be counted when we wish to determine the limits of our capacity for knowledge. What we demand in this case, then, are propositions that state what kind of truths we can arrive at in a regular way, not merely through the accidental mediation of a mistake. Such definitions of limits can still be very different. They can be *complete*, if they define the entire set of truths we are capable of knowing and the entire set of those we are incapable of knowing fully. And they are *incomplete* if this is not the case. Furthermore, they can define the set named by specifying this characteristic on one occasion or that characteristic on another; by virtue of this they can have very diverse uses for certain purposes. Finally, it is also quite obvious of itself that these specifications must turn out quite differently depending on whether what we understand by our capacity for knowledge is only each person's own capacity or the general human capacity for knowledge; whether we are speaking of specifications that are supposed to hold only for the present, or of specifications supposed to hold for our entire life on earth, for all of mankind throughout all of the centuries to come, or for all finite spirits in the universe.

(2) If we are seeking to determine the limits of our own capacity for knowledge or of all mankind, it is obvious that this can not be done by enumerating the truths that are unattainable to us as individuals or to all men. For in order for this to be possible, they would have to be known to us, not unknown. In general, when we attempt to make quite clear to ourselves the purpose of wishing to set forth a definition of the limits of our capacity for knowledge in a treatise on logic, what we discover (it seems to me) is that just about the only reason for this is the following: in order to be able to point out to anyone who wants to engage in an investigation of this or that sort a means by which he could judge in advance whether answering the question before him would not be beyond his cognitive powers. In which case, we want not only to spare him the fruitless effort the investigation would have caused him but also to free him from the risk of going wrong in that investigation.

(3) It follows from this particular purpose that many a definition that is quite correct in itself and perhaps not without usefulness for purposes of another kind is useless here. For example, it would be very true and worth taking to heart to define the limits of our knowledge by saying that we are only capable of knowing that is useful to us. But how unsuitable this definition would be for our present purpose!

(4) It also follows the announced purpose that no matter how exact it was, a definition of the limits of our capacity for knowledge which held true merely for the present or the past, i.e. which informed us merely of what we were not capable of knowing at this moment, would be of no use. For it does not follow from the fact that we have not been able to know a certain truth up to this moment that it will be impossible for us in the future as well. A definition we are supposed to be able to use for the purpose mentioned must hold true for the future; it must delineate a class of truths such that we shall not be in a position to know them not only at present but for all our future existence, or at least not so long as we are human beings.

(5) But since the truths themselves can not be enumerated individually, the *first* thing we could hit upon is probably this, that nothing but the object with which these truths unknowable to us are concerned would be individually specified. Limits defined in this way would have to consist of propositions of the following form: "Every object falling under idea *A* is incapable of being known by us." Or (what comes to the same thing)

"We are not capable of knowing any truth of the form: *A* is *X*." – But who does not see that such a claim would be absurd? For do we not know something of every object, at least what it has generally as an object in common with every other object? And is not the assertion that we are incapable of knowing anything about this object itself a judgment about it, so that at bottom there would be a contradiction in saying of any object that it can not be known at all, i.e. that not even a single true judgment can be made about it? – But the rejoinder will be made that when someone says of an object that he does not know it, it is by no means intended that he does not know a single one of its properties and so is incapable of expressing a single judgment about it. That suits me; now all I ask is that the properties it must be impossible for us to know when we are supposed to call the object unknown be indicated more specifically. If one chose to say that an object should be described as unknown to us if we can not cite any of the properties that distinguish it, then I call to mind the fact that according to this explanation there would be no objects completely unknown to us. For idea *A*, by which we identify the object, already express one of its distinguishing properties, in that property *a* is certainly a property such that it pertains only to objects falling under *A*. One would have to understand the explanation given still more narrowly, therefore. One would have to say, possibly, that an object is said to be *unknown* to us when we are not able to ascribe to it any distinguishing property of a sort not already implicit in the mere *idea of it*. In order to define the limits of our capacity for knowledge in this fashion, then, we would have to be able to set up and prove one or more propositions of the following form: "We human beings are not capable of knowing a property of any of the objects falling under *A* which belongs to it exclusively and would not already be implicit in the idea of *A* itself." – It is nothing impossible in itself, however, that an object should have properties not implicit in the idea of it (§64). We humans also have the capacity to know such properties, i.e to make synthetic judgments that are true. This can not simply be denied (§§197, 305). Therefore, it is obvious that the truth of such a proposition would have to be proved from the distinctive character of idea *A* alone. Now for my own part I confess that I have no knowledge of any such peculiarity of certain ideas. Indeed it even seems to me that a proposition of this form would be self-contradictory. For provided that *A* should not be composed of concepts already implicit in what is asserted of it (in which case the proposition would be a trivial tautology), the judgment set up in this case would already be itself the kind of judgment to be denied us.

(6) The second thing one might yet hit upon is that the propositions of such a nature that our cognitive faculty is never supposed to be able to decide whether they are true or false be listed individually. But how difficult would it have to be to prove claims of this kind? From nothing but the fact what we have so far been unable to decide whether property *b* belongs to the objects falling under idea *A* or does not belong to them, we are not justified at all in concluding that grounds for making such a decision will never be offered us in the future. In order to see the difficulty that obtains here more clearly, let us consider the two distinct cases that can arise depending on whether *A* and *b* are pure concepts or mixed ideas. If the first, and it is a matter of knowing a pure conceptual truth, then no experiences of any kind whatsoever we should make in the future could be of any help to us in deciding the question before us, at least not insofar as what is called for is the sort of knowledge in which there is no possibility of error. But who wants to be so dogmatic and to claim that we shall never be so fortunate, by means of however lengthy a consideration of the two concepts *A* and *b*, by more and more exact and more and more extensive comparison of them not only with each other but with all related concepts, and so on, as to know whether the two concepts can be united in a judgment or not? How many questions belonging to the domain of pure conceptual truths could not be decided for centuries and yet means were finally found to decide them. One need only recall, for example, the question as to whether the ratio of the diameter of the circle to its circumference is rational, and a hundred like it. – But if ideas *A* and *b*, of which we are to decide whether they can be combined in a true proposition, are mixed ideas, even if it should be the case that mere reflection can decide nothing on the matter, but experience is absolutely necessary, further that this experience is unheard of so far, indeed that we may not even be able to conceive how it could be acquired at all, it could still be forthcoming hereafter and tell us how to decide the question. Thus, for example, today we surely do not know how to settle the question as to the form of the moon's inhabitants, but anyone who would conclude from this that we shall never be able to, would be committing the same fallacy that someone would have committed centuries ago if he had denied the possibility of

calculating the moon's weight, because it was not yet possible at that time. The precarious character of this inference from the present to the future becomes all the more evident when we ask for a more precise specification of the sort of future one really wants to spread it over. The immediate future, which each of us still hopes to experience himself? Or the entire duration (who knows how long) of our whole race on earth? Or even the future that begins after leaving this earthly life, to eternity itself? One will hardly choose to venture the last, i.e. to claim that because we are not now able to decide whether a certain proposition was true or false, we would not be in a position to do so in eternity. But if we can not go so far without being audacious to the highest degree, it seems to me that we are already being *venturesome* in extending our prediction even to a much shorter time.

Note: In this way I acknowledge it to be my opinion that it is scarcely possible to define the limits of our capacity for knowledge, at least in the way that is really desired. Then someone may say that there is an implicit contradiction in this claim much like the one in (5). For the impossibility assumed here already seems to be a kind of limit for our capacity for knowledge itself. My reply to this is in the first place that I am not maintaining precisely that such a limit can *never* be defined, only that none such is known to *me*. In the second place, asserting this impossibility can be regarded as a description of a truth eternally to remain unknown only on the presupposition that there is such a limit in reality whether we know how to identify it or not. But I am not saying this. Instead I am more inclined to suspect that we are not in a position to identify such a limit because in fact there *is* none, in that the sum total of human knowledge admits of being enlarged to an infinite degree.

VOLUME FOUR

PART FIVE

Theory of Science Proper

CHAPTER ONE

GENERAL THEORY

§ 395

The Supreme Principle of All Theory of Science

Whenever the instruction we impart has an art or even just some action or other for its object, i.e. whenever we give directions about how an end attainable by free human activity can and should be pursued, we hope in vain to discharge this task satisfactorily and to give *complete* guidance if we do not take into account the relationship between the rules of *morality* and if not precisely the given end itself, all of the means we recommend for attaining it, i.e. the rules we impart in our instructions. For merely specifying the *goal*, i.e. the effect that is supposed to be attained by a certain procedure, never completely determines this latter, or does so only in the rarest cases. There are almost always several ways of acting which, no matter matter how different they may be in many other respects, are nevertheless equivalent in terms of the effect they bring about. Thus, for example, mechanics teaches us that innumerable ways can be contrived for lifting a given load with a given force and everyday experience shows by what a variety of ways and means men not only strive for one and the same goal but also actually achieve it. For example, one person knows this way of getting another's attention or sympathy or directing his will this way or that; someone else knows another way, and so on. If in giving instruction for attaining a given goal we paid attention to nothing whatsoever except this goal itself, so that any means that is equally effective also seems equally to be recommended, we would have to leave a great deal wholly undetermined, then. But since, as no one denies, the laws of morality extend to everything a man does knowingly and willingly, it can be supposed from the start that the various methods we could all cite as effective for attaining the goal before us will not always be indifferent for those higher purposes prescribed by morality. Now we must regard it as a command that stands as a matter of course, even if it is not stated explicitly, that we should choose from among several means equally effective for attaining the desired end only those which also suit best the purposes morality prescribes. For even if the person who is requesting instructions from us is not asking on his own behalf whether the means we shall specify are compatible with the laws of virtue or not, even then we can in no way be permitted to recommend to him means that are exactly contrary to those laws, at least when there are others equally effective that are in conformity with them. From this it follows that even if we are not in a position to investigate whether the end for attaining which we are supposed to recommend means is already moral in and of itself, nevertheless in specifying these means the relationship between them and the laws of morality can surely never be left out of account.

From this it follows, however, that the following proposition can always be set out in connection with any directions supposed to be given for an art or a way of acting: "One would have to conduct oneself in this case so that along with the end originally imposed as much of good or of the ends imposed by the moral law can be achieved as can be combined with it." Indeed, this proposition, I maintain, is of such a kind that all of the rules our instructions must contain can be derived from it, as consequences can be derived from their ground (a partial ground at least). For this single truth puts us in a position to define the conduct necessary for achieving the desired end as exactly as it can be defined, if defining it is not supposed to depend on mere arbitrary choice but only on rational grounds. Or what other considerations could there be from which regulations for that conduct could be derived? What one might be most likely to hit upon is that there might be other enjoyable advantages for the agent himself or for other people that could be achieved by an apt application of existing powers. We would then also have to give instructions for achieving them, despite the fact that they are not mentioned at all in the assigned task. That is fit and proper, but who does not see that all of it can be derived from the former truth? The laws of morality themselves require that anything which benefits one person without harming others be done. So if it is possible with just the same powers the agent is prepared to devote to the attainment of the end originally specified also to obtain some further good either for himself or others, it follows from our proposition above that in directing him we should call his attention to them and exhort him to them. And so it is clear that that proposition actually defines everything that can reasonably be stipulated in the conduct we are supposed to give him directions about. What is still left undetermined can not be determined by any grounds in its very nature. For what neither furthers the end the agent has cited himself better than something else nor corresponds more perfectly to the laws of morality, and so provides no imaginable good either to the agent nor to any other being, is truly to be regarded as indifferent in all respects, no matter how it may be set up. It is not rational grounds, but mere arbitrary choice that must lead us if we want to reach a decision on these matters as well.

If what has just been said holds true of all instruction, it must also hold true of what we are supposed to impart in the theory of knowledge, and so it is permissible to set forth the following propositions as its *basic principle*: "In dividing the whole domain of truth into particular sciences and in presenting these sciences in their own scholarly treatises, our procedure must always be as the laws of morality demand and consequently so that the greatest possible sum total of goods (the greatest possible advancement of the general welfare) is produced by it."

§ 401

A Proper Scholarly Treatise Must also Indicate the Objective Connection between Truths, as far as Possible.

According to the definition given in §393, the concept of a scholarly treatise is satisfied if the theories of the intellectual discipline it is devoted to are only presented in such an order and connection with other truths so that they will be accepted by anyone with the pertinent prior knowledge who thinks them through. I have not asked that the objective ground on which each of them depends be given as well. And this was deliberate, for in the light of what we saw in §216 *et al.*, stating the objective ground of every truth is not to be required if only because there are truths that have no such ground. Furthermore, even where there is a ground, it is sometimes uncommonly difficult, often totally impossible for us human beings to discover it. On the other hand, learning of the objective connection between truths is such a useful and pleasant thing, indeed the mere search for it brings so many benefits, that we are certainly not being unreasonable in asking the author of any scholarly treatise to make the effort to learn how truths hang together and point it out to his readers whenever this is feasible and not barred by higher considerations (i.e. by more important benefits). The benefits yielded by such an effort could be approximately the following: (a) In the case of many propositions we thought true and had the idea of advancing in our treatise, just by the effort of seeking out its ground we can arrive at the discovery that this claim is false, because no tenable ground can be found for it, while one can be found for the opposite. (b) On the other hand, if by way of our reflection on the ground of a proposition held to be true we come to learn of the grounds that tell us *why* it is true along with the proofs *that* it is true, it is obvious that our confidence in it will have to be very much heightened thereby. Thus investigating the ultimate ground of our obligations can yield the important and beneficial result that we become so much the more convinced how true the judgments expressed by common sense in this connection are. (c) In many cases indicating the ground of a proposition can be used, if not as the shortest, nevertheless as the most instructive and convincing proof of its truth. Thus the shortest proofs that can be given for certain theorems of geometry about the similarity of spatial objects are precisely those which are learned by seeking their objective ground. (d) It is still more important that by discovering the objective ground of a truth we are often placed in a position to find out a number of other useful truths. This is primarily the case with empirical truths, where the discovery of the ground of a phenomenon is at the same time the discovery of its cause. But knowledge of causes puts us in a position to fulfill a great many of our wishes and intentions, to free ourselves of countless evils that oppress us, and to bring about situations beneficial to ourselves or to others. For thousands of years it has been known that lightning strikes high towers and buildings. But since Franklin taught us why this happens, we have had a means of making ourselves secure for the most part against the destructive power of lightning. Discovering the ground is also seldom entirely fruitless within the domain of pure *conceptual truths*. For example, investigating the ground of why the angles at the base of the isosceles triangle are equal to each other led to the discovery of the important truth that any two triangles that have two sides and the enclosed angle equal are identical with each other. And through this we have arrived at a thousand other truths that were entirely unknown before. Investigating the ground of why the forces on the lever at equilibrium are inversely proportionate to their distances from the fulcrum permitted most important discoveries in mechanics, and so on. (e) In any case, someone who knows not only that something is true, but why it is true, has one truth more. Searching this truth out exercises the intellectual powers and its discovery yields a satisfaction of its own.

CHAPTER FOUR

ON THE PROPOSITIONS WHICH SHOULD OCCUR IN A SCHOLARLY TREATISE

§ 525

Explaining a Truth's Objective Ground

Not infrequently we can go about proving *that* something is by calling attention to the reason *why* it is. Now since pointing out the objective ground is such a useful thing that we should communicate it whenever possible in our scholarly treatises (§401), there is no doubt it must be regarded as a virtue in a proof if it derives the truth to be demonstrated from its objective ground. To differentiate them from others, we can call proofs of this type *demonstrations* and the rest of them, which are only aimed at assurance, *confirmations*. In some intellectual disciplines, particularly in the pure conceptual sciences and in moral theory, almost every proposition that has to be established can be proved from its objective ground. So in this area almost all proofs should be demonstrations. In others, particularly the empirical sciences, chemistry, medicine, history, and the like, it is possible only rarely to carry out a truth's proof from its objective ground alone, since in some cases it is entirely unknown to us and in some not known completely enough. But insofar as we have even a partial knowledge of it, we can sometimes use it in support of our proof to that extent. Thus, for example, we can never prove that a man's murder was committed by another person only from the fact that the latter lived in enmity with the former, threatened to murder him, had provided himself with a murder weapon, and the like. However, by adding these intrinsic grounds to certain other grounds of knowledge (e.g. eye witnesses), the degree of the event's probability can be increased a great deal.

§ 530

Proofs by Reduction to Absurdity

(1) We have already seen in §329 that it is a very useful means of convincing ourselves of the truth of a proposition before us to try to derive

an obviously false proposition from the negation of it. For if this is the case, it is clear that the negation is false and so the proposition negated is itself true. Proofs in which such a procedure is applied are ordinarily called proofs by reduction to absurdity (*deductio ad absurdum*), also *apagogic* or *indirect* proofs. All the rest are called *direct, immediate* or *ostensive* proofs. The question now arises whether there is not something objectionable about using indirect proofs in a scholarly treatise. In such a proof essentially all we demand of our readers is that they infer the truth of a proposition *M* to be proved by us from the fact that a conclusion Neg. *A* which they know to be false can be derived from its negation, or the proposition Neg. *M*, either directly or in combination with certain other propositions *B*, *C*, *D*,... they hold to be true. What we want, then, is really that they should construct the following argument: "If *M* were a false proposition, then Neg. *M* would be a true proposition, and so Neg. *A* would also be a truth. But Neg. *A* is false. Therefore Neg. *M* is also false and consequently *M* is a truth." Now since proofs of this kind are transparently clear, it is understandable that the degree of certainty an apagogic proof can provide for proposition *M* depends entirely on the degree of the reader's confidence in his knowledge of its two premises. If he finds them trustworthy enough, he will also find proposition *M* quite trustworthy. If the proposition Neg. *A*, the *absurdity* to which the reductive proof leads, is a proposition the falsity of which is self-evident to anyone, i.e. if proposition *A* is an indubitable truth, then everything comes down to the degree of certainty with which we have proved to our readers the hypothetical major premise, i.e. the proposition: "If Neg. *M* were true, Neg. *A* would also have to be true." If this is also self-evident or the number of propositions *B*, *C*, *D*,... we use to prove it not too large, all of them are sufficiently certain, and the inferences by which we derive it from them not merely probable inferences, then the confidence generated in proposition *M* by our apagogic proof will be satisfactory. If we only look to the objective of *persuasion*, then, there is no reason why the apagogic form of proof would have to be avoided in a scholarly treatise. And the fact that such proofs often offer themselves spontaneously justifies a suspicion that they will also strike our readers as easy and natural. What permits us to reject such proofs, nonetheless, is, I believe, the following. If from proposition Neg. *M*, which is false, the false proposition Neg. *A* is supposed to be derivable by mediation of the true propositions

B, *C*, *D*,…, then from propositions *B*, *C*, *D*,… and *A*, all of them true it must be possible to derive the conclusion *M*, likewise true. This leads us immediately to conclude that it has to be a roundabout way to derive proposition *M* if instead of deriving it right off from propositions *B*, *C*, *D*,… and *A*, first proposition Neg. *A* is derived from propositions *B*, *C*, *D*, and Neg. *M*, then the conclusion is drawn from Neg. *A*'s obvious falsity that there must be a false proposition among the premises as well, and since *B*, *C*, *D*,… are all decidedly true that the false proposition must be Neg. *M*, and consequently that *M* itself must be true. If the views set forth in §221 about the way in which truths intrinsically hang together are not mistaken, it will be clear that the propositions by which an apagogic proof supports the proposition it is supposed to prove can never represent its objective ground. For surely the objective ground of a truth can not lie in the *larger* number of propositions from which it is derived in this type of proof, since it is also derivable from a smaller number.

§ 557

How to Prove a Statement Specifying the Composition of an Idea

If it is still a mere *idea* that we are supposed to define and we maintain in our definition that it is a compound idea, and composed of such and such parts combined in such and such a way, we have to prove that the concept brought forward by the combination we have specified is really one and the same with what we were supposed to define. We shall do so if we convince our readers that no distinction can be shown that would have to obtain between these two concepts if there really were two of them. To this end we ask our readers to look for such a distinction themselves. We also assure them, however, that we have done this ourselves, and many others before us, without having discovered anything. Besides, we deal with some of these attempts in the book itself and show thereby that not a single difference obtained where it was most likely to be suspected. In order to do this we then show that the concept we have formed not only has the same *extension* as the given concept, but also that there is no difference in *content* between the two. The former is accomplished when we call the reader's attention to the fact that every object he would be inclined to set under the one concept could with equal justice be placed under the other. The latter can only be effected by inviting the reader to turn his attention inward upon himself, in order to become aware of whether there is anything different going on in his mind in connection with the one idea or the other, apart from those differences that lie in the different symbols or other accessory ideas that do not belong to the concept itself. If even the closest attention reveals no differences other than those of the kind just mentioned, the conclusion must be drawn that the concept formed in accordance with our definition and the concept we wanted te define are one and the same concept, and consequently that our definition is correct. And the readers will be all the more sure of this if we finally take up many other definitions that have already occurred or could occur to someone and show that they are incorrect, by demonstrating in a way similar to the one in the preceding section that the concepts formed according to these definitions are different from the given concept. For example if we had laid out the definition that the concept of space is composed of the concepts, possibility and location, combined in such a way that what we conceive of by space is the possibility of a location; and if we explained further that the concept of location arises from the concepts: relation, time, power, action *et al.*, in such a way that what we conceive by the locations of things are the relationships they have which are the ground of their particular powers at such and such a time and of their actions on one another; then in order to prove the correctness of these definitions we would have to show first of all that the concepts cited surely have at least the same extension as the concepts we are supposed to be explicating. We must call attention to the fact, then, that wherever one speaks of a space there is the possibility of a location for some thing and that conversely wherever one speaks of the possibility of a location for a thing, there is a space. Furthermore, we must call attention to the fact that nothing else must be established along with the powers of things in order to have determined their mutual action on each other at any time except their locations, and that in the contrary case, if only the powers of things have been extablished, but not their locations, the mutual effects they have to have on one another at any time is not yet determined. It will become clear from this that the two concepts have the same extension and the only thing we shall still have to show is that they do not differ in their components either. But there is no way to do this except by calling upon what our readers themselves feel when they compare these two concepts.

We shall ask them whether in fact when they think of a space they think of anything except the possibility of an object being there, and whether whenever they speak of a location a body assumes they are thinking of anything but a certain relationship between this body and other bodies, which contains the ground of its exerting, with the powers it has, precisely these and no other influences on the neighboring bodies. And so on.

§ 558

How the Proof that a Definition of a Given Proposition Is Correct Must Be Carried out

If the object we are defining is an *entire proposition*, according to § 127, there are only two components it can be concerned with, because the third, namely the copula, is the same in all propositions, from that point of view. So all we can define is which idea we take to be the subject idea and which we take to be the predicate idea. If we go still further and define these proximate components themselves, the proof that our definition is correct will be carried out in accord with the rules just indicated, since these parts are mere ideas. So all I need to speak of now is how we can prove to our readers that those *proximate* components into which we analyze a proposition have not been stated incorrectly. But there is no way of doing this except to prove that the proposition to be defined is entirely the same as the one obtained by forming a proposition out of the components given in our definition. We must show then that any difference someone may believe he has noted between the two is only apparent and disappears upon a more precise investigation. For this purpose we must then ask our readers once again to spare themselves no effort in looking even for differences that are only apparent so that they can then test what they believe they have found without bias. We must tell how we have already done so on our own part, but have never found any difference that prevailed against examination. Not satisfied with this, we must draw upon some of these attempts and cite some that could be the most plausible ones, and show that they do not exist in a brighter light. But any difference between two propositions can make its appearance only in the following two ways, either by consideration of the consequences that follow from them, when a consequence is derivable from the one that is not derivable from the other, or by considering their components, when we become aware that something is present in the idea of the one proposition that is not present in the idea of the other. Therefore the objections we bring out against the correctness of our definition can only belong to one of these two types. First of all, then, we must prove of every consequence that does not obviously follow from both propositions that it is derivable from the one in just the same way in which it is derivable from the other. But this by itself only proves the equivalence (§156) of the two propositions, not their identity. Therefore we must on each occasion go on and ask our readers to conceive of both propositions with the utmost clarity and then judge for themselves whether they offer any difference between them besides the merely accidental differences that consist in different symbols or other accessory ideas that do not belong to the propositions themselves. If we have an idea that in answering this question they may encounter some doubts, we must remove them by showing by means of a more precise examination how little they belong to the essence of the propositions. Finally, insofar as there are some other forms of analysis of the propositions to be defined which we suspect will not be obviously incorrect to everyone, it is necessary to speak of them and show that they do not accomplish what they are supposed to accomplish.

PART B

EXCERPTS FROM BOLZANO'S CORRESPONDENCE

Letter to J. E. Seidel in Sulzbach, 26 January 1833:

Honored Sir:

It surely does not generate the most favorable impression of the quality of a work if its author openly confesses that he has had some trouble in finding a publisher for it. Nevertheless, the history of scholarship teaches us that the greatest men of learning have found themselves in this situation, and precisely with those of their writings by which they subsequently brought about the most beneficial changes. And so I am that much less ashamed to find myself in such a position too, since in my case a number of special circumstances combine to explain this state of affairs. I live in a country (Bohemia) which has produced no important work in the area of philosophy – and that is where the book for which I am seeking a publisher belongs – for centuries. I confess a faith (the Catholic) and belong to a station (the clerical) which both have an unfavorable prejudice against them. I was a professor (of theory of religion at the university here) and I was removed from this position because I displeased both the ecclesiastical and the secular authorities. And both of them have taken such an attitude against me since then that it is as good as impossible for me to have a book – however innocent its content may be – appear under my name and with the censor's approval (Mss submitted to the censorship are returned most unfairly, without even giving me an answer).

I am undaunted by these circumstances, however. Instead I just use the highly desirable leisure made available by my dismissal to continue working with that much less interruption on the intellectual disciplines to which I feel myself called. These are primarily mathematics, philosophy and theology. It seemed to me that a treatise on logic (as a theory of how the entire domain of human knowledge must be divided up into particular sciences and what rules must be followed in working these sciences out and presenting them) was one of the most important works I was capable of providing. I devoted ten full years almost exclusively to this effort and I believe a work has been achieved of which no one will claim that its title said too much if it were set up approximately as follows: "Logic as Theory of Science in a Presentation which Deserves the Attention of Every Scholar." – Certainly anyone who reads only a few sections of this book will come upon something new, a more exact definition of a previously vague and uncertain concept, a clear account of the components of which we unconsciously make up a concept known to us all, the correction of a more or less widespread prejudice. But since I took it to be my unavoidable duty not to set up so many views without citing the reasons that moved me to adopt them and explaining why I could not hold to the previous theory: the work grew – despite the most conscientious avoidance of all repetitions and of everything that did not belong to the matter – to such a considerable extent that it will fill *four* printed volumes. I understand very well that this latter fact would have to give any publisher some pause, but it is equally obvious to me that it would not be to the purpose to begin with publication of a selection from the whole work, although I do see that such a selection has to appear later and I wish to be helpful in making it up.

I now turn to you, Sir, as a man who by general reputation is praised for knowing the desires of the booksellers and for aiming at something higher than mere pecuniary gain in his business, with the question whether you might decide to become the publisher of this book, if you receive the Ms from my hands without any demand for compensation whatsoever? –

Should you believe that even then there would be too much risk in undertaking this publication, I ask whether you would not at least condescend to carry out the following proposition, on which you not only would be risking the least thing yourself but there could also be achieved a correspondingly larger public for the work. What I wish, namely, is that next Easter you would have sent out from your book company an invitation for subscriptions to this work. With this invitation, however, that there should also be distributed gratis a sample of the book consisting of its first 4–5 pages (the introduction, say). For this purpose I would be willing not only to bear the expenses of printing those 4–5 pages in an edition of 2000 copies, but also if necessary to pay in advance. At what time and at what price the four volumes will appear if there should be a sufficient number of subscribers would also have to be set in this announcement (which in particular would have to be sent to France also). And of course I wish the date to be set as soon as possible and the price to be as inexpensive as possible. The fact that my name can not be identified on either the announcement or the title page of the work can of course have some detrimental influence on the number of subscribers, but not on the sale the work can hope for later on. For after it has appeared both my former students and all who know me well from my earlier writings will have no doubt as to the author, and I would not be able to prevent my being subsequently identified as the author in many public newspapers even if I wanted to.

Prof. Heinroth in Leipzig, who was given the Ms to read without my knowledge, wrote a preface for it in which he tenders me a great deal of praise, as I hope not fully undeserved. I leave it entirely to your judgment, however, whether this foreword should be printed with the sample sheets or not. The Ms is now still there in Leipzig, and – as soon as you, Sir, give me leave to do so, it shall be sent to you for your perusal at my cost and by whatever means you have the goodness to specify.

Finally, I ask you to send me your answer by post sealed in an envelope addressed *to Herrn Karl Singer, Grosshändler in Prag*. I am most respectfully yours,

BERNARD BOLZANO

Prag, January 26, 1833

From letter to M. J. Fesl, 8 February 1834:

You will understand that I can not satisfy your wish for a sort of survey of the book, at least not at the moment, since I do not have the manuscript. I do not even remember the book's title; I'm not satisfied with it besides. It reads along these lines: "Essay Towards a Comprehensive Treatise on Logic as Theory of Science with Continual Reference to Previous Presentations of this Discipline." Recommend a better one to me! You know enough about the work's content to do so. You know that in it I treat logic merely as a theory of science (a theory about the way the entire domain of human knowledge should be divided into particular subjects – sciences – and how they should be worked out). It has become so long (4 large volumes) truly not because I repeat myself or bring in foreign matter (e.g. rules that apply only to a particular science), but merely because of the abundance of its content, because of the all too many new views (deviating from earlier ones) and the (compressed) statement of the reasons for which I have not been satisfied with the views of others. One will seek in vain for rules like these: "To see an object clearly, one must keep the eye neither too close nor too far away." – On the other hand, I can promise with a good conscience that no scholar, if he just has a head for thinking (which to be sure not all scholars do), will take the book into his hand without finding something on every page that draws him to it either by virtue of the clarity of its presentation or the novelty of its point of view (for the most part by both at the same time). If the views set forth in the book are correct, they will lead to revolutions in more than one science, particularly in metaphysics, in moral theory and jurisprudence, in esthetics, in mathematics, in the rational side of physics, in philosophical linguistics and (God be with us!) in theology. – But I can give you no guarantee at all that these views are correct, to be sure. For that they seem correct to me, that I have tested them over such a long course of years and tested them with all the impartiality, calm and caution men are capable of, that I have sought out every objection to them more eagerly than other points that confirm them, that so many of my former students who have become acquainted with these views, e.g. Slivka, Přihonský, Schneider *et al.* have embraced them with heart and soul, that even one or another who have never been students of mine, upon a partial acquaintance with these ideas, declare them to be correct – all that proves little or nothing! Presumably Troxler, Beneke can also make the same claim for their own views, to which mine are almost totally opposed.

From letter to F. Exner, 22 November 1834;

I will now try once again to make the concept I attach to the terms: *objective proposition,* or *proposition in itself* or *proposition in the objective sense,* as intelligible as I can. What I understand by a proposition in the objective sense of the word is anything and everything that can be comprehended under it when both of the following points are supposed to be assertible of every proposition:

(a) *that it is either true or false,*
(b) *that it is nothing existent.*

This is the sense in which we are taking the word *proposition*, for example, when we say, "God knows all propositions that are true and adopts them as His own judgments, or judges in accord with them; He is also acquainted with all false propositions, insofar as He has them before His mind and recognizes them to be false." –

Previous linguistic usage also takes the word proposition in many other senses, however. I do not wish to see the above sense confused with any of them. Consequently, I do not by any means understand by a proposition in the objective sense

(a) a *judgment* someone *makes* or even only *thinks of*; for at the time he is making it or thinking of it this has reality, and so it is something existent.

(b) Still less do I require of a proposition that it be expressed in *words* or *symbols*, for they too are existing things.

I call propositions that are true truths, and so I do not ascribe any *existence* to truths either, neither *eternal* nor *temporal*. Thus I hold it to be a mistake to say, for example, that the truth that there is a God, or that $2\times2=4$ is an eternal truth, whereas that a bushel of oats costs 1 Fr is a temporal truth. The former is not eternal because only what exists can be eternal. The latter is not temporal because the proposition cited is false as it stands and is true only with the more precise specification: a bushel of oats costs 1 Fr this year (or some other year) in this country or that. But with this more precise specification it is as eternally true as the first one. If what I mean by *propositions* in the objective sense is understood after this fashion, it is easy to understand what I call *ideas* in the objective sense when I note that propositions are all composed of certain parts and that I call such of their parts that do not constitute propositions themselves ideas. But if propositions are not existents, it is evident in and of itself that ideas can not be anything *existing* either. Thus by no means do I understand by *ideas*, taking the word in its objective sense, a certain *phenomenon* in the mind of a thinking being, or an *activity* of that being, or anything *produced* by that activity in the being; for this is all something *real*, but the idea has no reality.

Finally, it is very easy to explain what I want to have understood by the words *proposition* and *idea* when I am taking them not in the objective but in the *subjective* sense. Propositions and ideas in the subjective sense are something real, phenomena or products in the mind of a thinking being. The *idea* in the subjective sense needs a being which has it; the *proposition* in the subjective sense needs a being which *thinks of* it or makes it as a *judgment*.

I believe I can make the concepts I attach to the words, *propositions and ideas in the objective sense and propositions and ideas in the subjective sense,* intelligible in this way without starting out from a still higher concept, from some *generic concept* that embraces propositions and ideas in *both the objective and the subjective senses,* understood, say, by propositions and ideas *in general.* I will not deny that for any two concepts with the relationship of exclusion to each other, e.g. *wax candle* and *syllogism,* it must be possible to give a proximate generic concept. In the example cited, it would be the concept, "*of a thing which is either a wax candle or a syllogism,*" but I do deny that giving such *generis proximi* is always useful, to say nothing of being necessary. And so I believe in particular that it is not necessary to give the genus prox. for the two concepts, *proposition in the objective and proposition in the subjective sense.* They are too incomparable, these two things, for such a genus prox. for them to be of use. The proposition in the objective sense is nothing existent, the proposition in the subjective sense is something existent: what are the two supposed to have in common? – The idea in itself is related to the idea thought of more or less as some real man is related to a painting of him. Now who would want to have a Gen. prox. for the man and his picture? – Comparing the two senses I distinguish by the words *objective* and *subjective* with the senses long tied to the expressions, *religion* or *art* or *science* in the *obj.* and *subj.* sense, will be still more to the point, however. It is well known, for example, that what we understand by geometry in the obj. sense is nothing but the sum-total of all the essential characteristics having to do with space. But by geometry in the subj. sense, e.g. the geometry of Euclid, what we understand is the sum-total of all the *knowledge* concerning space that Euclid possessed. I have never heard of a genus which is supposed to encompass geometry in the obj. and subj. senses as its proximate species. Likewise, the word religion is always taken in either the obj. or the subj. sense, and there is no talk of a third sense encompassing both of them.

The fact that we designate both concepts (of the obj. and the subj. idea) by the same word is by no means sufficient to prove that both concepts must fall under a proximate genus from which they are derived by adding a diff. spec.; as thousands of examples show. Thus, for example, the two concepts, *tongue* in zoology and *tongue* in a shoe surely have no genus prox., from which they are derived by addition of a diff. spec. Instead, the tongue of the shoe gets its name only from some remote similarity to the tongue in an animal body. Likewise, we call the obj. idea and the subj. idea both *ideas* because they have an intimate relation to each other, in that for every subjective idea (for every thought) there is some objective idea *apprehended by* it.

We can directly conclude from this how your views and mine differ on this matter. You want to derive the concepts, *proposition and idea in the objective,* and *proposition and idea in the subjective sense,* from a pair of more secure concepts, *proposition and idea in general.* I hold that this is impossible, since what I claim is that neither pair of concepts has such a gen. prox. from which they are supposed to be derived by addition of a diff. spec. – You define proposition and idea in the objective sense by saying that they mean phenomena in the mind of a thinking being conceived of *in abstraction* from all of their relationships in the mind. Now I do not grant that. I say that a subjective idea (in thought) remains something *existent,* and therefore no less a subjective idea, even if we abstract it from all of the circumstances in which it exists, and so never becomes what I mean by the objective idea. More or less as a painted man (i.e. the picture of a man) is still only a picture and not a real man even if it is abstracted from all of the circumstances in which it exists (in which picture gallery, in what place, in the neighborhood of which other pictures).

But you say you want to abstract from the *existence* itself which the subjective idea has. I reply: by *abstracting* from the *existence* of something which has existence, i.e. by just not *thinking* of it, we do not take its existence away. Consequently the idea *in thought* remains something existent even if you do not choose to think of its being so. It is just as the angle sum of the triangle equals two right angles even if you choose to abstract from this property of it entirely.

But if *abstracting* is supposed to mean *denying*, then your definition of the objective idea acquires the sense that it is understood to be a *phenomenon* in the mind of a thinking being which at the same time has no *existence* and so should not be a phenomenon. In that case, the reproach of using a self-contradictory concept you want to make to me would fall on your own definition. "There are propositions and ideas no one is thinking of" – that is actually something absurd, if propositions and ideas are supposed to be phenomena in the mind of a thinking being, for it means much the same as "there are men who do not exist." – But if we understand by *propositions and ideas in the objective sense* what I understand by it, there is nothing contradictory at all in thinking of such a proposition or such an idea without presupposing that there is a being in whose mind that proposition or idea is present. You say, "The concept of an idea (in the objective sense) which not only *did* not exist or *does* not exist in any mind, but also *can* not exist in any mind is a contradictory concept." It seems exactly the other way around to me, that the concept of an objective idea which at the same time existed somewhere, occurred as a phenomenon in this mind or that, would be a self-contradictory concept. For, as I have said, objective and subjective idea are a pair of mutually exclusive concepts. The former designates something which has no reality, the latter something which does.

If in order to avoid that contradiction you interpret the proposition, "there are unknown truths," to mean, "Thinking beings have not yet discovered all of the relationships all ideas have to each other," you shift without being aware of it from your concept of a proposition in itself to my own. For now what you mean by a truth in itself is a certain *relationship* obtaining between ideas, which is nothing existing and so is a mere *proposition in itself* in my sense of the word.

Finally you even claim that everything which is subjectively true is also objectively true, but not *vice versa*. Accordingly, the two concepts, *objective* and *subjective* propositions (and ideas) would not exclude each other at all, but one would be subordinate to the other. But that not only contradicts everything you have said before (where you divided propositions generally into objective and subjective, obviously opposing the objective and subjective to each other), but also contradicts ordinary linguistic usage, which always opposes religion, mathematics, etc. in the objective and the subjective sense and in fact does so as I do.

We are in agreement, moreover, that what I said to make intelligible the concepts of a proposition and an idea in both the objective and subjective senses is not a proper *definition* of these concepts (not a specification of their components). But if you add that in a proper definition of the concept of a proposition it must be derived from (or composed of) that of an idea, because the proposition is composed of ideas, I must say in reply that this is not a certain inference. The concept of a compound object by no means always has to be composed of the concepts of its parts. Thus, for example, the concept of a clock may well be composed of the concepts of an instrument, time, measurement, but not of the concepts of dial, hand, and so on. It might well be, then, that the *concept* of a proposition in the objective sense, if not entirely simple, still consists of quite different parts from those you assume in thinking of the components of which the *proposition in itself* is composed *itself*. So far as the concept of an *idea* is concerned in particular, I should almost believe that the account I have given in order to make it intelligible, an idea is a part of a proposition that is not yet an entire proposition itself, is the proper *definition* of that concept. For it is quite the usual thing for the concept of a part to include that of the whole. Thus, for example, the concept of the entire clock is surely included in the concepts of dial, hand, etc. For certainly what you mean by a dial is nothing but "a surface with numbers written on it in such a way that it can be used on a *clock*, and so on." – The concepts of head, neck, heart, lung, etc. can surely not be defined either, without mentioning the relationships of these parts to the entire organism, i.e. the concept of the whole occurs in the concepts of these parts.

From letter to J. P. Romang, 1 May 1847:

To describe in a word the essential difference between my philosophical and theological concepts and others, I might say it is the fact that I take greater care to raise everything I think of to the highest possible level of clarity and distinctness than has commonly been aimed at earlier. It is only by that, on the path to the more precise definition of concepts, that I came to all of the distinctive doctrines and views you encounter in my writings (even the mathematical ones). But if one wants to understand the reasons that support so many distinctive claims, one must make a beginning with the Logic. Don't be afraid of the bulky book when you see it! You do not need to read everything in it, the entire *fourth* volume you can confidently lay aside. The entire book with the heading *art of discovery* in the third volume, the whole large chapter, *On Inferences,* in the second, are superfluous for you. It will be enough for you to study in the *first* volume, say, §§19, 25 and 26 (on the concept of *propositions and truths in themselves*), §§48, 49, 50 (on the concept of an *idea in itself*), §§55, 56, 57, 58, 63, 64, 66, 67, 68, 69, 70, 72, 73, 79 (time and space), §85 (sequence), §87 (where the concept of the *infinite,* which I put in the place of the self-contradictory *Hegelian Absolute*) [sic]; in the second volume, §§125, 127 133, 137, 148 (analytic and synthetic propositions), §§154–158 (propositions with varied components, where the relationship of derivability, If *ABC*... are true, *M* is also true, is explained), §170 (propositions of the form: A certain *A* has *B*), §179 (propositions with if-then), §182 (the important concepts of *necessity* and *possibility*), §183 (temporal specifications), §197 (analytic and synthetic truths), §198 (concept of the relation of ground and consequence between truths), §§201, 202, 214 (basic truths), §221. Finally, in the third volume you might study just what is said about the concepts of *clarity* and *distinctness* themselves (§§280, 281) and then confidently set the entire work aside, only taking care to look up one thing or another as the occasion arises. The index in the fourth volume will make that easier.

Now to the comments prompted by the rest of the contents of your prized letter.

What Hegel understands by the concept *in itself*, equally well supposed to be the *object* or *matter* (*Sache*) itself, is in my opinion a non-thing, and so I plead for it not to be muddled up with what I conceive of as *propositions* and their *components, ideas in themselves.*

To my mind, an *existent* in the proper sense is = something real, e.g. a power, a substance, even a mere *thought* which as such exists at a particular time in one (or more) thinking beings. Now a *proposition in itself* does not seem to me to be such an existent at all, for it neither exists at a definite *place* (which only holds true of substances, limited) nor at a definite *time* (like the *thought* of it, say); it is neither a creation nor, absolutely not – God Himself nor is it in God, for a thought of it may well be in God, but not the proposition itself. But of course it is *something,* since we distinguish one from another. The concept $\sqrt{-1}$ belongs to the class of *object less* concepts, like the concepts: nothing, least interval, largest number and others. Even I admit, then, that $\sqrt{-1}$ (the symbol) means something, but just the (objectless) concept of a quantity which when squared would give the opposite of 1. But it does not yet follow from this that the *concept* itself *signifies* something, i.e. *represents* something (an object).

I am very glad that Herr Prof. has already arrived at the conviction himself that the negation in the proposition, *A – has – not b,* belongs to the predicate. The individual's belief is strengthened by the belief of many, and it is so easy to go astray in philosophy that we can never feel sure our views are correct if we do not gradually succeed in finding more and more who think the same. It is true that when I was still professor (1805–1820) I had the pleasure of finding that most of my auditors (500–600 annually) attached themselves not to the Professor of Philosophy but to me, but what weight does the applause of those people have even if it were always entirely sincere?

I speak of propositions with *Es* [it], such as "Es sind Leute vor der Thüre" [there are people at the door] in §172 of the Logic. Concerning those with the subject, nothing, (§170) I will comment right here that I deny all truth to any in which this concept really takes the position of the *subject idea.* For I call a proposition *true* if it *ascribes* to its *object* a property that *belongs* to it. But if the subject idea is *objectless* (like the idea of nothing) there is no object at all to which anything would be ascribed! It would be another matter with the proposition, "Nothing is – an objectless idea," for in this proposition the subject idea is not the idea of nothing, but the *idea* of that idea, which is denotative, of course. The proposition is really to be interpreted: "The idea, nothing, - has the character of lacking an object."

As regards synthetic truths I must refer to the book; only at least §64 must be read in advance.

Since I read that you, Sir, are now occupied with working out a moral theory, I would be eager to know which highest moral law you incline toward in it, since it seems to me that the explanation of God's justice given in your theory of religion §82, pp. 286ff, could easily lead to acknowledging the ultimate ground of all moral good to be the greatest possible advancement of the general welfare (of physical and moral goods). This is the only one I regard as correct and it pleased me very much just recently to discover that Prof. Fechner in Leipzig has embraced this principle in his little book on the highest good without having the slightest knowledge of my *Religionswissenschaft*, in which I set it forth.

From letter to R. Zimmermann, 9 March 1848:

Wissenschaftslehre, Vol. 1, §102, 1. The matter is not only unclearly presented, but as I am just now beginning to see entirely false. If we symbolize the concept of *any arbitrarily selected natural number* by *n* or, what would put it better, if *n* is supposed to represent any arbitrarily selected natural number, that already settles the matter of *what* (infinite) set of objects this symbol represents. This is not altered in the slightest by adding an exponent like n^2, n^4, n^8, n^{16}..., so that each of these numbers is supposed to be raised now to the second, now to the fourth... power. The *set* of objects n represents is always exactly the same as before, even though the *objects themselves* that n^2 represents are not precisely the same as those n represents. The false result was produced only by the unwarranted inference from a *finite* set of numbers, namely those that do not exceed the number N, *to all*.

If I am not mistaken, there is also another blunder in the mathematical example in the second volume, §159. An attempt was made to correct the passage in the list of printing errors (which I do not have, however). But, if I remember correctly, I declared it to be equally unsuccessful later on (§159.4).

It is shocking that the title for Vol. I, §19" The Existence of Truths in Themselves" is also to be criticized as *misleading*.

The concept of *knowledge* is also laid down and defined quite contrary to linguistic usage and in quite the wrong place. For we do not say of whatever someone accepts with complete confidence that he *knows* it, but rather of what he accepts that it is in accord with the *truth*. Thus we say: The child *knows* very well that you love him; he *sees* it from your behavior. He does not yet *know* anything about it, etc. In these cases the degree of confidence is the least thing that we want to emphasize by using the word, know. As in the following case: I scarcely *know* what I am saying, where I am, etc.

There also seems to me much that is mistaken in §284 in vol. 3 on the association of ideas, etc.

I have already let you know that I am no longer satisfied with the explanation in the Logic* of the concept of *definition*, nor with that of the concept of the relationship of ground and consequence.

So right away you have a half dozen examples of my infallibility! How many more may there be? How many I would find myself, if I had the time to read through the Logic once again! –

From letter to F. Příhonský, 10 March 1848:

A few days ago I received a letter from our dear young friend**, in which

* The *Wissenschaftslehre* (ed.).
** Robert Zimmermann (ed.).

he asks me about a passage in the *Wissenschaftslehre* vol. I, § 102 under 1 that neither he nor his father could unriddle. I read over it and find that what is presented there is false from the ground up, so the responsibility for their not understanding it did not lie in my two readers' mental capacities but in my learned *supervidit*. I hastened to report this to them the very next day in order to give them the encouragement they well deserved and on this occasion I enumerated a half dozen similar blunders in the book which I became aware of in just as casual a fashion as this one. How many more there must be still, which I would notice if I found the time to read through the book once again, and how many that I would not recognize even if they are there! Therefore nothing suits me more than to remain nice and modest and ask all of you if you would be helpful to me in this connection by openly letting me know of anything that seems mistaken to you or at least does not clearly impress you as having been proved incontrovertibly.

BIBLIOGRAPHY

A. WORKS BY BOLZANO

1. *Works on Logic, Epistemology and Methodology of Science*

Dr. B. Bolzanos Wissenschaftslehre. Versuch einer ausführlichen und größtentheils neuen Darstellung der Logik mit steter Rücksicht auf deren bisherige Bearbeiter. Herausgegeben von mehren seiner Freunde. Mit einer Vorrede des Dr. J. Ch. A. Heinroth Sulzbach 1837. 4 Vols.

Dr. Bolzano und seine Gegner. Ein Beitrag zur neuesten Literaturgeschichte, Sulzbach 1839.

Prüfung der Philosophie des seligen Georg Hermes von einem Freunde der Ansichten Bolzano's, Sulzbach 1840.

Bolzano's Wissenschaftslehre und Religionswissenschaft in einer beurtheilenden Uebersicht. Eine Schrift für Alle, die dessen wichtigste Ansichten kennen zu lernen wünschen, Sulzbach 1841.

Aufsatz, worin eine von Hrn. Exner in seiner Abhandlung: "Über den Nominalismus und Realismus" angeregte logische Frage beantwortet wird [Abhandlungen der königlichen böhmischen Gesellschaft der Wissenschaften, Fünfte Folge, 2, Prag 1843, Berichte, S. 71–78].

Was ist Philosophie? Von Bernard Bolzano. Aus dessen handschriftlichem Nachlaß, Wien 1849.

2. *Works on Mathematics*

Betrachtungen über einige Gegenstände der Elementargeometrie, Prag 1801.

Beyträge zu einer begründeteren Darstellung der Mathematik. Erste Lieferung, Prag 1810.

Der binomische Lehrsatz, und als Folgerung aus ihm der polynomische, und die Reihen, die zur Berechnung der Logarithmen und Exponentialgrößen dienen, genauer als bisher erwiesen, Prag 1816.

Die drey Probleme der Rectification, der Complanation und der Cubirung, ohne Betrachtung des unendlich Kleinen, ohne die Annahmen des Archimedes, und ohne irgend eine nicht streng erweisliche Voraussetzung gelöst: zugleich als Probe einer gänzlichen Umstaltung der Raumwissenschaft, allen Mathematikern zur Prüfung vorgelegt, Leipzig 1817.

Rein analytischer Beweis des Lehrsatzes, daß zwischen je zwey Werthen, die ein entgegengesetztes Resultat gewähren, wenigstens eine reelle Wurzel der Gleichung liege, Prag 1817.

Versuch einer objectiven Begründung der Lehre von der Zusammensetzung der Kräfte [Abhandlungen der königlichen böhmischen Gessellschaft der Wissenschaften, Fünfte Folge, 2, Prag 1843, Abhandlungen, pp. 425–464].

Versuch einer objectiven Begründung der Lehre von den drei Dimensionen des Raumes [ibid., Fünfte Folge, 3, Prag 1845, Abhandlungen, S. 201–215].

Dr. Bernard Bolzano's Paradoxien des Unendlichen herausgegeben aus dem schriftlichen Nachlasse des Verfassers von Dr. Fr. Přihonský, Leipzig 1851.

Functionenlehre. Herausgegeben und mit Anmerkungen versehen von Dr Karel Rychlík, Prag 1930.
Zahlentheorie. Herausgegeben und mit Anmerkungen versehen von Dr Karel Rychlík, Prag 1931.
Über Haltung, Richtung, Krümmung und Schnörkelung bei Linien sowohl als Flächen sammt einigen verwandten Begriffen [Spisy Bernarda Bolzana 5, Prag 1948, pp. 139–181].
K. Rychlík: *Theorie der reellen Zahlen im Bolzanos handschriftlichen Nachlasse*, Prag 1962.
B. van Rootselaar: *Bolzano's Corrections to His Functionenlehre* [Janus 56, Leiden 1969, pp. 1–21].

Bolzano's bibliography also embraces biographical works and correspondence, works on the history of the Royal Bohemian Society of Sciences, works on physics, metaphysics, moral philosophy, theology, social philosophy and aesthetics. Furthermore, an enormous amount of manuscripts is to be found in Prague and Vienna.

B. WORKS ON BOLZANO

An extensive, subject-indexed bibliography of the literature on Bolzano appears in the second Introductory Volume of the *Bernard Bolzano-Gesamtausgabe*, Frommann-Holzboog, Stuttgart, 1972.

1. *General Works*

Bauer, R., *Der Idealismus und seine Gegner in Österreich*, Heidelberg 1966, pp. 37–60.
Bergmann, H., *Das philosophische Werk Bernard Bolzanos*, Halle 1909.
Fujita, I., *Borutsāno no tetsugaku*, Tōkyō 1963.
Kolman, A., *Bernard Bolzano*, Berlin 1963.

2. *Biographies*

Winter, E., *Bernard Bolzano. Ein Lebensbild*, Stuttgart 1969. The first Introductory Volume of the *Bernard Bolzano-Gesamtausgabe*.
Winter, E., 'Leben und geistige Entwicklung des Sozialethikers und Mathematikers Bernard Bolzano 1781–1848', *Hallische Monographien* **14**, Halle 1949.
Zeithammer, G., *Dr. Bernard Bolzano's Biographie* [Manuscript in the Literary Archive of the *Památník národního písemnictví* in Prague, written 1848–1850].

3. *Logic*

Bar-Hillel, Y., 'Bolzano's Definition of Analytic Propositions', *Theoria* **16**, Lund 1950, pp. 91–117, and *Methodos* **2**, Milano 1950, pp. 32–55.
Bar-Hillel, Y., 'Bolzano's Propositional Logic', *Archiv für mathematische Logik und Grundlagenforschung* **1**, Stuttgart 1952, pp. 305–338.
Buhl, G., 'Ableitbarkeit und Abfolge in der Wissenschaftstheorie Bolzanos', *Kant-studien, Ergänzungshefte* **83**, Köln 1961.
Kambartel, F., Introduction to "Bernard Bolzano's Grundlegung der Logik", *Philosophische Bibliothek* **259**, Hamburg 1963, pp. VII–LIV.
Morscher, E., *Das logische An-sich bei Bernard Bolzano* [Diss. Innsbruck, 1968].
Scholz, H., 'Die Wissenschaftslehre Bolzanos', *Abhandlungen der Fries'schen Schule N. F.* **6**, Berlin 1937, pp. 399–472.

4. *Mathematics*

Jarník, V., 'O funkci Bolzanově', *Časopis pro pěstování matematiky a fysiky* **51**, Praha 1922, pp. 248–264.
Jarník, V., 'Bolzanova "Functionenlehre" ' *ibid.*, **60**, Praha 1931, pp. 240–262.
Jarník, V., 'Bernard Bolzano a základy matematické analysy', *Zdeňku Nejedlému Československá akademie věd*, Praha 1953, pp. 450–458.
Laugwitz, D., 'Bemerkungen zu Bolzanos Größenlehre', *Archive for History of Exact Sciences* **2**, Berlin 1962–1966, pp. 398–409.
Rootselaar, B. van, 'Bolzano's theory of real numbers', *ibid.*, pp. 168–180.
Rootselaar, B. van, 'Bolzano, Bernard' *Dictionary of scientific biography* (ed by C. C. Gillispie), Vol. **2**, New York 1970, pp. 273–279.
Rychlík, K., 'Über eine Funktion aus Bolzanos handschriftlichem Nachlasse', *Věstník Královské české společnosti nauk, třída matematicko-přírodovědecká*, ročník 1921–1922, no. 4, Praha 1923.
Rychlík, K., 'La théorie des nombres réels dans un ouvrage posthume manuscrit de Bernard Bolzano', *Revue d'histoire des sciences* **14**, Paris 1961, pp. 313–327.
Steele, D., *Historical Introduction to "Paradoxes of the Infinite by Dr. Bernard Bolzano"* (London 1950).
Wußing, H., 'Bernard Bolzano und die Grundlegung der Infinitesimalrechnung', *Zeitschrift für Geschichte der Naturwissenschaft, Technik und Medizin* **1**, Nr. 3, Berlin 1961, pp. 57–73.

5. *Metaphysics*

Huonder, Q., *Das Unsterblichkeitsproblem in der abendländischen Philosophie*, Stuttgart 1970, pp. 85–90.
Neemann, U., 'Kausalität und Determinismus bei Bernard Bolzano', *Philosophia naturalis* **13**, Meisenheim 1972, pp. 353–370.
Příhonský, F., *Atomenlehre des sel. Bolzano, insbesondere wie Bolzano die Schwierigkeiten, die dem Denker bei dieser Lehre begegnen, glücklich beseitigt hat. Dargestellt aus den Schriften desselben*, Bautzen 1857.
Stähler, W., 'Die Frage nach der Unsterblichkeit der Seele und ihren Voraussetzungen in der Philosophie Bolzanos', *Philosophisches Jahrbuch der Görresgesellschaft* **42**, Fulda 1929, pp. 232–251, 369–383.
Winter, E., 'Die Entwicklung der Auffassung B. Bolzanos von der Willensfreiheit', *ibid.* **45**, Fulda 1932, pp. 483–499.

6. *Theology*

Winter, E., 'Religion und Offenbarung in der Religionsphilosophie Bernard Bolzanos, dargestellt mit erstmaliger Heranziehung des handschriftlichen Nachlasses Bolzanos', *Breslauer Studien zur historischen Theologie* **20**, Breslau 1932.

7. *Social Philosophy*

Horáček, C., 'Bernard Bolzano und seine Utopie "Vom besten Staat" ', *Archiv für die Geschichte des Sozialismus und der Arbeiterbewegung* **2**, Leipzig 1912, pp. 68–97.
Salz, A., 'Bernard Bolzanos Utopie "Vom besten Staate" ', *Archiv für Sozialwissenschaft und Sozialpolitik N. F.* **13**, Tübingen 1910, pp. 498–519.
Schindler, F. S., 'Bolzano als Sozialpolitiker', *Deutsche Arbeit* **8**, Prag 1908–1909, pp. 683–699.
Seidlerová, I., *Politické a sociální názory Bernarda Bolzana*, Praha 1963.

8. *Aesthetics*

Svoboda, K., 'Bolzanos Ästhetik', *Rozpravy Československé akademie věd, ročník* **64**, řada SV, sešit 2, Praha 1954.

C. WORKS REFERRED TO BY BOLZANO IN THE SELECTIONS FROM THE WISSENSCHAFTSLEHRE

Abicht, J. H., *Verbesserte Logik oder Wahrheitswissenschaft auf dem einzig gültigen Begriff der Wahrheit erbaut,* Fürth 1802.
Aristotle, *Analytica posteriora* [Aristotelis Stagiritae peripateticorum principis Organum: Hoc est libri omnes ad Logicam pertinentes, Graecè, & Latinè: Jul. Pacio à Beriga interprete. Cum triplici Indice. Editio secunda accuratè recognita & emendata. Frankfurt 1592].
Aristotle, *De sophisticis elenchis* (ibid.).
Aristotle, *Metaphysica. Editio stereotypa,* Leipzig 1832.
Arnauld, A. and Nicole, P., *Logica sive ars cogitandi. Editio novissima,* Basel 1749.
Baumgarten, A. G., *Acroasis logica, aucta et in systema redacta ab J. G. Töllnero,* Halle 1765.
Beck, J. S., *Lehrbuch der Logik,* Rostock 1820.
Beneke, F. E., *Lehrbuch der Logik als Kunstlehre des Denkens,* Berlin 1832.
Bouterwek, F., *Idee einer Apodiktik,* Halle 1799.
Calker, F. A. van, *Denklehre oder Logik und Dialektik, nebst einem Abriß der Geschichte und Literatur derselben,* Bamberg 1822.
Crusius, C. A., *Weg zur Gewißheit und Zuverläßigkeit der menschlichen Erkenntniß,* Leipzig 1747.
Eberhard, J. A. (ed.), *Philosophisches Magazin* **1–4**, Halle 1789–1792.
Fechner, G. T., *Über das höchste Gut,* Leipzig 1846.
Fichte, J. G., *Nachgelassene Werke,* I: *Einleitungsvorlesungen in die Wissenschaftslehre, die transscendentale Logik, und die Thatsachen des Bewußtseins: vorgetragen an der Universität zu Berlin in den Jahren 1812 und 1813,* Bonn 1834.
Fries, J. F., *System der Logik,* Heidelberg 1811.
Gerlach, G. W., *Grundriß der Logik, zum Gebrauch akademischer Vorlesungen,* Halle 1817.
Hegel, G. W. F., *Wissenschaft der Logik,* Nürnberg 1812–1813.
Hegel G. W. F., 'Vorlesungen über die Philosophie der Religion. Nebst einer Schrift über die Beweise vom Daseyn Gottes', *Werke* **11–12**, Berlin 1832.
Herbart, J. F., *Lehrbuch zur Einleitung in die Philosophie,* Königsberg 1813.
Hillebrand, J., *Grundriß der Logik und philosophischen Vorkenntnislehre zum Gebrauch bei Vorlesungen,* Heidelberg 1820.
Hollmann, S. C., *Philosophia rationalis, quae logica vulgo dicitur* [---] *paulo uberioris in universam philosophiam introductionis,* Göttingen 1746.
Jacob, L. H., *Grundriß der allgemeinen Logik und kritischen Anfangsgründe der Metaphysik,* Halle 1791.
Kant, I., *Critik der reinen Vernunft* (4th ed.), Riga 1794.
Kant, I., *Metaphysische Anfangsgründe der Naturwissenschaft* (3rd ed.), Leipzig 1800.
Kant, I., *Logik, ein Handbuch zu Vorlesungen herausgegeben von G. B. Jäsche,* Königsberg 1800.
Keckermann, B., *Systema Logicae. Editio quarta recognita & emendata,* Hannover 1612.
Kiesewetter, J. G. K. C., *Grundriß einer allgemeinen Logik nach Kantischen Grundsätzen. Erster Theil, welcher die reine allgemeine Logik enthält* (3rd ed.), Berlin 1802. *Zweiter Theil, welcher die angewandte allgemeine Logik enthält* (2nd ed.), Berlin 1806.

Klein, G. M., *Anschauungs- und Denklehre. Ein Handbuch zu Vorlesungen,* Bamberg 1818.
Knutzen, M., *Elementa philosophicae rationalis sive logica,* Königsberg and Leipzig 1771.
Krug, W. T., *Fundamental Philosophie,* Zülichau 1803.
Krug, W. T., *Denklehre oder Logik,* Königsberg 1806.
Krug, W. T., *Handbuch der Philosophie und der philosophischen Literatur,* Leipzig 1820–1821.
Lambert, J. H., *Neues Organon oder Gedanken über die Erfassung und Bezeichnung des Wahren und dessen Unterscheidung vom Irrthum und Schein,* Leipzig 1764.
Leibniz, G. W., 'Nouveaux essais sur l'entendement humain', *Oeuvres philosophiques latines et françoises* [---] publiées par Mr. Rud. Eric Raspe, Amsterdam and Leipzig 1765.
Leibniz, G. W., 'Dialogus de connexione inter res et verba, et veritatis realitate; scriptus anno 1677', *ibid.*.
Locke, J., *An Essay Concerning Humane Understanding,* London 1690.
Maaß, J. G. E., *Grundriß der Logik; zum Gebrauch für Vorlesungen,* Halle and Leipzig 1806.
Maimon, S., *Versuch einer neuern Logik oder Theorie des Denkens,* Berlin 1794.
Mehmel, G. E. A., *Versuch einer vollständigen analytischen Denklehre als Vorphilosophie,* Erlangen 1803.
Metz, A., *Handbuch der Logik zum Gebrauch akademischer Vorlesungen,* Würzburg 1802.
Platner, E., *Philosophische Aphorismen nebst einigen Anleitungen zur philosophischen Geschichte. Ganz neue Ausarbeitung,* 1–2, Leipzig 1793, 1800.
Reimarus, H. S., *Die Vernunftlehre als eine Anweisung zum richtigen Gebrauche der Vernunft in der Erkenntniß der Wahrheit, aus zwoen ganz natürlichen Regeln der Einstimmung und des Wiederspruchs hergeleitet,* Hamburg 1756.
Reinhold, C. E. G., *Die Logik oder die allgemeine Denkformenlehre,* Jena 1827.
Reusch, J. P., *Systema logicum antiquorum atque recentiorum, item propria praecepta exhibens,* Jena 1760.
Rösling, C. L., *Die Lehren der reinen Logik durch Beispiele und Verbesserungen leicht verständlich dargestellt, mit Hinweis auf eine Sammlung besonderer kritischer Bemerkungen über mancherlei Lehren der Logiker,* Ulm 1826.
Savonarola, G., *Compendium aureum totius logice,* Leipzig 1516.
Schultz, J., *Prüfung der Kantischen Critik der reinen Vernunft,* 1–2, Königsberg 1789, 1792.
Schulze, G. E., *Grundsätze der allgemeinen Logik* (2nd ed.), Helmstädt 1810.
Sextus Empiricus, *Adversus Logicos* [Sexti Empirici Opera graece et latine, edidit J. A. Fabricius], Leipzig 1718.
Troxler, I. P. V., *Logik, die Wissenschaft des Denkens und Kritik aller Erkenntniß,* 1–3, Stuttgart and Tübingen 1829–1830.
Ulrich, J. A. H., *Institutiones logicae et metaphysicae* (2nd ed.), Jena 1792.
Wolff, C., *Philosophia rationalis sive Logica, methodo scientifica pertractata et ad usum scientiarum atque vitae aptata. Editio tertia,* Frankfurt and Leipzig 1740.

NAME INDEX

SUBJECT INDEX

SYNTHESE HISTORICAL LIBRARY

Texts and Studies
in the History of Logic and Philsophy

Editors:

LEWIS WHITE BECK (ed.), *Proceedings of the Third International Kant Congress.* 1972, XI + 718 pp.

‡KARL WOLF and PAUL WEINGARTNER (eds.), *Ernst Mally: Logische Schriften.* 1971 X + 340 pp.

‡LEROY E. LOEMKER (ed.), *Gottfried Wilhelm Leibnitz: Philosophical Papers and Letters.* A Selection Translated and Edited, with an Introduction. 1969, XII + 736 pp.

‡M. T. BEONIO-BROCCHIERI FUMAGALLI, *The Logic of Abelard.* Translated from the Italian. 1969, IX + 101 pp.

Sole Distributors in the U.S.A. and Canada:

*GORDON & BREACH, INC., 440 Park Avenue South, New York, N.Y. 10016

‡HUMANITIES PRESS, INC., 303 Park Avenue South, New York, N.Y. 10010

SYNTHESE LIBRARY

Monographs on Epistemology, Logic, Methodology,
Philosophy of Science, Sociology of Science and of Knowledge, and on the
Mathematical Methods of Social and Behavioral Sciences

RADU J. BOGDAN and ILLKA NIINILUOTO (eds.), *Logic, Language, and Probability.* 1973, X + 323 pp.
GLENN PEARCE and PATRICK MAYNARD (eds.), *Conceptual Change.* XII + 282 pp.
M. BUNGE, *Exact Philosophy – Problems, Tools, and Goals.* 1973, X + 214 pp.
ROBERT S. COHEN and MARX W. WARTOFSKY (eds.), *Boston Studies in the Philosophy of Science.* Volume IX: *A. A. Zinov'ev: Foundations of the Logical Theory of Scientific Knowledge (Complex Logic).* Revised and Enlarged English Edition with an Appendix by G. A. Smirnov, E. A. Sidorenka, A. M. Fedina, and L. A. Bobrova. 1973, XXII + 301 pp. (Also in paperback.)
K. J. J. HINTIKKA, J. M. E. MORAVCSIK, and P. SUPPES (eds.), *Approaches to Natural Language. Proceedings of the 1970 Stanford Workshop on Grammar and Semantics.* 1973, VIII + 526 pp. (Also in paperback.)
WILLARD C. HUMPHREYS, JR. (ed.), *Norwood Russell Hanson: Constellations and Conjectures.* 1973, X + 282 pp.
MARIO BUNGE, *Method, Model and Matter.* 1973, VII + 196 pp.
MARIO BUNGE, *Philosophy of Physics.* 1973, IX + 248 pp.
LADISLAV TONDL, *Boston Studies in the Philosophy of Science.* Volume X: *Scientific Procedures.* 1973, XIII + 268 pp. (Also in paperback.)
SÖREN STENLUND, *Combinators, λ-Terms and Proof Theory.* 1972, 184 pp.
DONALD DAVIDSON and GILBERT HARMAN (eds.), *Semantics of Natural Language.* 1972, X + 769 pp. (Also in paperback.)
MARTIN STRAUSS, *Modern Physics and Its Philosophy. Selected Papers in the Logic, History, and Philosophy of Science.* 1973, X + 297 pp.
‡STEPHEN TOULMIN and HARRY WOOLF (eds.), *Norwood Russell Hanson: What I Do Not Believe, and Other Essays.* 1971, XII + 390 pp.
‡ROBERT S. COHEN and MARX W. WARTOFSKY (eds.), *Boston Studies in the Philosophy of Science.* Volume VIII: *PSA 1970. In Memory of Rudolf Carnap* (ed. by Roger C. Buck and Robert S. Cohen). 1971, LXVI + 615 pp. (Also in paperback.)
‡YEHOSUA BAR-HILLEL (ed.), *Pragmatics of Natural Languages.* 1971, VII + 231 pp.
‡ROBERT S. COHEN and MARX W. WARTOFSKY (eds.), *Boston Studies in the Philosophy of Science.* Volume VII: *Milič Čapek: Bergson and Modern Physics.* 1971, XV + 414 pp.

‡ CARL R. KORDIG, *The Justification of Scientific Change*. 1971, XIV + 119 pp.
‡ JOSEPH D. SNEED, *The Logical Structure of Mathematical Physics*. 1971, XV + 311 pp.
‡ JEAN-LOUIS KRIVINE, *Introduction to Axiomatic Set Theory*. 1971, VII + 98 pp.
‡ RISTO HILPINEN (ed.), *Deontic Logic: Introductory and Systematic Readings*. 1971, VII + 182 pp.
‡ EVERT W. BETH, *Aspects of Modern Logic*. 1970, XI + 176 pp.
‡ PAUL WEINGARTNER and GERHARD ZECHA, (eds.), *Induction, Physics, and Ethics, Proceedings and Discussions of the 1968 Salzburg Colloquium in the Philosophy of Science*. 1970, X + 382 pp.
‡ ROLF A. EBERLE, *Nominalistic Systems*. 1970, IX + 217 pp.
‡ JAAKKO HINTIKKA and PATRICK SUPPES, *Information and Inference*. 1970, X + 336 pp.
‡ KAREL LAMBERT, *Philosophical Problems in Logic. Some Recent Developments*. 1970, VII + 176 pp.
‡ P. V. TAVANEC (ed.), *Problems of the Logic of Scientific Knowledge*. 1969, XII + 429 pp.
ROBERT S. COHEN and RAYMOND J. SEEGER (eds.), *Boston Studies in the Philosophy of Science*. Volume VI: *Ernst Mach: Physicist and Philosopher*. 1970, VIII + 295 pp.
‡ MARSHALL SWAIN (ed.), *Induction, Acceptance, and Rational Belief*. 1970, VII + 232 pp.
‡ NICHOLAS RESCHER *et al.* (eds.), *Essays in Honor of Carl G. Hempel. A Tribute on the Occasion of his Sixty-Fifth Birthday*. 1969, VII + 272 pp.
‡ PATRICK SUPPES, *Studies in the Methodology and Foundations of Science. Selected Papers from 1911 to 1969*, 1969, XII + 473 pp.
‡ JAAKKO HINTIKKA, *Models for Modalities. Selected Essays*. 1969, IX + 220 pp.
‡ D. DAVIDSON and J. HINTIKKA (eds.), *Words and Objections: Essays on the Work of W. V. Quine*. 1969, VIII + 366 pp.
‡ J. W. DAVIS, D. J. HOCKNEY and W. K. WILSON (eds.), *Philosophical Logic*. 1969, VIII + 277 pp.
‡ ROBERT S. COHEN and MARX W. WARTOFSKY (eds.), *Boston Studies in the Philosophy of Science*, Volume V: *Proceedings of the Boston Colloquium for the Philosophy of Science 1966/1968*, VIII + 482 pp.
‡ ROBERT S. COHEN and MARX W. WARTOFSKY (eds.), *Boston Studies in the Philosophy of Science*. Volume IV: *Proceedings of the Boston Colloquium for the Philosophy of Science 1966/1968*. 1969, VIII + 537 pp.
‡ NICHOLAS RESCHER, *Topics in Philosophical Logic*. 1968, XIV + 347 pp.
‡ GÜNTHER PATZIG, *Aristotle's Theory of the Syllogism. A Logical-Philological Study of Book A of the Prior Analytics*. 1968, XVII + 215 pp.
‡ C. D. BROAD, *Induction, Probability, and Causation. Selected Papers*. 1968, XI + 296 pp.
‡ ROBERT S. COHEN and MARX W. WARTOFSKY (eds.), *Boston Studies in the Philosophy of Science*. Volume III: *Proceedings of the Boston Colloquium for the Philosophy of Science 1964/1966*. 1967, XLIX + 489 pp.
‡ GUIDO KÜNG, *Ontology and the Logistic Analysis of Language. An Enquiry into the Contemporary Views on Universals*. 1967, XI + 210 pp.
* EVERT W. BETH and JEAN PIAGET, *Mathematical Epistemology and Psychology*. 1966, XXII + 326 pp.
* EVERT W. BETH, *Mathematical Thought. An Introduction to the Philosophy of Mathematics*. 1965, XII + 208 pp.
‡ PAUL LORENZEN, *Formal Logic*. 1965, VIII + 123 pp.

‡GEORGES GURVITCH, *The Spectrum of Social Time*. 1964, XXVI + 152 pp.
‡A. A. ZINOV'EV, *Philosophical Problems of Many-Valued Logic*. 1963, XIV + 155 pp.
‡MARX W. WARTOFSKY (ed.), *Boston Studies in the Philosophy of Science*. Volume 1: *Proceedings of the Boston Colloquium for the Philosophy of Science, 1961–1962*. 1963, VIII + 212 pp.
‡B. H. KAZEMIER and D. VUYSJE (eds.), *Logic and Language. Studies dedicated to Professor Rudolf Carnap on the Occasion of this Seventieth Birthday*. 1962, VI + 256 pp.
*EVERT W. BETH, *Formal Methods. An Introduction to Symbolic Logic and to the Study of Effective Operations in Arithmetic and Logic*. 1962, XIV + 170 pp.
*HANS FREUDENTHAL (ed.), *The Concepts and the Role of the Model in Mathematics and Natural and Social Sciences. Proceedings of a Colloquium held at Utrecht, The Netherlands, January 1960*. 1961, VI + 194 pp.
‡P. L. GUIRAUD, *Problèmes et méthodes de la statistique linguistique*. 1960, VI + 146 pp.
*J. M. BOCHEŃSKI, *A Precis of Mathematical Logic*. 1959, X + 100 pp.